Life Cycle Assessment of Energy Systems

Scrivener Publishing
100 Cummings Center, Suite 541J
Beverly, MA 01915-6106
Publishers at Scrivener
Martin Scrivener (martin@scrivenerpublishing.com)
Phillip Carmical (pcarmical@scrivenerpublishing.com)

Life Cycle Assessment of Energy Systems

Closing the Ethical Loophole of Social Sustainability

Nicholas Sakellariou

Scrivener Publishing

WILEY

This edition first published 2018 by John Wiley & Sons, Inc., 111 River Street, Hoboken, NJ 07030, USA and Scrivener Publishing LLC, 100 Cummings Center, Suite 541J, Beverly, MA 01915, USA

For more information about Scrivener publications please visit www.scrivenerpublishing.com.

Wiley Global Headquarters
111 River Street, Hoboken, NJ 07030, USA

For details of our global editorial offices, customer services, and more information about Wiley products visit us at www.wiley.com.

Library of Congress Cataloging-in-Publication Data

ISBN 978-1-119-41816-0

Cover images: Energy Field, Amrobih | Dreamstime.com
Cover design by: Kris Hackerott

Set in size of 11pt and Minion Pro by Exeter Premedia Services Private Ltd., Chennai, India

10 9 8 7 6 5 4 3 2 1

To the Divine Mother

Om Bolo Shri Sat Guru Bhagavan ki Jai

To my Teachers

To Tilemachos Sakellariou, the one who never ceased to believe in me; to Rafaela Sakellariou, the one who continuously shows me the way of the Light; and to Rusty Wells, the one who opened my heart and stretched my imagination.

Contents

Acknowledgements

I have enjoyed a rare privilege over the last seven years.

Thank you, Professor Alastair Iles of the University of California at Berkeley, for making this project possible. Your presence in my life has been a true blessing! Thank you Professor Matthew Wisnioski and Professor Gary Downey for making such intellectual and professional trajectories imaginable. My thanks also go to Professor David Winickoff who chaired my orals examination committee in the Department of Environmental Science, Policy, and Management and who read this manuscript.

It was indeed a rare privilege to study under Professor Laura Nader, as it has been an honor to work under Professor Carolyn Merchant.

For editing and excellent commentary I am indebted to Alastair Iles, Matthew Wisnioski, Bettina Lewis, and Martin O'Brien. For inspiration I am grateful to Juan Lucena, Donna Riley, Caroline Baillie, and Dustin Mulvaney.

Jordan Willis, thank you for your hard work in transcribing hundreds of hours of interview material.

Δημήτρη Σακελλαρίου, Νιούσκα Σακελλαρίου, Νικόλα Λαρεντζάκη, Αγνή Ιωάννου, Δάφνη Λαρεντζάκη; mi familia Mexicana Casavantes y Ambrosius: I can see your love in every page of this manuscript.

This material is based upon work supported by the National Science Foundation (NSF) under Grant No. 1354545. Any opinions, findings, and conclusions or recommendations expressed in this material are those of the author and do not necessarily reflect the views of the National Science Foundation.

Last but not least, in the past five years, more than 150 engineers and lay experts have shared with me their stories, archives, personal and professional documents and their vision for a more "sustainable" or socially mindful engineering practitioner. This project is theirs, including Tom Munsey, Jim Evans, Greg McIsaac, John Peet, and Gerry Te Kapa Coates.

Part I

ENGINEERING AND SUSTAINABILITY

1

Engineering Sustainability, Sustaining Engineering

Engineers are the unacknowledged philosophers of the postmodern world

Carl Mitcham, "The Importance of Philosophy to Engineering"[1]

Introduction

In light of "undeniable realities of acid rain, reduction in the ozone layer, and (now) CO_2 emissions," wrote the New Zealander engineer David Thom, chairman of the World Federation of Engineering Organizations (WFEO) Committee on Engineering and Environment from 1991 to 1999, "we see the dangerous failure… [of the position that]… the engineer is the servant of political processes." Thom, echoing many past and present

---

[1] Mitcham, Carl. "The Importance of Philosophy to Engineering." *Teorema: Revista internacional de filosofía* 17, no. 3 (1998): 27–47, 28.

engineering leaders, suggested that political arrangements could hardly be expected to settle the social impacts of technology. In this regard it was "incumbent on the engineer (in professional self-interest, if no for other reason) to become fluent in the analysis of [such] consequences" through adopting the tools and fundamental precepts of sustainability.[2]

Six years after Thom distinguished between engineering service and political servitude, he asserted that the profession had "a choice between two paths." Engineers, he elaborated, could either "trail behind the accelerating pace of events…until… [they] *are no longer relevant*," or they could "accept challenge, change, trauma and travail and march in the vein of the new Industrial Revolution."[3] This preoccupation with meeting the sustainability challenge so that engineers are not "left behind in the decision-making process that will influence the future shape of this world" not only prompted Thom's article, but also the 1993 American Association of Engineering Societies (AAES) statement on the "Role of the Engineer in Sustainable Development."[4]

This realization is how sustainability became an engineering ideology of assessing environmental impacts throughout a product or project's life cycle. At the same time, this ideology converged with the vision of engineering professional transformation conceived by practitioners as a response to a perceived societal demand for conservation. In the larger society of the late 1980s, there were growing concerns that technologies were causing massive environmental damage—this was an environmental crisis. At one level, because engineers work with technologies, their very work and worldviews were suddenly being endangered and questioned from the outside. At another, due to their professional practices and cultures, some engineers identified themselves as culturally and politically "invisible." A small minority of creative engineer-philosophers thus sought to rescue their profession's technology crisis through integrating "sustainability" into their principles and work.

In the late 1980s, engineers were confronted anew with the dominant image of a shrinking environment. Spokesmen for the profession suggested that engineering work was admired for creating an urban, technological civilization using the world's natural resources, while it was simultaneously blamed for exploiting such resources to the verge of extinction.

---

[2] Thom, David. "The WFEO Code of Ethics." *New Zealand Engineering*, June 1, 1988.

[3] Thom, David. "Engineering to Sustain the Environment." In *The Role of Engineering in Sustainable Development*, edited by Monica D. Ellis, 62–79. Washington, DC: AAES, 1994, 78, emphasis added.

[4] AAES. "Statement of the American Association of Engineering Societies on the Role of the Engineer in Sustainable Development." Washington, DC: AAES, 1993.

By the 1990s, some engineers were writing about both ongoing evidence regarding environmental constraints and resource deficiency, and a need to apply root engineering values, expertise, and practices in a process of transformation. The prevailing image of growth-driven change gone awry, and development fraught with ecological disaster, substantially mobilized international and US engineering organizations and elite practitioners, who wanted to keep pace in the race to a technological future.

Responding to the economic-environmental challenge, the 1990s produced two distinct engineering ideologies of sustainability—one emphasizing engineering innovation, and the other emphasizing socio-cultural change. The first ideology, based on creativity, resembles an *ideology of technological change,* as characterized by engineering historian Matt Wisnioski in his analysis of American engineering in the 1960s. The technological change ideology of sustainability refers to *engineering reform* controlled and directed by engineers themselves—in other words, technological practices can be improved through the application of expertise. In this book I am building on Wisnioski's dialectical framework adding to it another dimension for the 21st century; I highlight how the dialectic between sustainability and engineering has been defined largely by the ideology of technological change.[5]

Wisnioski's compelling argument is that an intellectual crisis of technology within American society (between 1957 and 1973) presented a conceptual lens through which engineers could interpret technology as modernity. He shows that an ideology of technological change served as the counter-paradigm to an ideology of *technopolitics* while positing that technology was neither good, nor bad, nor was it neutral. Since the 1970s, Wisnioski contends, the solution that American engineers have favored for the dilemmas of technology and social progress has been that "[t]hrough rational management,... technology's unintended consequences could be minimized and its positive capacities maximized."[6]

The second and less influential ideology of engineering sustainability, with its emphasis on socio-cultural change, stems from a minority of practitioners and academics during the 1980s and 1990s who self-identified with the conceptual framework of social responsibility. Engineers associated with organizations like Engineers for Social Responsibility (ESR), the subaltern US group

---

[5] Wisnioski, Matthew H. *Engineers for Change: Competing Visions of Technology in 1960's America.* Cambridge, MA: The MIT Press, 2012.

[6] Wisnioski, Matthew. "How Engineers Contextualize Themselves." In *Engineering in Context,* edited by Steen Hyldgaard Christensen, Bernard Delahousse, and Martin Meganck, 403-416. Aarhus: Academica, 2009.

of American Engineers for Social Responsibility (AESR) discussed in chapter 3, and later the International Network of Engineers and Scientists for Global Responsibility (INES) mindfully suggested a more culturally and politically sensitive vision for engineering sustainability. The technopolitics ideology of sustainability is about *engineering challenge*: it places more emphasis on the devolution of expertise from the existing model of engineering and society, and it questions the dominant values of engineering practice.

Ideology, then, is important for understanding the current problem with how sustainability is defined in engineering. It is defined predominantly in a narrow way, such that a particular type of scientific investigation is considered valid to answer questions of sustainability. And the way sustainability is framed bears resemblance to other cultural patterns in engineering—it gets stripped of power issues, of people, of alternative ways of thinking about the topic in general, including environmental justice, class issues and a free-market critique. A sustainability engineer who is not paying attention to power relations is likely to reproduce social injustice; we see that, for example, in terms of who becomes an engineer and in terms of the entire experience of technical education as one that delivers a certain conformity to a set of values and a set of applications in engineering. It is my hope that some engineers who do not or cannot identify with an alternative professional culture will start to feel as though they *have* a relationship with non-traditional philosophies into discussions about sustainability. Indeed one reason for examining the history and politics of sustainability engineering in ideological terms is that it extends an understanding of the current co-existence of corporate system approaches along with a reformist movement in considering a redefinition of the profession and its practitioners.

Three points need to be emphasized regarding the growth of sustainability identity in engineering. First, as I will show in detail in chapter 2, "sustainability engineering" did not come about naturally, but required substantial ideological and institutional transformation. Technological change as the dominant engineering ideology is largely confined to the narrow limits of technical problem solving. Advocating for apolitical expertise, most engineers conceptualize themselves as mathematical problem solvers and society as a set of discrete problems, to be solved through the application of scientific principles and mathematical theorems.[7] In

[7] Seely, Bruce E. *Building the American Highway System: Engineers as Policy Makers*. Philadelphia: Temple University Press, 1987. Keniston, Kenneth. "The crisis of the engineering algorithm." Paper presented at the Institute of Advanced Studies in Humanities, Politecnico di Torino, Torino, Italy, 1996, manuscript.

other words, many engineers have an identity based on technical problem solving within narrow horizons.

The backbone of Life Cycle Assessment (LCA), for example, is a computational puzzle: a cradle to grave (i.e., life cycle) inventory analysis that construes an accounting balance of material, energy and chemical flows for the entirety of an industrial process or product system. The very term "system," as discussed in chapter 2, derives from a sustainability engineering-specific ontological assumption and worldview that is inherent in the design and execution of LCAs. Life cycle thinking is used by technical experts—primarily engineers—to draw path models of raw material extraction, to component or commodity transportation, manufacturing, use, to end-of-life for different production processes. Depending on the "scope" of a typical study undertaken, different researchers may draw *different* models that correspond to the *same* production process or system. LCAs identify environmental—and more recently social—"footprints," thus establishing industrial benchmarks against which engineering progress can be imagined and quantified. Corporations, governmental agencies and other technical expert constituencies supported LCAs as means to counteract environmental criticism and to substantiate "responsibility" and "transparency" to critics, stakeholders, consumers and the society at large. The scope (the choice of boundaries for an LCA) and the methodology (an engineering exercise of collecting and reporting about the "impacts" and "benefits" of industrial activity) are examples of why a conventional LCA might embody ideological assumptions and societal/political boundaries.

Here I show that technopolitics and technological change have co-evolved and have challenged each other; yet stories like that of the AESR told in chapter 3 reveal the power of the dominant engineering ideology. The co-evolution of engineering ideologies of sustainability has shaped not only technical methodologies like LCA, but also the way engineers experienced societal politics and conceptualized themselves as technical professionals.

Importantly, the efforts to foster sustainability identity in the engineering profession reflected a politics of alliance-making and connections with the corporate world. The 1980s—especially in the US and the UK—were dominated by a philosophy of voluntary industry action and neo-liberal economics that introduced the idea of public-private partnerships, the Trojan horse of "sustainable development" premising a synergy between communities in the Global South and private corporations overseen by the public sector. Contemporary developments like the emergence of Social Life Cycle Assessment (SLCA), detailed in chapter 4, attest to the importance of the larger political economy of the 1980s and 1990s as the economic (neo-liberalism) and institutional bases of the co-evolution of the technological change and technopolitics ideologies of sustainability.

Second, the contributors to the sustainability engineering discourse in the 1990s represented only a very small fraction of the engineering community in the US. Set forth by a minority of vocal proponents—a handful of enlightened environmental consulting managers, members of powerful professional elites and the US Corps of Engineers, some engineering educators, and a few radical professional voices including a subaltern engineering group like the AESR—the ideas of "sustainability engineers" and their relation to the larger profession were not common or widely shared by their rank-and-file colleagues. In some part, this is explained by the fact that the sustainability engineering luminaries possessed either one or a combination of qualities: degrees or expertise beyond engineering, background in professional organization leadership, political activism, and conduct of intellectual life beyond engineering interests or research. But while sustainability engineering in the 1990s was grounded in the historical conditions of voluntary industry action and neo-liberalism and in the support of professional elites, it simultaneously coincided with the efforts of more visionary engineers to provoke change in a very static and conservative profession. Thus in 1999 Edgar Woolard, former CEO and chairman of DuPont, spoke for many in the engineering profession when he wrote that "environmentalists" have been viewed as promoting "very radical change—based on what many in industry perceived to be philosophical or ideological grounds at best and pure emotionalism at worst."[8] The development of the sustainability identity in engineering, then, has been contentious. On the one side, during most of the 1990s, the engineering minority who expressed concerns regarding the environmental crisis received either skepticism or was rejected on the ground that it produced a deviant culture. On the other side, the 1990s also generated debates between those engineers who advocated for sustainability but differed in how they approached it. As these debates between sustainability enthusiasts—regardless of ideological orientation—unfolded, the identity of the sustainable engineer underwent integration between technological change and technopolitics.

Third, that the dialectic between sustainability has created various sites where engineers are struggling to reshape professional identity. These sites, I show in chapter 3, include the development of technical methodologies and tools, professional codes of practice, and educational reforms. These

---

[8] Woolard, Edgar S. "Creating Corporate Environmental Change." *The Bridge* 29, no. 1 (1999): 8-11.

sites are important because they make the politics of sustainability visible: they reveal, for example, how some practitioners resisted the idea that engineers have a responsibility to lead as politicizing their work.

Entangled with the politics of sustainability is the question of what epistemologies are valued in engineering. When, for example, there is the presumption that engineering suggests an evaluation of technical expert knowledge, which is seen as objective over "lay knowledge," there is a political choice being made about what counts as legitimate sustainability expertise. The case studies presented in chapters 6 and 7 illustrate how a reductionist sustainability epistemology of greenhouse gas emissions held by renewable energy engineers has resulted in a tendency to downplay the environmental and social justice considerations of solar and wind project development in California's Western Antelope Valley.

The idea of sustainability officially entered the US engineering profession with the then Chief of Engineers' Hank Hatch's keynote speech at the American Society of Civil Engineers (ASCE) convention on October 8, 1989—by 2004-2005 the term had acquired a substantial stake in engineering cultures around the world.[9] The sustainable technology crisis (1989-2003), I found, was similar to previous technology crises within the US engineering profession in that it comprised a renegotiation of engineering's cultural and epistemological confines and a reconsideration of the scope of engineering servitude. The very idea of engineering being *ipso facto* a potent force to enhance human welfare was yet again in doubt, in light of the manifest negative effects of engineering projects on the environment.[10] In this book I argue that sustainability has become meaningful in an *engineering context* through a historical process of inserting considerations into engineering cultures and technical methodologies, particularly LCA, that have progressively challenged dominant views of what engineering practice should be. Thus I recount how the growth of "sustainability" concerns has driven an expanded engineering discourse.

Engineering contextualizations of sustainability together with calls for professional leadership emerged as a response to the emerging dominant

---

[9] Hatch, Hank. "Keynote speech at the ASCE Convention on October 8, 1989," manuscript, courtesy of Hank Hatch.

[10] See respectively, Layton, Edwin T. Jr. *The Revolt of the Engineers: Social Responsibility and the American Engineering Profession*. Cleveland, PA: Press of Case Western Reserve University, 1971, and Wisnioski, Matthew H. *Engineers for Change: Competing Visions of Technology in 1960s America*. Cambridge, MA: MIT Press, 2012. In all the aspects in which engineering leaders in the 1990s deviated from their predecessors they are closer to "the challenge of reorienting technology" (Thom, 1994, 78).

image of a collapsing planet from the late 1980s onwards. From the perspective of engineering elites and their organizations, and often from the viewpoint of anonymous professionals, there was a solution to the perceived environmental crisis of the late 1980s and early 1990s: they would demonstrate to themselves and to everyone else that—through sustainability engineers' diagnoses—nature's recovery was achievable. They saw the environmental crisis as an opportunity to create a unifying force around the concept of sustainability and to invest more engineering expertise into public policy.[11] Engineering contextualizations of sustainability were developed so that professionals could "attain the proper status they deserve."[12] In short, sustainability ensured a functioning of professional power; the narrative of sustainable technology deeply grounded in the 1980s political economy and cultural order became essential to the relevancy and social image of engineering.[13] Thus, as the case study of the US AESR group illustrates, engineering sustainability bore resemblance to interpretations of "social responsibility" and "humane technology" in the 1930s and 1960s, respectively.[14]

---

[11] "[Engineering] constitutes a powerful force that is presently fragmented into various subdisciplines and areas of specialty." Hence, "[t]here is the need to unify the global engineering community, particularly around the issue of sustainability." Carter, Archie N. "Editorial." *Journal of Professional Issues in Engineering Education and Practice* 119, no. 3 (1993): 213. See also, Sanio, Michael. "The Role of Engineers in Sustainable Development." Paper presented at the 1997 AAAS CAIP Meeting, AAAS Headquarters, Washington, DC, 1997. The same year Thom argued that "[i]t is the wider concept of sustainable development itself that embraces all engineering activity" (Thom, David. "The Role of Engineers to Promote Cleaner Production." In WFEO, *The Engineer's response to sustainable development*, 36-45. Washington, DC: WFEO, 1997. Another 1996 report by the Civil Engineering Research Foundation (CERF) read: "The American Society of Civil Engineers (ASCE) recently reported that over the past two decades there has been a marked decline in the number of civil engineers who have held leadership positions in public policy development in the United States...Policy in this context is taken to include the establishment of infrastructures to support sustainable development." Civil Engineering Research Foundation (CERF). "Engineering and Construction for Sustainable Development in the 21st Century: Assessing Global Research Needs." Washington, DC: American Society of Civil Engineers, 1996.

[12] Wiggins, John H. "Challenge for Engineers." *Journal of Professional Issues in Engineering Education and Practice* 121, no. 3 (1995): 199.

[13] See, for example, Hank Hatch, Sustainable Development, excerpts from an address to the Presidents' Circle of the National Academies—quoted in ASCE, *Sustainable Engineering Practice: An Introduction*. Reston, VA: American Society of Civil Engineers, 2004, v.

[14] See footnote 10 above.

By monitoring the interplaying ideologies of technological change and technopolitics in the period between 1989 and 2003, this book continues a reflection on engineers as the "unacknowledged philosophers" of postmodernity. Engineers "build" postmodernity, we are customarily told by technology scholars; Wisnioski's thesis indicates that equally important is the creation of collective culture in a postmodern world and how engineers express themselves through their developing philosophies, intellectual histories, and social theories.[15] We generally assume that engineers rally around technical facts and methods to avoid political or philosophical gridlock. In reality, engineer-leaders have appropriated non-engineering discourse to renegotiate engineering knowledge and identity in addressing some of the most pressing existential dilemmas facing their discipline. To sustain engineering and reclaim normative control over technological matters, visionary engineers have transcended their traditional intellectual province and in the process "engineered" social and political theories of sustainability into terms that speak to engineers.

The Sustainable, Yet Invisible, Engineer

In the late 1980s and early 1990s, a period in which "sustainable development" became the defining conceptual framework in global environmental policy, articles were written independently of each other by an engineer-historian and an engineer-anthropologist. They featured the same alarming title: "The Invisible Engineer."[16] Henry Petroski, professor of civil engineering and history at Duke University, was deeply concerned about engineers remaining all but invisible *in both the cultural sphere and the social consciousness*. To combat engineering invisibility, fueled by professional specialization and "independent agendas" that have "robbed the profession of a single voice," Petroski contended that engineers "must seize every opportunity" to interact directly with the public. Others shared his concerns, such as Gary Downey, professor of science and technology studies at Virginia Tech: his view was that the isolation of engineers and engineering work *from most social studies of technology* has been due to the perceived division between the social and knowledge contents of engineering, as well

---

[15] Wisnioski (2012).

[16] Petroski, Henry. "The Invisible Engineer." *Civil Engineering—ASCE*, 60, no. 11 (1990): 46-49; Downey, Gary L., Arthur Donovan, and J. Timothy Elliott. "The Invisible Engineer: How Engineering Ceased to Be a Problem in Science and Technology Studies." *Knowledge and Society* 8, (1989): 189-216.

as an apparent division between structures (e.g., corporate, scientific) and professionals' everyday, contingent practices. Both authors invoked the invisibility metaphor to warn that there was a real danger of the "essential features" of engineering being missed by the public and by non-engineers.

Around that time, and twelve years after the publication of the environmental classic *Our Common Future*, the National Academy of Sciences (NAS) Board on Sustainable Development lamented that the relatively little input on sustainability by the engineering community contributed to their invisibility. In their 1999 report *Our Common Journey*, the NAS described an interdisciplinary overview of scientific scholarship and engineering practice at the intersections of environment and development; they lauded how "a broader systems perspective" has become the centerfold of "research programs on global change" in the US and abroad. The NAS therefore embarked on the seemingly urgent task of reinvigorating the engineering dimensions of sustainability amidst an internal debate on why development discourse had "increasingly" drifted "from its scientific and technological base". "In the last decade," they noted, sustainable development had been influenced more "by political than by scientific perspectives."[17]

Whether or not prior to 1999 "political" perspectives on sustainability were ignored or suppressed in favor of "scientific" perspectives, many engineers were championing explicitly social, cultural and political perspectives on sustainability, beginning around 2004-2005. To some extent, this was due to initiatives promoted by reputable engineering institutions such as the National Academy of Engineering (NAE) and the ASCE. The lamentation, then, that extra-engineering elements have influenced the sustainable development trajectory fails to acknowledge that many engineers, indeed, have not only engaged in sustainability work but have linked together the political and scientific perspectives of sustainability.

Far more a philosopher of sustainability engineering than an invisible engineer, the then vice president of international environmental engineering firm CH2M Hill Don V. Roberts was "getting tired" of reading about scientists' contributions toward the sustainable future. "As engineers," he cautioned, "we have poor visibility in the environmental community."[18]

---

[17] World Commission on Environment and Development (WCED). *Our Common Future*. Oxford: Oxford University Press, 1987. National Research Council. *Our Common Journey: A Transition Toward Sustainability*. Washington, DC: National Academy Press, 1999, 275 and 283.

[18] Roberts, Don V. "Sustainable Development—A Challenge for the Engineering Profession." In *The Role of Engineering in Sustainable Development*, edited by Monica D. Ellis, 44-61. Washington, DC: AAES, 1994, 46.

Indeed, another American practitioner in the mid-1990s bemoaned in the *Journal of Professional Issues in Engineering Education and Practice* that the Rio Conference consigned to oblivion everything engineering.[19] Their failure to engage with environmental politics made engineers claim they were deemed to remain poised on the outskirts of relevant expertise.[20] In response to such fears, Maurice Strong, the Canadian businessman and former United Nations (UN) environmental official, promised that adopting sustainable development ideas would "result in a profound change in the public perception of the engineer, as well as for the individual engineer's perception of his or her own professional role."[21]

In sum, Petroski's argument fits the circumstances of sustainability engineering in the 1990s. The politics of engineering identity formation in the period between 1989 and 2003 relied on, and helped to produce, the sustainable, yet invisible engineer. On one level, the critique from within the profession has conceded the perceived decline of engineers' status and their influence in setting policy. At another level, engineers' cultural and political marginalization may be a direct result of their resistance against what some engineering professionals disdainfully describe as the "political perspectives" of sustainability.

But how have issues of engineering identity and visions of sustainability interfaced? How did professionals condemning the social dimensions of sustainability as "unrealistic" still find socio-political philosophies to legitimize their views of sustainable development? And how have opposing assumptions about the role of technology and engineering professionals in society fed on each other to determine the main stakes in sustainability

---

[19] Wiggins (1995): 199. Along similar lines, Hatch remarked to a 1998 World Bank audience that "[s]cientists have played a significant role in helping us to understand the fundamental impact of human interactions on the global environment and developing policy to support the decision-making debate. But it is engineers who use that science to plan, build, and operate the infrastructure that will directly contribute to-or detract from-the goals of sustainable development." Hatch, Henry. "Panelist Remarks." In *Partnerships for Global Ecosystem Management Science, Economics and Law. Proceedings and Reference Readings from the Fifth Annual World Bank Conference on Environmentally and Socially Sustainable Development*, edited by Ismail Serageldin and Joan Martin-Brown, 70-73. Washington, DC: The World Bank, 1998.

[20] Rubin, Debra K. "FIDIC delegates debate new ecological activism." *Engineering News Record* June 28 (1990): 15–16.

[21] Strong, Maurice. "The Engineer as Agent of Global Change." Speech delivered at the American Association of Engineering Societies and Engineering Foundation Conference, entitled "Sustainable Development: Creating Agents of Change," held at Snowbird, Utah, August 4, 1995.

engineering? To understand these questions one need to turn first to the ideologies of technological change and technopolitics.

What Prompts the Sustainability Engineer? The Ideologies of Technological Change and Technopolitics

Recasting Engineering Progress as the Golem-Like View of Sustainability

The first of a series of books published after 1989 by the NAE's Program on Technology and Sustainable Development (TSD) addressed "the paradox of technology."[22] The "paradox of our time," argued Massachusetts Institute of Technology (MIT) president Paul Gray, was best captured in the metaphoric tale of the Golem of Prague. Hailed as proof of engineering progress, environmental technology was really "[a]n artificial creature *created to serve* [;]..., [one which] exhibited *a mind of its own*, acting in mischievous ways *unanticipated* by its maker."[23]

This conception underlay all technological change articulations of sustainability. It represents technology as a product of engineering servitude to society that, alas, has acquired the properties of a self-governing force, the application of which has often had second-order consequences that were neither anticipated nor understood by its designers. The Golem-like view of sustainability engineering also regards effective engineering management as the force needed to maintain the global growth machine running without social costs. Sustainable development à la technological change thus became the principal engineering ideology for imagining both society and self.[24] Top engineering functionaries, engineering society leaders, and executives of environmental consulting companies propagated the notion that sustainability and the ideology of technological change were complementary. What is more, sustainability *meant* traversing the unintended consequences of technical change.[25]

Sustainability and technological change, combined, became the working model for making engineering progress in the 1990s. Accepting that

---

[22] The same year Hatch talked about the "irony" that "further development is needed to handle both the problems of growing population and the problems of past development." Hatch, Hank. "Keynote speech at the ASCE Convention on October 8, 1989," manuscript.
[23] Jesse Ausubel H., and E. Hedy Sladovich, eds. *Technology and Environment*: National Academy of Engineering, 1989, 192.
[24] Ibid., 201.
[25] See for example, Frosch, Robert. A. "Sustainability Engineering (editorial)." *The Bridge* 29, no. 1 (1999): 2–3.

sustainable technology engenders a dialectic between unintended consequences and effective management of environmental impacts "free[s] us [engineers] from the futility of searching for magic bullets… It allows us to embrace progress and take steps to improve the quality of life of humans and the environment."[26]

The Rationalities of Engineering Ideologies of Sustainability

The more we understand the rationalities of technological change and technopolitics, respectively, the more we recognize the ideological underpinnings of sustainability in an engineering context. Despite their shared concerns with systemic interdependence, the technological change view of sustainability thrives on the assumption that it is both rational and objective, while the technopolitics view is based on challenging such claims to objectivity and questioning the value of sustainability's political ends. The technological change's twin claim to rationality and objectivity is directly linked to the engineering identity of the "doer" and her ability to depict and manipulate material balances—"engineers must focus on the what and the how tos," contended Hatch in a 1992 speech entitled "Relevant Engineering in the 21st Century."[27] And whatever their differences, for the most part, both ideologies of sustainability assume a certain level of engineer autonomy in the operation of sustainable technology.[28] The typology in Table 1.1 classifies the rational bases of the engineering community's ideologies of sustainability by listing them under two headings: a) Premises, and b) Core assumption.

Throughout the 1990s, sustainability engineering was consistently discussed on the assumption of allegedly definable technological foundations.[29] The technological change ideology of sustainability was the

---

[26] National Research Council, 1999.

[27] Hatch, Henry J. "Relevant Engineering in the 21st Century." *Journal of Professional Issues in Engineering Education and Practice* 119, no. 3 (1993): 216–219.

[28] For example, a conference entitled "Preparing for a Sustainable Society," co-sponsored by the IEEE Society on Social Implications of Technology and IEEE Toronto Section, took place in Canada (Ryerson Polytechnical Institute, Toronto, Ontario, Canada, June 21–22, 1991). The conference's Call for Papers read: What is a sustainable society? How will the relationship between technology and society change if a strategy of sustainable development is adopted? Can society control and redirect the technological system it has created or is this system now controlling society? Burkhardt, Helmut, and H. Willem Vanderburg. "Preparing for a Sustainable Society." *IEEE Technology and Society* 10, no. 4 (1991): 6–8.

[29] For example, see Rajagopalan, Visvanathan. "The Engineer's role in sustainable development." *Civil Engineering* 62, no. 8 (1992): 6.

Table 1.1 Rational bases for engineering ideologies of sustainability

Engineering ideologies of sustainability	Premises	Core assumption
Technological Change	Engineering creates prosperity—yet engineers have unintentionally contributed to environmental problems; "Sustainable," means "environmentally sustainable"; Environmental problems can be eliminated by technological means without requiring a sacrifice of prosperity.	The reorientation of environmental technology is autonomous, thus the exploitation of natural resources should neither be decreased nor increased, but ought to be effectively managed.
Technopolitics	The design and social integration of engineered systems reflect normative, though not readily recognizable, assumptions and values; therefore, "development's" effects on the social order are equivalent to any other form of political action; Engineering and science play a supportive, not central, role in the quest for sustainability; The logic of competitive productivism must be reconsidered.	Provided that technology is autonomous, the urgent focus of sustainability engineering is reconceptualizing and redirecting democratic control of technical means and technical expertise.

dominant influence in the sustainable development debate at least during the fourteen years from 1989 to 2003—a period I define here as the sustainable technology crisis. This position, advocated by the most vocal and powerful engineering constituencies, maintained that engineering and technology had created prosperity and improved the quality of life but that engineers, as Don Roberts put it, "unintentionally contributed to global environmental problems."[30] In celebrating innovation while defending the dominant business and economic culture of the 1990s US sustainability engineering leaders like Roberts and Hatch referenced the profession's well-intentioned obliviousness, thus paving the way for the rise and legitimation of environmental metrics.[31]

At least until 1998, most spokesmen of professional societies, engineering leaders in general, or codes of ethics in engineering argued that sustainable development signified environmentally sustainable development.[32] "On a scale of one to ten, with one signifying 'strictly development with modest modifications' and ten signifying 'strictly environmental protection,' I am probably about a three," said Hank Hatch in a 1992 interview.[33] Engineering leaders, like Hatch or ASCE's executive director, James E. Davis, embraced the oxymoronic predicament of "sustainable growth"

---

[30] Don Roberts quoted in Rubin (1990).

[31] Hatch, Henry J. "Sustainable Development." Manuscript from talk given at during a mini-symposium on sustainable development, Tufts University, March 8, 2000. Edgar Woolard, who spent most of his career working for General Motors and DuPont, recalled: "Environmental metrics were not much of an issue when I started as a young engineer at General Motors 36 years ago. The metrics we used were the number of cars we produced and how good their quality was…It is not that we were not responsible. I think we were…. Environmental performance was not a key factor in whether or not we met our business objectives." Woolard (1999).

[32] "That is, our goal is clearly development, but is heavily modified by the expression 'environmentally sustainable,'" ibid. See also Ausubel and Sladovich (1989); Leonard, Raymond S. "Information Systems for Engineering Sustainable Development." Paper presented at the Workshop on Engineering Partnership for Sustainable Development: A Workshop in Conjunction with Prep Con IV of the United Nations Conference on Environment and Development. March 1-3, 1992, New York, NY; IEEE. "White Paper on Sustainable Development and Industrial Ecology." Hoboken, NJ: IEEE Electronics and the Environment Committee, 1995; Council of Academies of Engineering and Technological Sciences (CAETS). *The Role of Technology in Environmentally Sustainable Development: A Declaration of the Council of Academies of Engineering and Technological Sciences.* Washington DC: CAETS, 1995.

[33] Atkisson, Alan. "Green Engineering and National Security: The US Army Corps of Engineers looks to the future and embraces the concept of sustainable development, an interview with Lt. General Henry J. Hatch." *In Context* 32 (1992): 40.

which was, in fact, part and parcel of the professional discourse of sustainability in the late 1990s.[34]

Hence when the late Roy Weston, pioneering environmental consulting engineer and founder of US environmental company Roy F. Weston Inc., first articulated an engineering vision of sustainable development as "the economic model of the future" in 1994 he also assumed that sustainability implied unbounded prosperity.[35] And engineering elites projected this view onto the developing world casting growth as an international management responsibility: "We simply must address the needs of the developing world, or sustainability will be impossible. The market is the best way to do this," remarked the industrial engineer and NAE member Chad Holliday, DuPont's then CEO, in 1999.[36]

On a broader scale, the technological change axis maintains that the reorientation of technology is autonomous.[37] Parallel to the assumption that technology is neither good, bad, nor neutral, runs the idea that exploitation of natural resources should neither be decreased nor increased, but ought to be effectively managed. Thus the technological change advocate's argument for sustainability is: although development "inevitably cause[s] some harm to the environment," halting or diminishing it is "clearly unrealistic."[38] Continued development is needed, argued the Council of Academies of Engineering and Technological Sciences (CAETS) in the mid-1990s, to mitigate the impacts of past engineering projects, and economic growth is inevitable to protect natural resources.[39]

The dominating feature of the ideology of technological change, this section has showed, is that sustainability engineering essentially combines

---

[34] Charles, Michael. "Sustainable growth: administration pursues 'livability agenda.'" *Civil Engineering* 69, no. 3 (1999): 10. "Reducing wealth (or rather consumption) appeals to some as a way to encourage sustainability," wrote Deanna Richards, then director of NAE's program on Technology and Sustainable Development, "but it is an unlikely outcome." "It may even prove to be foolhardy." Richards, Deanna J. "Harnessing Ingenuity for Sustainable Outcomes." *The Bridge* 29, no. 1 (1999): 16–22.

[35] Weston, Roy F. "Sustainable Development: The Economic Model of the Future." Paper presented at the New Mexico Conference on the Environment, Albuquerque, NM, United States, April 25, 1994.

[36] Reisch, Mark S. "Striving for Sustainability: Chemical industry leaders wrestle with sustainable development, Responsible Care." *Chemical and Engineering News* 79, no. 36 (2001): 17–22.

[37] See, for example, Thom, David. "Engineering Education and the New Industrial Revolution." *International Journal of Engineering Education* 14, no. 2 (1998): 89-94.

[38] Coates, Geoffrey H. "Facilitating Sustainable Development: Role of Engineer." *Journal of Professional Issues in Engineering Education and Practice* 119, no. 3 (1993): 225–229.

[39] CAETS (1995).

business as usual and environmental stewardship.[40] A different picture emerges, however, when a closer investigation is made of the qualities of engineering expertise and the end results of innovation. In the rest of this section I demonstrate how the sustainability ideology of technological change emphasized certain qualities to the exclusion of others and how advocates of technopolitics warned that the unchallenged adoption of economic growth has led the engineer to serve a contestable set of social and political values.

Nowhere is the interlocking relationship between sociopolitical change and sustainable technology manifested more than in the writings of technopolitics proponents. The ideology of technopolitics assumes that sustainability decision making is a normative model for making engineering choices regarding social and political futures. In addition, technopolitics theorists argue that the emphasis on efficiency—spelling out essentially political priorities on the basis of metrics of input and output—masks profound questions about the compatibility of human life, institutional structures, and nature. This orientation is evident in the work of UK-born chemical engineer John Peet, who has been active with ESR for almost three decades. In 2002, Peet thought that "most applications of science and technology are not governed by a philosophy of science, but by political economic philosophies." Consequently, he argued, "sustainability is at base the reflection of a social and moral problem that can only be solved by addressing the dominant values of society, especially materialism and growth."[41]

Robert Hudspith, a professor of mechanical engineering at MacMaster University, has furnished a clear expression of the precepts of a technopolitics ideology of sustainability:

> There are several barriers that hinder an understanding of the role of technology in sustainability. The first barrier concerns the concept of neutrality...For example, the automobile is considered to be inconsistent with sustainability only if it is used without adequate pollution controls or is used inefficiently...It is pointed out that *technologies have characteristics that reflect certain underlying assumptions about what is valuable in life*. However, the barrier to working this through is the inability to be specific about these characteristics; vague

[40] This feature was described by an engineer as "[l]ike the Wall Street Journal meets Mother Earth News." Tracey, Dennis. "Sustainability and Environment: Art vs. Science." *EFS Newsletter*, December 2001.

[41] Peet, John. "Chemical Engineering & Sustainability: Is Green Processing Enough?" Paper presented at the APCChE (Asia-Pacific Conference on Chemical Engineering) Conference, Christchurch, NZ, September 30 - October 3, 2002, Proceedings paper #235.

> generalizations do not get translated into new design criteria. For example,...*little is done to show how this tendency gets built-in to the technology.* A third barrier concerns our *tendency to evaluate technologies as isolated entities* without acknowledging overall trends or how technologies are linked in systems. The effect of a technology often depends on the degree to which it has become systematized.[42]

These themes have been repeated by other technopolitics ideologues. The Canadian engineer academic Helmut Burkhardt, John Peet, and Sharon Beder—the prolific Australian engineer-writer, educator and ESR member—all subordinated technological and economic means to cultural and political ends. As Peet put it in 2000, "technology and economics can and must contribute to its [sustainability's] resolution, but are unlikely to assist in its identification."[43] For "neither Science nor Economics can tell us what should be. That is the key issue of sustainability," he declared two years later.[44] Technopolitics ideologues advanced the idea that sustainability begged a more radical reexamination of engineering cultural worldviews. In 1991, while directing a firm called the Altruistic Engineering Consultancy, the female engineer Chantal Toporow cautioned that sustainable development had become coercion: engineering apolitical mindsets had led to the growth and diffusion of a monolithic culture. [45] A year earlier she had been the lead author along fellow members of the Los Angeles Institute of Electrical and Electronic Engineers (IEEE) of a volume entitled "Delicate balance: technics, culture and consequences."[46]

In many instances, technopolitics advocates of sustainability engineering reconsidered the very logic of competitive productivism: i.e., the assumption that there is an inherent economic determinism embodied in sustainable technology. "[T]he extent to which economic democracy has

---

[42] Hudspith, Robert. "A teaching tool that exposes the non-neutrality of technology as it relates to sustainable development." In *Preparing for a sustainable society*, edited by Helmut Burkhardt and H. Willem Vanderburg, 294-301. Piscataway: Institute of Electrical and Electronic Engineers, 1991.

[43] Peet, John. "Being Fully Human and Creating a Better Future: Sustainable Development from an Integrated Systems Perspective." Paper Presented at the Workshop on Sustainable Development, Sigtuna Foundation, Sweden, 8-9 June 2000.

[44] Peet (2002).

[45] Toporow, Chantal C.M. "Values-led Technologies." In *Preparing for a sustainable society*, edited by Helmut Burkhardt and H. Willem Vanderburg, 228-235. Piscataway: Institute of Electrical and Electronic Engineers, 1991.

[46] Toporow, Chantal, McCagie Rogers, Nik Warren, and Justin Biddle. "A Delicate Balance: Technics, Culture, and Consequences." California State University, Los Angeles, October 20-21, 1989. Torrance, California: Los Angeles Chapter IEEE SSIT-30, 1990.

been weakened," noted Helmut Burkhardt and Willem Vanderburg in *Technology and Society*, is directly related to corporate decision making guided by international competition. In other words, "development begins unquestioningly out of the fear that if one corporation does not develop the technology, their competitors will."[47] Through their writings, engineering sustainability advocates of an ideology of technopolitics presented a societal vision that made direct links between the dominance of the modern economic system and the proliferation of engineering identities based on growth driven by technical expertise: they recognized, for example, the role engineering education has played in mystifying sustainability by normalizing environmental and social decline as unintended consequences of industrialization and engineering progress.[48]

These ideas were intended to provide support for the fundamental assumption of a technopolitics view of sustainability engineering. The "real issue," declared Richard Devon of AESR, "is exploring options in the social relations of expertise, not just exploring the moral dilemmas of individuals."[49] In other words, technopolitics ideologues linked environmental protection to broader questions concerning the politics of technology and engineering identity formation, especially the need for a participatory, deliberative design of engineered systems. In their arguments, sufficiency-based definitions of sustainability mingled with discussion of engineering tools, methodologies and structures that would corroborate participatory modes of technological governance. Much technopolitics work has explored "expert and stakeholder inputs into technology choice decisions."[50] By conscious, self-reflective engineering effort, these proponents believed, professionals could accomplish the "both possible and desirable goal" of integrating quantitative and qualitative approaches in sustainability engineering decisions.[51] In accomplishing this task, Peet postulated that engineers are challenged by the fact that notions of sustainability expertise are continuously expanded beyond engineering professionalism.[52] His view was in stark

---

[47] Burkhardt and Vanderburg (1991). See also Sharon Beder. "Engineers, Ethics and Sustainable Development." Paper presented at the 10th International Congress of Logic, Methodology and Philosophy of Science, Florence, Italy, 1995.

[48] Bazan-Arias, Cathy. "Readers respond: Ethics and sustainable development." *CENews* January 6, 2009.

[49] Devon, Richard. "Towards a Social Ethics of Engineering: The Norms of Engagement." *Journal of Engineering Education* 88, 1 (1999): 87-92.

[50] Herkert, Joseph R., Alex Farrell, and J. James Winebrake. "Technology Choice for Sustainable Development." *IEEE Technology and Society* 15, no. 2 (1996): 12–20.

[51] Ibid.

[52] Peet (2000): 8.

contrast to some technological change ideologues' perspective that "so much [development] needs to be done that some countries cannot afford the luxury of democracy and public debate."[53]

Conclusion

In this chapter I defined analytically the politics of sustainability engineering in terms of a dialectic relationship between two ideologies: technological change and technopolitics. I grounded the rational bases of the engineering ideologies of sustainability in a set of premises and core assumptions. I then anchored historically the development of a sustainability discourse in American engineering in the metaphor of the "invisible engineer" to introduce the identity politics on which that discourse depended. This chapter's analysis of rational bases of sustainability engineering situated the ideology of technological change—the dominant engineering ideology of sustainability—in the capacity of engineering professionals and their organizations to recast engineering progress in the 1990s. Sustainability as effective management of natural resources was thus located in the notion of technology's unintended consequences.

Chapter 1 argued that to the extent that technological change and technopolitics suggest the foundation of the identity of the "sustainable engineer," the various visions of sustainable technology become contingent on the dialectic between these two ideologies—thus setting limits on the engineer's capacity to challenge the boundaries (ideological, cultural, professional, and methodological) of her discipline. The next chapter goes deeper into the history and the evolving politics of sustainability engineering beginning not with the conflicting aspects of technological change and technopolitics, but with their common departures.

---

[53] Cottell, Michael N.T. "Facilitating Sustainable Development: Is Our Approach Correct?" *Journal of Professional Issues in Engineering Education and Practice* 119, no. 3 (1993): 220–224.

2

A Critical History of Sustainability Engineering

Common Departures

"The central theme of our age is interdependence," Maurice Strong declared in 1972 when he was appointed the first executive director of the United Nations Environment Program (UNEP) as well as secretary-general of the UN Conference on the Human Environment, which took place in Stockholm in June 1972.[1] The Stockholm conference expressed a particular worldview that had taken shape amongst UN officials. It asserted that "there can be no fundamental conflict between environment and development."[2] This stance evolved through a series of UN events, such

[1] Strong Maurice, "Introduction." In Rowland, Wade. The Plot to Save the World. Toronto: Clarke, Irwin & Company Limited, 1972, x.

[2] Ibid., x. In 1969 secretary-general of the UN, U Thant, was advocating that a "global partnership" was needed to "improve the human environment, to defuse the population explosion, and to supply the required momentum to development efforts." Quoted in Meadows Donella H., L. Dennis Meadows, Jørgen, Randers, and W. William Behrens

as the UN General Assembly adopting the World Charter for Nature in 1982, where it recognized that "due account shall be taken of the long-term capacity of natural systems...; [that]... impact[s] on nature shall be controlled, and [that] the best technologies that minimize significant risks to nature or other adverse effects shall be used."[3] These events culminated in the 1987 report put out by the World Commission on Environment and Development (WCED). In this report, sustainable development was defined as "meet[ing] the needs of the present without compromising the ability of future generations to meet their own needs."[4]

Three months before the UN's Economic and Social Council recommended that its General Assembly convene a conference on the problems of the human environment, an interdisciplinary group of thirty industrialists, economists and academics were summoned independently in the Academia dei Lincei in Rome. The "Club of Rome"—as the group's initiators decided to call themselves—was organized by the Italian economist Aurelio Peccei, at the time manager of Italconsult, an engineering consulting firm active in developing countries.[5] Dedicated to shedding light on "the global system in which we all live," the Club of Rome operated on the

---

III. *The Limits to Growth: A Report for the Club of Rome's Project on the Predicament of Mankind*. New York: Universe Books, 1972, 17.

[3] UN General Assembly, *Draft World Charter for Nature*, October 30, 1980, A/RES/35/7. <http://www.refworld.org/docid/3b00f1a938.html>
Accessed November 12, 2013.
Two years earlier, in 1980, the International Union for Conservation of Nature and Natural Resources (IUCN) published the World Conservation Strategy in collaboration with the UNEP and the (then) World Wildlife Fund (WWF). According to the Strategy, "Two features characterize our time. The first is the almost limitless capacity of human beings for building and creation, matched by equally great powers of destruction and annihilation...The second is the global interrelatedness of actions, with its corollary of global responsibility. This is turn gives rise to the need for global strategies both for development and for conservation of nature and natural resources." The Strategy also includes a definition of sustainable development that engineers and non-engineers have more recently referred to as the "triple bottom line": "For development to be sustainable it must take account of social and ecological factors, as well as economic ones..."
<https://portals.iucn.org/library/efiles/edocs/WCS-004.pdf>
Accessed November 14, 2013.

[4] WCED (1987): 8.

[5] The meeting's roots are traced in a 1965 speech Peccei gave promoting industrialization in Latin America, which caught the attention of Alexander Kind, at the time Director General for Scientific Affairs for the Organization for Economic Co-operation and Development (OECD). Peccei's talk was given during the first meeting of the Atlantic Community Development Group for Latin America (ADELA), an international pro-industrialization institute focusing on Latin America.

premise that "the major problems facing mankind are of such complexity and are so interrelated that traditional institutions and policies are no longer able to cope with."[6]

Two years after the first Rome gathering, an academic from MIT named Carroll Wilson, who was a founding member of the Club of Rome, suggested to Peccei that a colleague's methodology of "system dynamics" may be appropriate to address the "world problematique."[7] Peccei agreed, and Wilson's fellow professor Jay W. Forrester of the Alfred P. Sloan School of Management flew to Switzerland to attend a June 1970 meeting organized by the Club in Bern, where he first presented his ideas to the group. The prospect of applying his computer model-based method to the Club's "Project on the Predicament of Mankind" excited Forrester, who started working on the first equations of a world system model on his flight back to the US. Though he did publish that work as *World Dynamics* in 1971, Forrester himself did not undertake the investigation set forth by Peccei and his colleagues. This assignment was passed on to his closest disciple, Dennis L. Meadows. Under the auspices of a $250,000 fund from the Volkswagen Foundation, Meadows pushed the accelerator of sustainability engineering.[8] His "MIT Project team" conducted the research published just a few weeks before the Stockholm conference under the title *Limits to Growth* (1972), a book popularly recognized as launching the sustainable development movement.[9]

Meadows *et al.* explained that system dynamics begins with the "recognition that the structure of any system—the many circular, interlocking...relationships among its components—is often just as important

[6] Meadows *et al.* (1972): 9–10.

[7] Wilson was the first general manager of the Atomic Energy Commission.
Cook, Joan. "Carroll L. Wilson, Science and Energy Expert." *New York Times*, January 13, 1983.
<http://www.nytimes.com/1983/01/13/obituaries/carroll-l-wilson-science-and-energy-expert.html>
Accessed November 25, 2013.

[8] Simmons, Harvey. "Systems Dynamics and Technocracy." In *Models of Doom: A Critique of the Limits to Growth*, edited by H.S. D. Cole, Christopher Freeman, Marie Jahoda, and K.L.R. Pavitt, 192–208. New York: Universe Books, 1973.

[9] Parenti, Christian. "'The Limits to Growth': A Book that Launched a Movement." www.thenation.com, December 5, 2012.
<http://www.thenation.com/article/171610/limits-growth-book-launched-movement#>
Accessed December 6, 2014.

in determining its behavior as the individual components themselves."[10] Furthermore, the book's authors set themselves the task of understanding the physical conditions—"let us suspend the requirement of political feasibility"—under which the interaction of five "accelerating trends" (world population, industrialization, pollution, food production, and resource depletion) could be placed under control so that a "self-imposed limitation to growth" or a "state of global equilibrium" could be attained.[11]

In the decades following the publication of *Limits to Growth*, a range of scholars expanded on systems dynamics to form a new paradigm for thinking economically about the environment. This paradigm substantially influenced engineers' social theories of sustainability. In particular, along with the idea of limits to growth, Kenneth Boulding's metaphor of the planet Earth as a spaceship system closed in terms of materials yet manageable through the relationships between its elements, struck home with engineers.[12] In 1971, Herman Daly added, the stationary state would depend less on natural capital than upon moral capital.[13] Also in 1971, the Romanian economist (and Daly's mentor) Nicholas Georgescu-Roegen, published his manifesto on the entropic nature of the economy, in which he postulated that any attempt to describe reality in mathematical terms—a phenomenon he called "arithmomorphism"—is unable to capture the qualitative change that Georgescu considered an inherent element of economic processes.[14] By the end of the 1970s, the Russian-born Belgian, Ilya Prigogine, had won the Nobel Prize for mathematically describing "non-equilibrium states," and the self-taught economist Hazel Henderson was proclaiming the "End of Economics."[15] Both advocated an evolutionary

[10] Meadows *et al.* (1972): 21 and 31. The authors dedicated the book to Peccei, "whose profound concern for humanity has inspired us and many others to think about the world's long term problems."

[11] Ibid. 159 and 156; 168–169.

[12] Boulding, Kenneth E. "The Economics of the Coming Spaceship Earth." In *Environmental Quality in a Growing Economy*, edited by Henry Jarrett, 3–14. Baltimore, MD: Resources for the Future/Johns Hopkins University Press, 1966. In his *The Meaning of the 20th Century* (New York, Harper and Row, 1964) Boulding has termed the open economy of the past as the "cowboy economy," as opposed to the "spaceman economy" that describes the closed economy of the future.

[13] Daly, Herman E. "Toward a Stationary-State Economy." In *The Patient Earth*, edited by John Harte and Robert Socolow, 236–237. New York: Holt, Rinehart, and Winston, 1971.

[14] Georgescu-Roegen, Nicholas. *The Entropy Law and the Economic Process*, Harvard, 1971.

[15] Nicolis, Gregoire, and Ilya Prigogine. *Self-Organization in Non-equilibrium Systems*. New York: John Wiley & Sons, 1977; Henderson, Hazel. *Creating Alternative Futures: The End of Economics*. New York: The Berkeley Windhover, 1978.

nature of economic development. Such systemic and evolutionary theories excited the imagination of engineers: "many believe [Georgescu and Prigogine's] work will establish more *realistic* underpinnings to both technological and economic policy during the coming era of scarcity," reported *Chemical and Engineering News* in 1979.[16]

That sustainability engineering intellectuals have assumed certain common epistemological assumptions is illustrated by the following example: "To me as an engineer," maintained Don Roberts of CH2M Hill "a sustainable system is one that is either in equilibrium, operating at a steady state, or a system that changes at a rate considered to be acceptable."[17] For Roberts—whose contribution in "bringing sustainability into the lexicon of the engineering profession" was so notable that the ASCE awarded him the society's highest accolade—the model for sustainability is exemplified by "natural ecosystems" operating in non-linear "closed loops."[18] Yet a technopolitics thinker, John Peet, shared the idea that the sustainability of natural systems depends on a "web of communication and control" mechanisms that contribute to maintain their balance.[19]

---

[16] Lepkowski, Will. "The social thermodynamics of Ilya Prigogine." *Chemical and Engineering News*, April 16 (1979): 30–33, emphasis added. Another engineer, at the time professor of industrial and systems engineering at Ohio State University, wrote: "Some contemporary economists and others, who are still not heard very loudly by their economics colleagues, are attempting to call our attention to the entropic nature of our economic processes…The economic process is entropic in that we extract high quality (low entropy) nonrenewable resources from nature and diffuse them in extraction, production, consumption, and disposal into irretrievable waste products (high entropy) in the air, water, and land. This process cannot go on forever." Overby, Chuck M. "Product Design for a Sustainable Future: A Matter of Ethics?" Paper presented at the Annual Conference of the American Society for Engineering Education, University of Massachusetts, MA, 1980.

[17] Don Roberts quoted in Prendergast, John. "Engineering Sustainable Development." *Civil Engineering-ASCE* 63, no. 10 (1993): 39–42. Roberts' interpretation became the standard definition of sustainability for engineers and was propagated throughout the programmatic texts of sustainability engineering of the era studied in this chapter.

[18] Unknown. "ASCE Names 10 Distinguished Members for 2009. *Engineering News Record*, ENR.com, July 23, 2009.
<http://enr.construction.com/people/awards/2009/0723-ASCENamesMembers.asp>
Accessed January 10, 2014.
See also, Roberts, Don V. "Sustainable Development-A Challenge for the Engineering Profession." Paper presented at the FIDIC Annual Conference, Oslo, Norway, June 18, 1990.

[19] Peet, John. *Energy and the Ecological Economics of Sustainability*. Washington, DC: Island Press, 1992, 77.

Whatever means were ultimately suggested by proponents of either technological change or technopolitics, both ideologies of sustainability presumed a fundamental role for "systems thinking" as an initiating conceptual mechanism; they also presumed that natural, economic, and social realities can be better penetrated through interlocking relationships, implying that engineers can generate understanding by the theories and models they produced. Recovering the historic continuity between ideas of systemic interdependence and growing ecological concerns in the 1970s is important because in succeeding decades such common departures facilitated the convergence of engineering ideologies of sustainability.

Sustainability's Technological Change Roots

The parallels between the principal framework for conceiving and operationalizing sustainable development provided by the MIT system dynamics group, and the technological change ideology of sustainability, are not perfect. In addition, a careful reading of *Limits to Growth* suggests that discursive expressions of both engineering ideologies of sustainability have co-existed and have in fact nourished each other since at least the early 1970s.[20]

Nevertheless, a closer look at the compatibility between system dynamics and the technological change view of sustainability is illuminating. First, beginning in the early 1980s, the assumption about the existence of resource scarcity was relied upon as an unquestionable maxim by the US and international engineering communities. Second is the underlying suspicion that humanism is anti-technology. And third is the recognition that engineers must engage proactively in the political realm.[21]

To ensure the continuation of growth in a limited world, technological change ideologues of sustainability have relied on the effective and efficient use of technology. Management of the Earth's resources was thought

---

[20] "Every new technology has side effects, of course, and one of the main purposes of model building is to anticipate those effects…Unfortunately the model does not indicate, at this stage, the *social* side-effects of new technologies" (emphasis in the original). This implies that "social side effects must be anticipated and forestalled *before* the large-scale introduction of a new technology" (emphasis in the original). But also, "[w]e have felt it necessary to dwell so long on an analysis of technology here because we have found that technological optimism is the most common and most dangerous reaction to our findings from the world model. Technology can relieve the symptoms of a problem without affecting the underlying causes." Meadows *et al.* (1972): 146, 148 and 154.

[21] See, for example, FIDIC. *Consulting engineers and the environment: guide for actions.* Lausanne, Switzerland: FIDIC, 1994.

by key sustainability engineering advocates like environmental consultant Roy Weston to be dependent on the mechanisms of capitalism and the free market. Like the technological change ideologues, however, proponents of a socially, politically and culturally informed vision of sustainability in an engineering context invariably articulated their theories around the same notion: driven beyond its "limits to tolerance"—"the world system is, and will likely remain, highly unstable and uncontrollable."[22]

The debate over the assault on nature and the widespread criticism against environmentally damaging technology increased the professional anxiety and defensiveness of engineers. Practitioners and leaders of the profession devoted attention to greening and humanizing engineering education. Yet they simultaneously argued that "[i]f engineers must be humanized, why should not liberal arts students be exposed to technology?" The rationale was that non-engineering students shall become more appreciative of technology's fine powers to mold modern life. Thus, technological literacy was perceived as a medium to communicate not only the assets of engineering progress, but the "dangers of environmental overkill" as well.[23]

And if, as the political scientist Harvey Simmons convincingly indicates, Forrester was certain "of the desirability of educating more and more people to accept the system dynamics approach," so too were leading proponents of sustainability engineering in the 1990s repeatedly called upon by their fellow professionals to become policy leaders in educating non-engineers about the need to aggressively apply sustainable development.[24] The standpoint that urged practitioners to be "policy leaders" in educating "the public" on the engineering implications of sustainability tended to frame

---

[22] Toporow (1991); Duncan, Richard C. "The Evolution of Social Control: Is a World Society Governable?" In *Preparing for a sustainable society*, edited by Helmut Burkhardt and H. Willem Vanderburg, 256–263. Piscataway: Institute of Electrical and Electronic Engineers, 1991.

[23] Chadderton, Ronald A. "Should Engineers Counteract Environmental Extremism?" *Journal of Professional Issues in Engineering Education and Practice* 121, no. 2 (1995): 79–84.

[24] Such calls appear as early as 1991: WFEO. *Arusha Declaration*—The UNCED Conference 1992 Statement by the World Federation of Engineering Organizations on Environment and Development, 1991. In the beginning of the engineering sustainability movement, it was very common for practitioners to maintain that an engineering professional must "…learn to converse more effectively with those around [him/her]." Rajagopalan, Visvanathan. "The Engineer's role in sustainable development." *Civil Engineering* 62, no. 8 (1992): 6. See also, Thompson, Frederick A. "Sustainable Development: 'Depoliticking' the Environment and Making It a Matter of Natural Economics." WESTON WAY April/May 1995.

the engineer's development role in rather polarizing terms (i.e., engineer versus community, or company versus liability claim). "If we have to persuade by public debate that the positive value [of sustainable technologies] outweigh the negative we need to do so perceptively, not just from facts," argued a leading UK civil engineer. As a consequence, "the value of female engineers" must be recognized for "they are usually better than men at gauging the feelings of the public audience and reacting sympathetically."[25]

Whether technopolitics advocates of sustainability had the same adversarial view about the role of practitioners in development—or about essentializing women engineers—is doubtful. At least one subtle critic of sustainability engineering, the educator and engineering ethicist Joseph Herkert, opposed sustainable development if the concept was "merely a vehicle for promoting the views of the engineering community in the public policy arena." While at best an incomplete conceptualization of sustainability, the considerable neglect of issues of justice, he wrote, could "at worst...be perceived as self-serving, even technocratic visions of how public policy issues should be managed."[26]

Familiar Ideological Bases for Engineering Theories of Sustainability

By drawing parallels between system dynamics and visions of technological change the common ideological bases for engineering theories of sustainability can be demonstrated and appreciated. The links between the Club of Rome and the UNEP, the previous section argued, gave rise to a conception of the world as an interdependent network system facing global challenges due to a number of common constraints for which broad-scale—primarily technological—solutions must be developed.[27] This was often translated by UN bureaucrats, World Bank officials and elite engineers alike into a vision

[25] Cottell (1993). Cottell was then president of the Institute of Civil Engineers (ICE), UK.
[26] Herkert, Joseph R. "Sustainable development and engineering: Ethical and public policy implications." Paper presented at the 1997 International Symposium on Technology and Society, Glasgow, UK, 20–21, June 1997, 175–180.
[27] The WCED—commonly referred to as the Brundtland report in honor of the Commission's chairman, and Norway's prime minister at the time, Gro Harlem Brundtland—(1987:8) reads: "[t]he concept of sustainable development does imply limits—not absolute limits but limitations imposed by the present state of technology...", while at the same time "... technology and social organization can be both managed and improved to make way for a new era of economic growth."

for creating worldwide and multidisciplinary partnerships by tapping the spirit of innovation to engineer and improve systemic efficiency.[28]

In this section I discuss how extra-engineering concepts such as "wholeness" became part of the sustainable engineer's definition of reality—a normative prerequisite for sustainability engineering intellectuals. This is evident in the writings of Roy Weston who, along with Roberts and Hatch, was at the forefront of sustainability engineering for more than fifteen years: "My challenge is to strive to be holistic and as objective and scientifically sound as I can in presenting what I believe is reality."[29]

A standard cognitive feature of engineering contextualizations of sustainability derives, then, from ecosystems theory and acknowledges the physics, chemistry, biology and ecology of interconnected "subsystems" such as population, natural resources, and technological development. Engineers' efforts to cast sustainability in terms legible to fellow professionals were hampered by conflicting meanings of ecosystem complexity—as a result, engineers' theories were rife with normative speculations about nature and society. Practitioners' theories of sustainability became part and parcel of the dominant professional discourse, using terms like "holistic," "organism," "mutuality," "metabolism," "integration," or "equilibrium"—traditionally non-engineering terms. Semi-abstract concepts from the systems-orientation of sustainability engineering, such as consumption and production, were treated as energetic flows. Thus such theories, undergirded by the *mechanical* metaphor of self-regulatory "feedback loops," incorporated both extra-engineering and rational elements.

As criticism increased against the negative environmental consequences of engineering, the same engineers who had articulated theories of sustainability that stressed technological innovation began referring to the interdependence of society, economy and nature in an ecological sense. This hybrid synthesis further blurred the boundaries between technological change and technopolitics. For it was open to interpretation whether the interrelationship among the parts of a complex evolving system required

[28] See, for example, the statement of the International Council of Academies of Engineering and Technological Sciences (CAETS): "For some environmental issues—for example, those involving oceans and atmosphere—the planet must be regarded as a single system best served by coordinated international action." In WFEO (1997): 59.
[29] Weston, Roy F. "Sustainable Development: To Better Understand the Concept." WESTON WAY April/May 1995.

management through technology, or whether they required a different approach based on the values of equity and reciprocity.[30]

Concepts such as "holism" or "feedback loops" stretched the engineer's confines of discursive style and creative imagination. For example, a practitioner from Canada wrote about "holistic engineering" that "it also relates to healing." The idea "implies treating the works of man and the natural environment in which they are embedded as a single system..., so that the whole remains healthy." Holistic engineering, he postulated, is based on the principle that "[a]ll life is sacred and should be treated with reverence."[31] In the same spirit, the Canadian engineering educator Willem Vanderburg proposed in 1991 "a new kind of technological development...guided by negative feedback about its contextual implications."[32] Technological change and technopolitics were built into the engineer's discursive experimentation with the vocabulary of systems theory: in blaming "technological euphoria" for downplaying environmental protection, an engineer could exhibit not only his adherence to a "holistic approach to economic development," but his faith to engineering management as well.[33]

For engineers, sustainability filled a cognitive gap that allowed them to cope with the bewildering systemic complexity they had previously overlooked. Technological change and technopolitics engineer-theorists of sustainability both drew upon the heritage of ecology and ecosystems theory when investigating the interrelationships between subsystems in their wholeness as well as in their particularity. Both perspectives linked their attempts to transform the world and the engineering profession with the finite availability of natural resources. Regardless of ideological proclivity, sustainability engineer-theorists regarded the "system" as representation *and* validation of reality. Similarly to ecosystem scholars, the MIT engineers who articulated the cognitive framework now fully incorporated in

---

[30] In the most programmatic publication on sustainability engineering of the American engineering community in the 1990s—AAES (1994): 3—"[e]nvironment, technology and economic development [were] seen as interdependent concepts in which industrial competitiveness and ecological sustainability [would] be addressed together as a common goal."

[31] Khanna, Jiti. "Holistic Engineering: Beyond the Code of Ethics." *Engineering-ASCE* 63, no. 10 (1993): 6.

[32] Vanderburg, Willem H. "Sustainable Development and the Practice of Engineering." In *Preparing for a sustainable society*, edited by Helmut Burkhardt and H. Willem Vanderburg, 9–16. Piscataway: Institute of Electrical and Electronic Engineers, 1991.

[33] Haimes, Yacob Y. "Sustainable Development: A Holistic Approach to Natural Resource Management." *IEEE Transactions on Systems, Man, and Cybernetics* 22, no. 3 (1992): 413–417.

the development policies of the UN, favored a "systemic," "holistic" and "self-organizing" approach to bringing "order out of chaos," explicitly claiming a non-dominating, non "Baconian view of the world as a resource to be exploited."[34]

Engineers' philosophical aptitude is thus of more relevance for the nature of engineering work they produce than it is commonly admitted. Such aptitude not only influences the magnitude of conceptual resources they can draw upon, but also their comfort levels in challenging the ideological and methodological boundaries of their discipline.

"Resource scarcity, environmental constraints, and new ethical ideas are combining to challenge our [engineering] foundation values," wrote mechanical engineering professor Charles Overby of Ohio University. Further, his overview of the main philosophical ideas of John Locke, Adam Smith, and Jeremy Bentham indicates that the "quantification of morality with a 'calculus of pleasure and pain' readily lends itself to a society in which economic values reign supreme." This in turn necessitates the pursuit of new engineering design metrics and criteria, according to Overby. Ten years ahead of its time, his presentation at the 1980 American Society for Engineering Education (ASEE) Annual Conference defined "product design for sustainability." "We know today that there may not be enough and as good resources left in common for others, therefore one response that we might make, from an engineering perspective, is to more creatively address ourselves to the design of products that are less consumptive of resources—product design for a sustainable future." With enough discrimination, he maintained, engineers could thus view societal and regulatory calls for environmental compliance—not as "horrible example[s] of excessive governmental involvement," but as opportunities for the professional to draw connections "with more of the *life-cycle of the product* and the residuals from its production."[35] Likewise, sustaining all three realms of society, nature and economy, in the words of Hank Hatch, implied that engineers become "aware of the life-cycle effects of [their] actions."[36]

Systems analysis, this section demonstrated, became very popular in the early 1970s—this is what the Club of Rome and the authors of *Limits to Growth* were pioneering. Having lapsed briefly in mainstream engineering

---

[34] Evans, Roger J. "Commentary on the Code of Ethics." *Journal of Professional Issues in Engineering* 114, no. 2 (1988): 148–156, 154.

[35] Overby (1980): 8. "Paradoxically," Overby (1980): 3 noted, "it is the very fruitfulness of science and technology in our domination of nature that is helping to bring about the challenge."

[36] Hatch (1989).

discourse, system analysis was revived by ideologically unlike engineers like Overby, Roberts and Hatch as a response to the sustainability crisis. In this way, systems analysis becomes a new cognitive mode for representing the world. It both allows for greater holism and continuity in engineering methods. Clearly, the concept of "systems" is essential to LCA, the idea of assessing impacts throughout a product or project's life cycle—without having a system, LCA is not conceptually possible. The notion of the system in an engineering context—semi-non-analytical and semi-mechanical—facilitated the conceptual link between sustainability and assessment of the "aggregate long term consequences of [technological] decisions, in terms both of time and place."[37]

As such, "the system" became the chief conceptual platform upon which technological change and technopolitics ideologues of sustainability found common ground, and could debate the conflicting nuances of the definition of sustainable development and the motives of technological practice. More than other engineers, sustainability ideologue John Peet stressed the application of the system metaphor as a discursive view for understanding engineering itself: "I am sure we [engineers]... accept...that many of the words we use to describe this complex system are not analytically definable...We can, however, clarify their meaning [in our profession] through discourse—a dialectical process."[38] Peet and other engineers who developed social theories of sustainability view the links between "sustainability" and "engineering" as not only technological, ecological, social, economic or political in nature, but imaginative and cultural as well.

The compelling internal logic of the systemic perspective inspired by natural and social ecologies had—and still has—a liberating effect on both the engineer and layperson's intellectual mood: whatever insight or consequence each realm (nature, society, the economy) had failed to address individually could be recaptured through the dynamic interdependence of an integrated whole, divided into interacting subsystems. This does not contradict the commonly stated assumption that sustainability's widespread support is due to "the fact that different groups can read competing meanings" into it.[39] Rather, I suggest that in an engineering context,

---

[37] AAES (1993). [The document—AAES's Policy on Sustainable Development—includes the "Action Principles for the Engineering Profession," and was republished in *The Role of Engineering in Sustainable Development*, edited by Monica D. Ellis, 3–6. Washington, DC: AAES, 1994.

[38] Peet (2000).

[39] Fischer, Frank, and Michael Black, eds. *Greening Environmental Policy: The Politics of A Sustainable Future*, St. Martin's Press, New York, 1995, xiv.

"sustainability" is multi-ideological and has cognitive connotations that resonate equally with ideologies of technological change and technopolitics.

Building on the work of ecologists and economists, this section argued, advocates of sustainability engineering adopted a fuzzier way of thinking through "systems." Such engineers were not social theorists or ecologists by training but—in contrast with their quantitative analytical precision—borrowed loose metaphors that they thought captured the nature of the sustainability fellow engineers should aim for. For example, one question is whether the "global economy" (granting the existence of such a thing, in the first instance) is an "organic" or "cybernetic" entity. But in doing so, they were actively reshaping the epistemic bases of their profession. From this standpoint, they bridged the disciplinary and ideological spaces between philosophy and engineering; they became the amateur philosophers of postmodernity. Yet the dominant vision of sustainability engineering, this section showed, is grounded in its technological change roots. Even if an engineer wants to become more holistic, "he" must be grounded in engineering methods and believe in the power of engineers to achieve technological reform.

Speaking to the Converted: Institutional and Ideological Alliances

In the aftermath of the WCED in 1987, engineering elites pondered over the challenges and opportunities that environmental protection was introducing into their professional agendas. Meanwhile, international development institutions, NGOs, and national agencies further sculpted the institutional and ideological territory of engineering for sustainable development. In summer 1989, UNESCO turned its attention to providing countries of the Global South with sustainable development-related technological expertise at the engineering education level.[40] A leading environmental economist, Robert Repetto of the World Resources Institute (WRI), identified an "urgent need" for \$20-\$50 billion annually to support private sector engineering projects and "technology transfer" in the developing world.[41] This discourse quickly found its way into US government policies. In 1995, the US National Science and Technology Council

[40] UNESCO sponsored the First International Symposium for Engineering Deans and Industry Leaders at the Fawcett Center for Tomorrow at Ohio University.

[41] Repetto, Robert. "Economic Aid and the Environment." *EPA Journal* 16, no. 4 (1990): 20–23. [Repetto (1990) (a)].

(NSTC) defined the "sustainable vision" as a transition away from command-and-control regulatory mechanisms.[42] This was a period of neoliberal political/economic hegemony. Thus, voluntary industry action was favored—including technical innovation. For example, the government's role in supporting sustainability, according to the NSTC and many other engineering and technology constituencies, was to fund private research on market-driven technologies.[43] In 1996, the President's Council on Sustainable Development identified "[i]ncreased US exports or transfers of cost-effective and environmentally sound technologies to developing countries" as one of the "indicators of progress" of the nation's international responsibility.[44]

As American engineers then organized themselves into partnerships, formed "sustainability task forces" and held engineering sustainability conferences, they also found allies in the most strategically important institutional settings. Maurice Strong felt it "refreshing to be speaking to the converted"[45] as he was addressing an AAES and Engineering Foundation (EF) crowd in the summer of 1995. This was small wonder, in light of his attitude that "the best route to environmental sustainability is industrial

---

[42] NSTC. *Bridge to a Sustainable Future: A National Environmental Technology Strategy.* Washington, DC: Executive Office of the President, 1995. The report mentions that "[o] rganizations such as the American Association of Engineering Societies [AAES] and the Civil Engineering Research Foundation [CERF] will continue to play an important role in achieving a consensus within the engineering community on approaches to sustainable development." A few months after the publication of NSTC's document, *ASCE News* reported that the AAES Task Force on sustainable development met with representatives of the Office of Science and Technology Policy (OSTP) to "aid the National Science and Technology Council in its sustainable development efforts." Unknown "ASCE leads world's engineers in advocating sustainable development." *ASCE News* 20, no. 1 (1995): 1 and 6.

[43] Such constituencies included the IEEE. See for example, IEEE (1995). The NSTC report argued that, "[o]rganizations such as the American Association of Engineering Societies and the Civil Engineering Research Foundation will continue to play an important role in achieving a consensus within the engineering community on approaches to attaining sustainable development…" NSTC, 1995.

[44] The White House, The President's Council for Sustainable Development. "Sustainable America: A New Consensus for Prosperity, Opportunity, and A Healthy Environment for the Future." The White House, 1996.
<http://clinton2.nara.gov/PCSD/Publications/>
Accessed November 21, 2013.
David T. Buzzelli, vice president and corporate director, health and safety and public affairs for Dow Chemical, and Jonathan Lash, then president of the World Resources Institute (WRI), were the first co-chairs of PCSD. Hileman, Bette. "Clinton forms sustainable development council." *Chemical and Engineering News* 71, no. 25 (1993): 7–8.

[45] Strong, Maurice (1995).

efficiency" and that "the principal engines of the transition...will be capital and technology."[46]

The early 1990s were a period when American engineering leaders joined their prominent international colleagues in a synchronized attempt to cast sustainability in engineering terms, and marked the birth of the greening of industry in the US and elsewhere.[47] The idea of an interconnected global whole constrained by common limits to growth was assumed by key development institutions such as the UN, the World Bank, the International Monetary Fund (IMF) but also NGOs like the WRI.[48] The UNEP and the World Bank, in particular, expressed their "tremendous support" toward engineering alliances for sustainability.[49] In the US, the Environmental Protection Agency (EPA) applauded the US Corps of Engineers and EPA administrator, William Reilly's good friend Hank Hatch for "developing...[a] new generation of environmental law and the new culture of environmental protection."[50] Reilly and Hatch's "new culture" reflected the evolution of modern direct regulation systems in the US

[46] Strong, Maurice F. "The Road from Rio." *The Bridge* 23, no. 2 (1993): 3–7.

[47] Ember, Lois R. "Environment protection: Global companies set new endeavor." *Chemical and Engineering News* 69, no. 14 (1991): 4. In April 1991, 150 companies, including some major US corporations, adopted the "Business Charter for Sustainable Development," which was based on 16 principles developed by the International Chamber of Commerce (ICC). The Charter was launched during the Second World Industry Conference on Environmental Management (April 10–12, 1991, Rotterdam, The Netherlands).

[48] Addressing an engineering audience in 1991, the US member of the WCED, William P. Ruckelshaus, argued that "if poor nations seek to advance using the same methods pioneered by rich countries, the world ecological system will suffer further damage," quoted in Rubin (1990).

[49] Carroll, William J. "Environmental Challenge for Engineers." *Journal of Professional Issues in Engineering Education and Practice* 121, no. 2 (1995): 126–129, 127.

[50] William Reilly, then administrator of the U.S. EPA, delivering a speech at the U.S. Corps of Engineers Law Day, 1990 in Washington, DC, on May 2, 1990. Reilly, William K. "Generational Change in Environmental Law." *Natural Resources Journal* 31, no. 1 (1991): 1–6. Earlier the same year, the UN Economic Commission for Europe and the US EPA held a workshop on "The Economics of Sustainable Development." UN Economic Commission for Europe & U.S. Environmental Protection Agency Workshop, The Economics of Sustainable Development. Washington, DC: Smithsonian Institution, January 23, 1990. Examples of significant environmental collaboration between the EPA and the US Corps in the 1980s include the 1982 and 1984 agreements between the two agencies stating that "upon EPA request, the Corps will manage design and construction contracts and provide technological assistance to EPA in support of remedial response cleanup of hazardous waste sites," and the January, 1989 "Memorandum of Agreement on Wetlands," signed to initiate co-administration of the 404 provisions of the Clean Water Act.

where most of the burden was shifted to the private sector and where agencies like the EPA began addressing environmental protection by designing and promoting impact assessment methodologies like LCA that industries could voluntarily apply in their operations. This was therefore the time of environmental standards and design for "clean technologies"[51]; it was also the era of engineering professional organization policies and engineering ethical codes for sustainability.

Development Institutions and Corporations Ratifying Sustainability Engineering

The very first engineering conference on sustainability, the annual meeting of the International Federation of Consulting Engineers (FIDIC), held in Oslo, Norway, June 17–21, 1990, was aptly titled "Sustainable Development: A Challenge for the Engineering Profession." It hosted 300 delegates from more than forty countries.[52] FIDIC's conference featured presentations by Don Roberts, who represented the US environmental consulting sector, and Kenneth Piddington, then head of the Environment Department of the World Bank.[53] Indeed, Piddington's presence at FIDIC's annual conference gave substance to the Bank's message of integrating environmental concerns into its development policy, since he announced that in 1990 the Bank had offered $430 million in freestanding loans for environmental

---

EPA. "Memorandum of Agreement Between the Department of the Army and the Environmental Protection Agency Concerning the Determination of the Section 404 Program and the Application of the Exemptions Under Section 404(F) of the Clean Water Act." January 1989.
<http://water.epa.gov/lawsregs/guidance/wetlands/404f.cfm>
Accessed November 20, 2013.
EPA. "U.S. Army Corps of Engineers Superfund Partnership with EPA." Undated.
<http://www.epa.gov/oig/catalog/programs/213.html>
Accessed November 20, 2013.

[51] See for example: Risk Reduction Engineering Laboratory, US EPA and Pollution Prevention Office. *The Environmental Challenge of the 1990s: Proceedings of the International Conference on Pollution Prevention, Clean Technologies and Clean Products, Washington, DC, June 10–13, 1990*. Cincinnati, Ohio: Risk Reduction Engineering Laboratory, 1990.

[52] Rubin (1990).

[53] David Thom, who is mentioned in the introduction of this book, then chairman of the Institution of Professional Engineers, New Zealand (IPENZ), viewed his compatriot's (Piddington) appointment as head of the Environment Department of the World Bank as proof that a "hard headed and powerful organization has acknowledged at last that the twin issues of environment and development are inseparable" (Thom, 1988).

engineering projects around the globe.[54] According to Piddington, engineers "had the opportunity to identify with the urgent assault on global environmental problems." Six years before his Oslo appearance, Piddington had given an invited lecture at the Institution of Professional Engineers in New Zealand (IPENZ), in which he captured the essence of the technological change ideology of sustainability ideology: "The environment and the economy, if properly managed, are mutually reinforcing—and are supportive of and supported by technological innovation."[55]

Soon after the publication of the Bank's 1992 report on development and the environment, which surmised that for the first time in history urban populations would soon overtake rural populations, Roberts commenced building on the Bank's claims to reify his agenda on engineering projects for sustainable development. In 1994 he cited the WCED in arguing that it is "technologically possible" to sustain "roughly twice the present population" via new technologies.[56] And in 1996 he elaborated on the idea that "future megacities" must be reengineered to take advantage of underground space, while advising members of the International Tunneling Association (ITA) to start intervening in debates about sustainable development.[57]

Some environmental NGOs like the WRI authoritatively articulated the rationale most commonly used to support engineers' arguments that "in the absence of radical and unlikely redistribution of resources",[58] more

---

[54] A 1992 World Bank report argued that "[o]n the surface, there appears to be a tradeoff between…the central goal of development—and protecting the environment. This Report will argue that in every realm of economic activity, development can become more sustainable. The key is not to produce less, but to *produce differently*." The World Bank. *WDR 1992: Development and the Environment*. Oxford: World Bank and Oxford University Press, 1992, 25. With regard to the publication of the 1992 Report, Lawrence H. Summers, the bank's chief economist is quoted as saying that "[p]romoting development is the way to protect the environment. Protecting the environment *has a role to play* in promoting development" (emphasis added). Ember, Lois R. "Economic Growth Viewed as Engine for Sustainable Development." *Chemical and Engineering News* 70, no. 20 (1992): 19–20, 19.

[55] Unknown. "The Ethic of Environmental Engineering." *New Zealand Engineering* 39, no. 8 (1984): 19–21.

[56] Roberts (1994).

[57] Roberts, Don V. "Sustainable development and the use of underground space." *Tunneling and Underground Space Technology* 11, no. 4 (1996): 383–390.

[58] WRI's Bob Reppetto quoted by William K. Reilly in Reilly, William K. "Sustainable Development: From Paradox to Paradigm." Keynote Address—UN Economic Commission for Europe & US Environmental Protection Agency Workshop, The Economics of Sustainable Development. Washington, DC: Smithsonian Institution, January 23, 1990.

and better technology was needed to support the forecasts of economic growth. Intersecting elements of a rhetoric of sustainable development, as formulated in WCED documents, and the speeches of Maurice Strong and Hank Hatch, characterize the institutional and ideological convergence that occurred by the mid-1990s—a merger that enabled spokesmen of the engineering profession to claim that sustainability was a technological problem and that engineering knowledge was *the* prime vehicle for global social change.[59]

Only a few months before the FIDIC conference in Oslo discussed the role of the consulting engineer in sustainable development, EarthTech 90 was held on the Mall in Washington D.C., at the foot of the Capitol. This event, co-organized by the WRI, featured a technology fair, an international forum and more than 100 exhibitors. The invitation for the event, directed at companies and co-drafted by Al Gore Jr., emphasized "the twin imperatives of environmental protection and economic development."[60] *Chemical and Engineering News* described the event as proof of business' ecological awareness, an event where co-sponsors Roy F. Weston Inc. and

---

[59] "We have in the past been concerned about the impacts of economic growth upon the environment," argued the WCED. "We are now forced to concern ourselves with the impacts of ecological stress...upon our economic prospects" (1987, 5). "The ability of the global economy to grow," wrote Hatch (1993), "could be limited by changes in our already stressed environment [particularly for the] poor and redeveloping world of Central Europe and the former Soviet Union." Thus it is important that in these countries "the rebuilding job be done on an environmentally sound and sustainable basis as they move to market economies" (Strong, 1993).

[60] EarthTech 90 archival material. The EarthTech '90 events were marked by protests from Greenpeace activists who maintained that "Earthtech '90 is the corporate version of Disneyland. These corporations are trying to sell the American public a kind of fantasy, where the largest polluters turn into the biggest advocates [of sustainability]." EarthTech organizers responded that "...you'd have to be naive not to think [that companies use the fair to advance their public image]...[b]ut that doesn't detract from the benefit of the ideas and products they're displaying."

Mehle Michael. "Greenpeace Upstages Corporations At Earthtech — Environment Fair Produces Protests." *Seattle Times*, Thursday, April 5, 1990. <http://community.seattletimes.nwsource.com/archive/?date=19900405&slug=1064891> Accessed November 19, 2013.

A couple of weeks later, *Chemical and Engineering News* editor Michael Heylin was commenting on the April 22, 1990, Earth Day: "The basic questions are no longer if we should clean up and if we can afford it. The only question today is how we can clean up and, especially how it can be done without crippling economic progress for the large bulk of mankind that still lives in poverty." Heylin, Michael. "Earth Day Revisited." *Chemical and Engineering News* 68, no. 16 (1990): 3.

other companies, including DuPont, "displayed technologies, products and concepts for environmentally sustainable development."[61]

Very powerful corporations and institutions came together to ratify a particular view of engineering. Driven by conferences and initiatives for communicating and escalating "sustainable development" up through the hierarchy of corporate staff and managers, this convergence was also facilitated at this very historical moment by the neoliberal model that permeated and underpinned the entire economic system in the 1990s.

Sustainability and Technological Change in the 1995 US National Security Strategy

To engineering managers of high-profile environmental consulting companies like Roy Weston and Don Roberts, American engineering was simultaneously the embodiment of, and legitimizing force for, universal democracy and the open, market-system economy. Upper echelon government leaders (and engineers by training) John H. Sununu, Hank Hatch, and Brent Scowcroft—who in 1991 served as White House chief of staff, chief of engineers, and assistant to the president for national security affairs, respectively—fostered regulations that explicitly linked engineering with the reduction of government barriers to private enterprise and with the unleashing of innovative and productive forces in society.[62] Sununu, Hatch and Scowcroft became involved with the 1991 National Security Strategy of the United States, where sustainability was applied in the context of promoting market principles and economic reform through structural arrangements with the IMF and the World Bank.[63] At the same time, "in the competition for scarce resources" the 1991 Strategy saw technology

---

[61] Unknown. "Firms show ecological awareness at tech fair." *Chemical and Engineering News* 68, no. 16 (1990): 5.

[62] Sununu, who has three degrees in mechanical engineering from MIT, was the first White House chief of staff for George H. W. Bush, serving from 1989 to 1991. Scowcroft, who received a BS degree in engineering at West Point, and MA and PhD. degrees in political science from Columbia University, served as assistant to the president for national security affairs for President George H. W. Bush from January 1989 until January 1993. Hatch graduated from West Point in 1957 and also holds an MS in Geodetic Science from Ohio State University.

[63] The White House. *The National Strategy of the United States*. Washington DC, The White House, 1991.
<http://nssarchive.us/NSSR/1991.pdf >
Accessed November 18, 2013.

development—for peace and for war—as imperative.[64] The WRI, again, stressed the importance of developing novel, more efficient, technology as part of both US domestic and foreign policy in its 1992 report "Back to Futures."[65] Such organizations favored the "twin" objectives of economic growth and environmental protection.[66]

It was in this political milieu that engineers and technology-focused NGOs began creating economic evaluation methodologies designed to appraise and legitimate pollution prevention technology. Total Cost Assessment (TCA)—first systematized in 1991 by the Tellus Institute for the US EPA, based on research performed by General Electric (GE), and further developed by the American Institute of Chemical Engineers (AIChE)—is an example of an early accounting approach to engineering sustainability.[67] Application of these economic instruments, according to FIDIC, should occur "during the whole life-cycle of products from source through production, distribution use, and final disposal…"[68]

This section has touched on the ideological and institutional alliances from which engineers' contextualizations of sustainability derived and enhanced social and political support. In the early 1990s, the operating

---

[64] Ibid.

[65] Heaton, George R. Jr., Robert Repetto, and Rodney Sobin. *Backs to the Future: U.S. Government Policy Toward Environmentally Critical Technology*. Baltimore: World Resources Institute Publications, 1992.

[66] Collaborations between e.g., the WRI and the UNEP, which took place in the early 1990s, were also instrumental in setting up the agenda for sustainability engineering. The WRI, for example, was commissioned by the UNEP to author a report entitled "Natural Endowment: Financing Resource Conservation for Development," in which financing instruments such "green" or "Ecovest" investment funds were explored (Repetto, 1990). Another report, entitled "Great Lakes, Great Legacy?" was produced by the Institute for Research on Public Policy of Ottawa and the Conservation Foundation of Washington in 1990. The report is one of the first attempts to "…focus on regulation and enforcement toward the concept of sustainable development" in the context of a developed country. Bob Repetto, of the World Resources Institute, and US EPA administrator William Reilly were both involved in preparing the report. Jim MacNeill, lead author of the Brundtland Report, was at the time running the Institute for Research on Public Policy. Colborn, Theo E., *et al.* "Great Lakes, Great Legacy?" Washington, DC, and Ottawa, Ontario: The Conservation Foundation and the Institute for Research on Public Policy, 1990, 226.

[67] American Institute of Chemical Engineers (AIChE). Total Cost Assessment Methodology.
<http://www.aiche.org/ifs/resources/total-cost-assessment>
Accessed November 20, 2013.

[68] FIDIC (1994).

philosophy of sustainability within engineering and non-engineering institutions alike was firmly rooted in an ideology of technological change.

Engineers Going for Sustainability (1985–1997)

Engineering Groundwork (1985–1991): The Institutions and Metrics of Sustainability

Official recognition of the relevance of sustainability to the engineering profession in the US began with Hank Hatch's keynote speech to ASCE's national convention in New Orleans, Louisiana, on October 8, 1989.[69] "It is we engineers who hold most of the keys to the solutions of the world's environmental problems," Hatch declared, simultaneously inviting a new nation-building mission for the Army Corps of Engineers.[70] During his second career as chief executive officer and chairman of major US environmental engineering consulting companies, Hatch sought, more actively than any of his engineering colleagues, to convince fellow practitioners that the future of the world—as well as that of the profession—required new ways of executing technology and being an engineer. This imperative to active transformation, the previous section showed, had been built on the supposition that ecosystems science "had identified the *real* problems."[71]

Yet Hatch's and others' goal to appropriate sustainability as "THE engineering issue for the 'nineties and beyond'" had begun gaining institutional momentum within at least one major international engineering professional organization, WFEO, already in the mid-1980s—*prior* to the publication of *Our Common Future*.[72] During its Annual Plenary Session in the context of the "Second World Congress on Engineering and Environment," in November 1985, the WFEO—which represented over eight million professionals in the 1990s—approved the first ever "Code of Environmental Ethics for Engineers." The Code was an ambitious and progressive document exhibiting commitment to "social equity," the "local system of values," and negotiating "the best possible social and political solution" in

[69] Ironically enough a site of one of the Corps' biggest blunders (the poor engineering of the city's water defenses).
[70] Hatch (1989).
[71] Carroll, William J. "World Engineering Partnership for Sustainable Development." *Journal of Professional Issues in Engineering Education and Practice* 119, no. 3 (1993): 238–240.
[72] Hatch (1989), emphasis in the original.

engineering project development. It was also the first time that all three terms, "engineer", "sustainability" and "sustainable development" appeared together under the same banner. The Code read as follows:

> "When you carry out any professional activity:
> Study thoroughly the environment that will be affected, assess all the impacts that might arise in the state, dynamics and aesthetics of the ecosystems involved, urbanized or rural, as well as in the pertinent socio-economic systems, and select the best alternative for an environmentally sound and sustainable development...
> Be aware that there are certain essentials for our continued existence—the interdependence of ecosystems, the need to maintain the diversity of species, the need for resource recovery and for harmony in all our many relationships, one with the other—and that each of them poses a threshold of sustainability that should not be exceeded."[73]

Prior to publishing the Code, the WFEO, whose "Standing Committee on Engineering and Environment" was formed in 1979, had had an eight-year history of involvement with natural resource management, non-waste technologies, and environmental education for engineers.[74] In 1985 the WFEO began to express a clear vision for the engineer's role in development. Moreover, the WFEO Code argued for both a *logic* and *a technique* ("assess all the impacts") of engineering project evaluation. And with the inclusion of terms like "aesthetics" and "harmony"—a result of WFEO's aspiration to achieve the "interdependence of ecosystems"—the Code clearly moved away from engineering codes of ethics' traditional focus on contractual relationships between engineers, and between engineers and their clients.

Meanwhile, as the WFEO was gearing up to address global sustainability, preoccupation with maintaining engineering control over technological affairs remained essential to their organizational narrative. Indeed, a call for engineering leadership and active involvement in public decision making served as the common denominator of shared values among the various national and international engineering societies in the 1990s. For example, FIDIC's vision of reconstructing the public's attitude on engineering in the age of sustainable development is most apparent in the federation's efforts to influence the direction of the UN Conference on Environment and Development (UNCED) held in Rio de Janeiro in June 1992 as a follow up to the Stockholm Conference twenty years before.

---

[73] Thom (1988).

[74] Ibid.

Dissatisfied that no engineering constituency was engaging in preparations for the Rio conference, FIDIC formed its own Environmental Task Committee (ENVTEC) in 1988.[75] Alas, led by Norwegian consulting engineer Knut Ekenberg, the committee's policy statement became highly controversial during the society's annual meeting in Oslo in 1990 because delegates fought the idea that engineers have a responsibility to lead as politicizing the nature of their work. The contemporary identity of the "sustainable engineer" is rooted in a similar kind of professional politics: FIDIC's 1990 annual conference introduced engineering professionals into a dialectics between engineering ideologies and what it means to do "sustainability engineering" work. As reports from that conference indicate, the identity of the sustainable engineer was negotiated among participants over a spectrum which ranged from advocating the environmental assessment of impacts throughout a project's life to refusing to "be associated" with either clients or projects that are evading environmental protection. Such dialectic between technological change and technopolitics ideologies of sustainability challenged dominant values of the profession. For some, FIDIC may have been a professional organization that "should" be considering the political implications of engineering design; but, as Ekenberg put it "we are not used to thinking that way."[76]

In June 1990, FIDIC approved ENVTEC's statement entitled "Consulting Engineers and the Environment" which recognized the importance of adverse environmental effects in all the phases of an engineering project—from initial planning to decommissioning. The policy, which had been under development for almost two years, was first published in a special issue of *Independent Consulting Engineer*. It asserted that "engineers should provide leadership in sustainable development" while casting such leadership in terms of quantifying environmental benefits and adverse impacts of projects.[77]

---

[75] WFEO (1997). After familiarizing themselves with the work of the Brundltand Commission, ENVTEC members reviewed drafts of UN documents intended for the Rio Conference, including the 310-page Agenda 21 plan, only to discover that "Engineers were not appropriately recognized for their contributions to both environmental and developmental projects."

[76] Rubin (1990). See also: Unknown. "If we could just channel that passion elsewhere." *Engineering News Record* June 28 (1990): 130.

[77] FIDIC. "FIDIC on the Environment: Special Issue of Independent Consulting Engineer." *Journal of International Federation of Consulting Engineers*, Autumn, 1990. Lausanne, Switzerland.

In the meantime, the WFEO became independently involved in reviewing drafts of Agenda 21, and they articulated, almost identically to the ENVTEC remarks, that the engineering profession's troubles with sustainability began with its lack of input: "Engineering contributions to both developmental and environmental projects were omitted [from the Agenda] and... these omissions weakened the report." By 1997, the enactment of Agenda 21 principles and programs "applicable to engineering" had reached the status of "long term objectives" in WFEO's mission statement.[78]

WFEO and FIDIC therefore set the tone for how national engineering societies would later incorporate sustainable development into their own policies.[79] Under the guidance of David Thom, leaders of professional engineering societies in Australia and New Zealand agreed upon a draft environmental policy, which was then adopted by the WFEO as the "Arusha Declaration" during the society's September 1991 General Assembly in Tasmania. The Arusha Declaration was later transmitted to the organizers of the Rio Summit.

A non-binding policy, the Declaration offered the engineering profession one of the first blueprints for sustainable engineering action. It also called upon engineers to adopt a "new role" and to "realize [their] contributory potential within the short time frame required for the development and implementation of [sustainable] policy."[80] The reorientation of technology as a reaction to the image of global systemic problems à la the Brundtland Commission is exemplified in the Declaration as a new model of combined assessment of environmental and economic impacts. A "sustainability ethic" discourse exists in the Declaration too, but this ethics is primarily understood as "the impact of individual choices on nature and people." The Declaration also reflects Thom's perspective that engineering expertise must become an integral component of national and international strategic planning. In line with the engineering profession's tendency to evade

---

[78] Ibid and WFEO-ComTech. "Engineers and Sustainable Development." August 2002. Manuscript.

[79] Sustainability and engineering leadership was not limited to the US. The Institution of Engineers, Australia (IEAust), for example, during its Annual General Meeting in 1993 passed a motion directing that the "Council acknowledge the leadership role the engineering profession must provide in attainment of sustainable development and that Council develop special plans to achieve this leadership role and report progress regularly to the members," Jones, Lorie. "Acceptance Speech." Engineers Australia, WA Division, AGM, December 10, 2007.

[80] Thom David quoted in WFEO (1997) and WFEO (1991), respectively.

politics, it calls for the creation of centers for "capacity building" which can then prescribe engineering-specific guidelines for the manufacturing of "sustainable technologies." The declaration called on practitioners to commit to the "the 6 Rs": reduce consumption, replace activities, reuse materials, recycle, focus on renewable resource, and restructure institutions.[81]

Between 1985 and 1991 elite engineering professionals and organizations set the groundwork for the dialectic between engineering ideologies of sustainability. An appreciation of how "sustainability" challenged the ideological boundaries of engineering was thus heightened by seeing the 1985 WFEO Code of Ethics featuring elements of both technological change and technopolitics. This section further illustrated how anchoring historically the development of a sustainability discourse in engineering in the metaphor of the "invisible engineer" may be useful to appreciate the identity politics that underpinned that discourse. The sustainability engineering groundwork presumes a struggle of engineers to make themselves relevant—challenging for example, the fact that while setting the global sustainability agenda, the UN was not taking engineering work into account.

From Invisible Engineers to Facilitators of Global Stewardship

In the course of the 1990s, Hank Hatch and other prominent engineers were programmatically emphasizing the relevance of sustainability to increase the visibility of the engineering profession. Elite practitioners could build on a conglomeration of socio-political and economic ideas to claim that sustainable development is, in essence, the engineering challenge of realizing technology in the global environment. Thus a new engineering identity as facilitator of global stewardship emerged. For leaders of the American consulting engineering community such as CH2M Hill managers, who happened to be some of the most vocal proponents of a technological change ideology of sustainability, this identity was important because it legitimized the expansion of professional roles to include environmental engineering work in the US and abroad.[82] The idea of engineers as "global stewards" also legitimized that process where engineering leaders in the 1990s sought a convincing discourse to reinvigorate America's declining engineering school enrollment. The identity of facilitating global

---

[81] WFEO (1991).

[82] Poirot, James W. "Consulting engineers need to go public. Heightened public awareness will help profession embark upon a new age of innovation." *Engineering News Record*, January-March (1989): 3–6, special advertising section.

stewardship fit well with the dominant engineering culture in the 1990s in which establishing a visible and socially beneficial image for the profession by emphasizing the importance of engineering innovation and practitioners' technical challenges in addressing the environmental crisis was central. Particularly in the US, this identity was tied to the neoliberal economic idea of "public-private" partnerships and its assumed relationship with the protection of national security from the threats of environmental degradation. In an interview that Hatch gave in the year he retired from the Corps of Engineers, he said, "I can clearly anchor my concerns for environmentally sustainable development into this broader concept of national security."[83]

Nonetheless, playing the role of "facilitator of global stewardship" could also challenge the traditional conditioning of engineers to direct technological development. "The incongruity between the ideal of sustainable development and the way it is typically characterized by the engineering community," wrote Joe Herkert, "has significant implications for...the potential role of engineers as facilitators... [in] a sustainable society."[84]

The engineering community thus hardly converged on the meaning of facilitating environmental stewardship—much less on whether such stewardship should become the backbone of a sustainable engineering identity. In context, practitioners also debated the perceived need for sustainability metrics, approaches and evaluation practices. Indeed, as the next section will show, such metrics were seen, on the one hand, as an arena from which professionals would ground sustainability in traditional engineering expertise and, on the other hand, as a basis to challenge such notions of expertise altogether.

The Turn to Quantifying Sustainability

While some elite professionals and their organizations were trying to integrate "sustainability" into a new normative vision—engineers as facilitators of global environmental stewardship—they contended that engineering

---

[83] Atkisson (1992): 40. Another engineer and member of the Corps, James M. Waddell, championed the "overall role [where] engineers [would] become facilitators of sustainable development...[i]n doing this, they rightly assume a leadership position. Waddell, James. "Engineering for Sustainable Development" manuscript, 1991. In recent years we see a similar discourse in the Pentagon's fears about climate change impacts.

[84] Herkert further asserted that "we [engineers] must avoid the temptation to embrace technocracy by proclaiming ourselves to be the gatekeepers of all relevant insights on sustainable development" (1997): 179.

tools must be developed to accomplish this. In many ways, this rise of tool-making is a very technical turn: it is taming the sustainability spirit through specifying that only engineering tools will be able to make it work. Quantitative methods like LCA are far more compatible with the dominant engineering culture, practices, and cognitive traditions. As a consequence, those who develop the tools could de-politicize sustainability again.

At the same time the WFEO was adopting the "Arusha Declaration" during the society's September 1991 meeting emphasizing reduction of consumption, the reuse of materials, and recycling, a major shift on the perception regarding the role that materials accounting techniques should have was occurring with the US EPA. Having concluded in 1975 that the application of so-called resource and environmental profile analyses (REPAs) on recycling was not conducive to the attainment of the agency's regulatory objectives due to the associated "micro-managing of private business," for the next fifteen years, the EPA kept its policy-making focus away from LCA-type of studies. In 1972 the EPA viewed REPAs as means to establish a mandatory regulatory mechanism to reduce solid waste. In 1991, however, engineers in the agency decided that it would be more "realistic" to try developing tools and then letting industry use them.[85] The technical turn in sustainability engineering is thus part of a broader political and ideological debate regarding the role of the government and the market in managing environmental problems. The idea, for example, that the government should proceed like a corporation and that a rationale be incorporated into the EPA's framework to operate like a private sector actor is characteristic of the G. H. W. Bush administration (1989–1993). On the one hand, during that time, the sustainability discourse of US EPA, the WRI, and key technological change advocate Roy Weston were thus alike—they commonly promoted the development of voluntary guidance sustainability methodologies for industry, they praised the market-based incentive turn in policy, and argued that the future of sustainability engineering depended on World Bank and newly formed sustainable development NGOs' and industrial coalitions' funding for advanced technologies.[86] On the other hand, this discourse also echoed the different perspective between engineer-public officials, like Sununu, Hatch

---

[85] Interview data and Hunt, Robert G., and E. William Franklin. "LCA—How it Came About—Personal Reflections on the Origin and Development of LCA in the USA." *International Journal of Life Cycle Assessment* 1, no. 1 (1996): 4–7.

[86] See, for example, Repetto (1990); Reilly, William K. "A perspective from EPA." *EPA Journal* 16, no. 4 (1990): 12; Weston, Roy F. "Towards Understanding the Idea of Sustainable Development." Roy F. Weston, Inc., 1992. Manuscript.

and Scowcroft serving at the time President G. H. W. Bush, and technopolitics advocates of sustainability.

Discourse on sustainability engineering metrics thus flourished in parallel with the evolution of a political, economic and institutional system that operationalized the vision of a sustainable industry. In 1991, in preparation for the Earth Summit, a group of executives at major corporations, the Business Council for Sustainable Development (BCSD), founded by Swiss billionaire Stephan Schmidheiny, began debating with International Standardization Organization (ISO) and the International Electrotechnological Commission (IEC) on the elaboration of environmental standards.[87] Schmidheiny, who had established himself as a global sustainable development guru with the publication of his 1992 bestseller *Changing Course: A Global business perspective on development and the environment*, extended the sustainability discourse into the context of "eco-efficiency" and "corporate environmental responsibility," for which a product's life cycle is considered the main conceptual and analytical tool. In 1994 he helped set up a collaboration between the University of Tokyo, the Swiss Federal Institutes of Technology, and MIT. Among the "Alliance for Global Sustainability" projects was the development of "analytical tools...to help manufacturers predict environmental impacts of products throughout their life-cycles."[88]

Central to this dual interest in conceptualizing environmental policy in the context of market principles and sustainability as LCA guidelines and databases that corporate actors can voluntarily use were four documents published between 1991 and 1993 by the US EPA and the Society of Environmental Toxicology and Chemistry (SETAC), respectively. SETAC's "technological" (1991) and "conceptual" (1993) frameworks for LCA, as well as its "Code of Practice" (1993) were developed under the leadership of LCA world expert James Fava who worked for key sustainability engineering advocate Roy Weston in the early 1990s and more recently served on committees at the WRI and the World Business Council on Sustainable Development (WBCSD). The practitioner's role in crystallizing sustainability is manifested in chemical engineer Mary Ann Curran's work. Her tenure at the EPA coincided with the ascendancy of LCA over other environmental management approaches in the Agency's intellectual trajectory. American engineers working for Franklin Associates Ltd.—a consultancy

---

[87] Sentenced in summer, 2013, to eighteen years in prison for one of his former companies' (Eternit) involvement in Italy's biggest asbestos scandal.
[88] EFS Newsletter, February 1999.

that designed the early REPAs for EPA in the 1970s—also produced the EPA's 1993 "Life-cycle assessment: Inventory guidelines and principles" which, in conjunction with SETAC's publications, is considered to have officially launched the LCA field.[89]

Some of the first industrial sustainability metrics also emerged in 1994 (the British Standard 7750 as well as the European Commission's Voluntary Community Eco Management and Audit Scheme (EMAS), and the ISO 14000-series). Yet another translation of sustainability metrics into engineering practice was the US National Institute of Standards and Technology (NIST) Building for Environmental and Economic Sustainability (BEES) project founded in 1994. Developed primarily through the work of Barbara Lippiat—whose background is in economics—BEES uses an ISO 14000-based "multidimensional life-cycle approach" to integrate environmental and economic performance data in the decision making of "green building" materials.[90] In the mid- and late 1990s the International Chamber of Commerce (ICC)—which technological change engineer advocates of sustainability considered instrumental—worked together with UNEP and FIDIC to develop the so-called Environmental Management System (EMS) Training Resource Kits based on the ISO 14001 and the EMAS.[91]

Like Total Cost Assessment, this first generation of technological change sustainability engineering metrics—"quality systems" such as BS 7750 and ISO 9000—were attempts to quantify corporate performance in terms of environmental performance, and to draw connections between

---

[89] SETAC. *A technological framework for life-cycle assessment.* Edited by J. Fava, R. Denison, B. Jones, M. Curran, B. Vigon, S. Selke, and J. Barnum. Proceedings of a workshop in Smugglers Notch, Vermont. Society of Environmental Toxicology and Chemistry, August 18–23, 1990, in Pensacola, Florida, 1991.
SETAC. *A conceptual framework for lifecycle impact assessment.* Edited by J. Fava, F. Consoli, R. Denison, K. Dickson, T. Mohin, and B. Vigon. Proceedings of a workshop in Sandestin, Florida. Society of Environmental Toxicology and Chemistry, February 1–7, 1992, in Pensacola, Florida, 1993.
SETAC. *Guidelines for life-cycle assessment: A 'Code of Practice.'* Edited by E Consoli, D. Allen, I. Boustead, J. Fava, W. Franklin, A. A. Jensen, N. de Oude, R. Parrish, R. Perriman, D. Postlethwaite, B. Quay, J. Skguin, and B. Vigon. Pensacola: Society of Environmental Toxicology and Chemistry Press, 1993.
US EPA. "Life-cycle assessment: Inventory guidelines and principles." EPA/600/R-92/245. Prepared by Battelle and Franklin Associates for the US EPA. Cincinnati: Risk Reduction Laboratory, Office of Research and Development, 1993.
[90] Lippiatt, Barbara C. "Selecting Cost-Effective Green Building Products: BEES Approach." *Journal of Construction Engineering and Management* 125, no. 6 (1999): 448–455.
[91] Coates (1993).

environmental compliance and economic factors in the context of a product or process profile.[92] A characteristic example of technological change engineering involvement in sustainability metrics research in the early 1990s is the establishment of the Center for Waste Reduction Technologies (CWRT), an affiliate of AIChE. CWRT was created in 1991 to "benefit industrial sponsors and society... [and] to identify, develop, and share technology and management tools that measurably enhance the economic value of industrial sponsor organizations while benefiting the environment or addressing issues of sustainability."[93]

The identity of the sustainable engineer and the wave of metrics development described above were both responses to the perceived problematic nature of "engineering work" in the 1990s. Technopolitics advocates of sustainability also sought to define sustainability, no less than to reposition the engineer's identity. Yet as they began to address the point of what knowledge comes to be integrated into engineering they also challenged quantitative metrics as the best way to enact sustainability. Technopolitics proponent Joe Herkert and the late Alex Farrell detailed one of the first attempts to design engineering-specific knowledge tools to operationalize sustainability. Building explicitly on the approach and vocabulary of ecological economist Herman Daly—but in striking contrast with technological change approaches to sustainability metrics—Farrell conceptualized sustainability indicators and decision models as a means to incorporate non-expert knowledge. The basis for sustainability engineering, he elaborated, was expertise; nevertheless, he countenanced that "knowledge tools

---

[92] Many of the most prominent projects that furthered the sustainability metrics discourse in engineering in the 1990s faced significant technological and data challenges. For example, the proceedings of a 1997 ASCE conference entitled "Innovative Civil Engineering for Sustainable Development" noted that engineering-specific "QA [quality assurance] guidelines need to be developed for civil engineering programs with flexibility for their use in sustainable development programs." And at the Civil Engineering Research Foundation's (CERF) 1996 symposium participants maintained that "life-cycle cost goes hand-in hand with environmental impact and its effects," while identifying the current lack of systematic methodologies and support tools for collaborative decision making, the fragmented nature of the industry, and the lack of historical data and information on product maintenance and repair as critical barriers to sustainability engineering. See Nivargikar, Rao and Debra Reinhart, eds. *Quality Assurance-A National Commitment*: ASCE, 1997 and CERF (1996).

[93] ASCE (2004).

which link cultural and political factors to the ecological and economic objectives of sustainability are especially needed."[94]

In 1991, while Roy F. Weston Inc.'s James Fava was leading a team of practitioners in the first attempt to standardize the LCA methodology, a group of engineers from Canada and the US gathered in a IEEE Toronto-sponsored conference to debate the extent to which sustainability could provide a grounding for shifting values in engineering. In that conference there were some few who saw the challenge in equating sustainability engineering with LCA. Among them Steven B. Young, then a doctoral student at the University of Toronto, noted that there is, perhaps paradoxically, an opportunity for a deeply reflective turn in the attempt to operationalize sustainability via LCA. For the very definition of boundary conditions in LCAs—the rationale determining the level to which "impacts" of a life cycle are to be considered and the basis for including/excluding processes in the system—will always be enmeshed with extra-engineering values predetermining the LCA's "comprehensiveness and meaningfulness."[95] As chapters 4 and 5 illustrate, by the beginning of the 21st century, such consolidations of engineering and extra-engineering assumptions—an amalgam of quantitative sophistication and socio-political valuation—would be called "social LCA" and would be perceived as legitimate engineering professional territory.

For example, "[i]nstead of [a] specific quantitative assessment," argued MacMaster University's mechanical engineering professor Robert Hudspith in 1991, the sustainability practitioner must be engaged in practically addressing the question of how engineering work may operationalize the insight that technology embeds social and political values. In linking engineering work with engaging users in design and tapping into local environment and community expertise—which he defined as "internal," engineering, characteristics of technology—Hudspith suggested that sustainability's potency lay not in LCA per se but in its ability to subject professionals to a new standard of knowledge infrastructure.[96] That is, if engineers "are to be leaders and collaborators in sustainable development," one member of AESR wrote, they must also be willing to consider and promote...approaches that are *not necessarily consistent with their own*

[94] Farrell, Alex. "Sustainability and the Design of Knowledge Tools." *IEEE Technology and Society* 15, no. 4 (1996): 11–20.

[95] Young, Steven B. "The Materials Life-Cycle Approach." In *Preparing for a sustainable society*, edited by Helmut Burkhardt and H. Willem Vanderburg, 318–325. Piscataway: Institute of Electrical and Electronic Engineers, 1991.

[96] Hudspith, Robert (1991).

TABLE 1. Proposed Current and Complementary Goals, Strategies, and Values of U.S. Engineering Culture

Current (1)	Complementary (2)
(a) Goals	
Individual achievement and career advancement	Team, community, and social development
Creative invention and manipulation of sophisticated technology	Resolution of complex social and environmental problems
(b) Strategies	
Short-term, practical focus	Long-term, systemic focus
Manipulation of objects (I-it relationships)	Interaction with living beings (I-Thou relationships)
Analyze parts in isolation	Analyze whole systems
Deference to managerial and social authority	Deference to sustainability interests, ecological and carrying capacity limits
Competition	Cooperation
Quantitative rigor	Qualitative understanding
Rational, deductive logic	Inductive intuition
(c) Values	
Political conservatism	Political liberalism
Order	Chaos, spontaneity
Control	Adaptation, partnership
Utility and practicality of means	Existential value and appropriateness of ends
Efficient use of resources	Preservation of resources
Predictable reliability	Spontaneous adaptability
Determinant systems	Indeterminate systems
Standardization, uniformity	Diversity
Economic growth	Steady state economy and qualitative improvement
Manipulation of nature	Cooperation with nature
Acceptance of status quo	Questioning of status quo

Figure 2.1 "Proposed Current and Complementary Goals, Strategies and Values of US Engineering Culture." Adapted from McIsaac and Morey (1998): 114.

values and culture." Among the strategies they proposed to reinvigorate the dominant US engineering culture were improvement of "qualitative understanding" as opposed to "quantitative rigor" and deference to "sustainability interests" rather than "managerial authority."[97] (Figure 2.1).

---

[97] McIsaac, Gregory F., and C., Nancy Morey. "Engineers' Role in Sustainable Development: Considering Cultural Dynamics." *Journal of Professional Issues in Engineering Education and Practice* 124, no. 4 (1998): 110–119, 118, emphasis added.

Hank Hatch envisioned the technological and ethical infrastructures of the sustainable practitioner, yet he perceived her knowledge infrastructure in strictly engineering terms. When cofounder of AESR Thomas (Tom) Munsey cited Hatch's definition of environmentally sustainable development in his paper presented at the same 1991 conference in Toronto he was critical not so much about sustaining the environment, but about the implications of the Corps' technological change ideology to "building and securing this Nation." Munsey observed that sustainability was being construed in terms of LCA that remained highly technical and was not allowed to interrogate what sort of projects and development paths are being pursued. Persuaded that technological plethora has been a false promise, and that implementing sustainability lies not with technical methodologies but with societal restructuring per se, Munsey opted for a parameter that quantitatively minded engineers tend to neglect; a "sociological risk assessment" founded on spiritual dimensions discussed by fellow technopolitics authors like Italian visionary architect Paolo Soleri and Bulgarian hydraulic engineer Eduard Naudascher. He also advocated for a sustainability engineering identity as a model of socio-political "crusader" which also required that engineers move away from the dogmatic standardization of a method like LCA managed by a group of academic experts hoping to "establish a name, and therefore a meal ticket." Munsey's technopolitics position, then, to the question of sustainability metrics premised stepping into "unfamiliar [qualitative] territory…[where engineers needed] to redefine quality of life."[98]

Munsey's position, along with the AESR case study presented in the next chapter, offer an invaluable contrast with the dominant LCA focus—rather than trying to quantify sustainability, engineers need to understand what people want from their lives. This may be crucial for the development of a sustainability engineering identity for it opens space for more participatory approaches in engineering project design. Many engineers would have called Munsey and the AESR "radical" or "real crack-pot"—as they did—but this is only because of what is a dominant economistic philosophy in industrialized economies.

It was thus during the technical turn to sustainability that some practitioners questioned metrics consonant with the engineering's insistence

---

[98] Munsey, Thomas E. "Sociological risk assessment. Likely confrontations and conflicts in the war for sustainability." In *Preparing for a sustainable society*, edited by Helmut Burkhardt and H. Willem Vanderburg, 402–409. Piscataway: Institute of Electrical and Electronic Engineers, 1991.

on rigorous quantification—which they perceived as a sign of the engineer's irreflexive self and the profession's irreflexive culture. It was then also in the midst of competing engineering ideologies that LCA experts presented their subject matter as ideological referee providing a technique and an intellectual/professional space within which the coevolution of technological change and technopolitics would become possible. That is, the integration of technological change and technopolitics ideologies of sustainability informed the criteria for deciding on appropriate engineering metrics, therefore channeling engineering expertise into unchartered waters. Sustainable development, it was argued by a Canadian engineering professor in 1996, "seeks to include social and environmental impacts in the processes and accounting of economics. Therefore, risk assessments for sustainable development must consider all impacts whether quantitative or qualitative."[99]

The US-Led World Engineering Partnership for Sustainable Development

The ambitious spirit of sustainability engineering in the 1990s is perhaps best captured by the creation of the US-based World Engineering Partnership for Sustainable Development (WEPSD). Born in the traditional gentlemen's Cosmos Club in Washington DC, the Partnership was solidly grounded in the legacy of technological change characteristic of such engineering and development elites like the Corps of Engineers, WFEO, FIDIC, US NAE, the National Science Foundation, and the World Bank (Figure 2.2.). The WEPSD case study is illuminating because while sustainability engineering was appearing in many places, a set of US actors were especially influential. Funded by corporations and corporate foundations, this constituency connected with the powerful corporate and government elites that ratified the professional contract of the sustainable engineer; they lined up in support of conservative, free market think-tanks for sustainable development politics. At the same time, the Partnership was a new entity, a "global coalition" of engineers, businesspeople and scientists explicitly dedicated to the engineering profession's unification through a new common identity as facilitator of sustainability. The organization's mission thus resonated with the familiar attitude that to save engineering one must repurpose it.

---

[99] Kroeger, Heidelore I., and P. Simonovic Slobodan. "Risk as a sustainable development Criteria." In *North American Water and Environment Congress & Destructive Water*, edited by Chenchayya Bathala, 1531–1536. New York: ASCE, 1996.

Leading lights at the May meeting of the newly formed World Engineering Partnership for Sustained Development, *from left, are Henry Hatch of Atlanta-based Law Engineering; Maurice Strong, secretary general of the U.N. Conference on Environment and Development; William J. Carroll, one of the partnership founders (and a former ASCE president); and Don Roberts of CH2M Hill, who was also instrumental in the partnership's founding.*

Figure 2.2 WEPSD leadership: Hatch, Strong, Carroll, and Roberts. Adapted from Unknown. "Momentum Gathers for Sustainable Development." *ASCE News* 18, no. 6 (1993): 6.

In line with the technical turn to sustainability engineering, the WEPSD became involved in a quest to provide practical content to the idea of sustainable development, as Partnership leaders like Weston and Roberts gave expression to their social theories of sustainability. Established to foster effective relationships between engineering societies and practitioners, the WEPSD reached out to government officials, expert administrators and environmental elites. The Partnership was aiming fundraising initiatives at the World Bank, the UN and other private entities in order to inaugurate its first round of projects.[100] A year later, in 1994–1995, in collaboration with the Center for Field Research (CEF) and the environmental research group Earthwatch, WEPSD issued a call for proposals, offering grants from \$10,000 to \$100,000, and requiring volunteer work to perform environmental impact assessment in engineering projects for development. According to the call for proposals, priority was given to projects conforming with Agenda 21, with suggested areas of exploration—the "*Metrics of Sustainability*"—including: "environmental and technological impact assessment," "projects which establish metrics for sustainability in a given culture," and "programs that contemplate systems of life cycle analysis of project impacts..."[101] The Partnership thus rose to the challenge presented by international development institutions like the World

[100] Prendergast, John (1993).
[101] AAES (1994): 11.

Bank, describing how a sustainable future could be run by engineering task forces "to advise all Nations on the preparation and implementation" of sustainability.[102]

The initial WEPSD US group—Hatch, Roberts, Weston and a couple of others—was formed in December 1991 in Washington D.C., but officially the Partnership was created in March 1992 during a two-day convention that took place concurrently with a UN preparatory workshop for the Rio Summit (a summit attended by UNCED secretary general Maurice Strong). The same US sustainability engineering constituency that organized the convention drafted a vision statement they then shared with UN delegates during an open-house reception at the United Engineering Center in New York. Such a combined inventive and problem-solving potential, indicating the future triumph of international engineering for environmental protection, prompted Strong to suggest that "the concept of sustainable development would be impossible without the full input by engineers."[103]

The Partnership, which in 1995 represented approximately ten million engineers from eighty countries, was joined by WFEO and FIDIC during the March 1992 gathering; its membership was later enhanced to include the International Union of Technological Associations (IUTA), the Consortium for International Earth Science Information Network (CIESIN) and the ASCE.[104] At a time when Hatch was chairman of the Atlanta, GA-headquartered engineering and environmental services corporation Law Companies Group, Inc., the executive directorship of the WEPSD was assigned to William Robertson, who had been appointed associate chief of engineers by Hatch in the late 1980s. With Robertson directing them, a meeting of eminent engineers, scientists and administrators was held in the New York City headquarters of the UN in May 1993 to discuss the engineering implications of Agenda 21. Present at the meeting were also Mohamed El-Ashry, director of the environment for the World Bank, Gus Speth of the WRI, and Nitin Desai, UN undersecretary for policy development and sustainability.[105]

The Partnership's mission was the subject of another tellingly entitled conference, co-sponsored by the UK's Institute of Civil Engineers (ICE) and the ASCE: "Facilitating Sustainable Development—the role of the

---

[102] AAES (1994): 10.

[103] WFEO (2002).

[104] The IUTA and CIESIN joined in Fall 1992, whereas ASCE became member of the WEPSD in May 1993.

[105] WFEO (1997); Unknown. *ASCE News* (1993).

engineer." The agenda of the "Bermuda Summit," as this September 1992 conference later became known, was organized by William (Bud) Carroll, then president of the WFEO, and the ICE's Roger Dobson. The overall goal of the Partnership seemed to suggest a "very positive step," although ambivalence remained about whether "a mindset change" was required for professionals to carry out engineering with "an additional ethical requirement;" to collaborate with the International Chamber of Commerce (ICC) to reach out to "boardrooms of companies;" to assist communication with "varied" clients;[106] or to realize that engineers "must take a stronger stance on environmental protection while accomplishing economic growth."[107]

The WEPSD claimed that existing public and private technology decision making was not sustainable. This was particularly true, said the Partnership's mission statement, in the absence of "constructive relationships" with other stakeholders advocating common goals. The WEPSD argument, in effect, sought to cast "facilitation" of sustainability as calling for the development of global engineering communication networks. The result was that computerization and optimal use of information technology became one of the Partnership's main organizational foci. Engineers were moved by such technological visions: Raymond S. Leonard, for example, a consultant in nuclear technology for Los Alamos National Laboratory, challenged the WEPSD members during the 1992 meeting in New York to integrate information systems into the actual definition of engineering for sustainable development.[108]

A challenge, ambition and optimism is also evidenced in CERF's report to a 1996 symposium on "Engineering and Construction for Sustainable Development in the 21st Century," which featured over 700 attendees from thirty-five different countries:

> There is a need for information technologies—including methods, techniques, software and hardware—to create seamless decision support environments so that information about all aspects of an organization, and at different levels of detail, can be available to all decision makers.... Decision support tools must be in place to regulate logical and clear methods of decision making that allow for the meeting of a variety of institutional goals: productivity, bottom line, and sustainability.[109]

---

[106] Coates (1993); Prendergast (1993).

[107] Prendergast (1993).

[108] Leonard (1992).

[109] CERF (1996).

The dual calls for technological leadership and operationalization of sustainability in engineering projects contained in the Partnership's statements become more explicit in other WEPSD enterprises, funders and collaborators. In 1993 for example, WEPSD assigned to the Global Environment and Technology Foundation (GETF) the task of creating the World Engineering Network (WENet), "a global communication and information exchange system focused on sustainable development."[110] GETF had been created in 1988 by West Point graduate Tom Harvey, who was at the time chairman of Roy F. Weston Inc. The WENet project, for which development funding was initially provided by the NSF, was viewed by sustainability proponents as a "major step in linking all engineers and sharing technology with developed and developing nations."[111] Through its "Standing Committee on Information and Communication," WFEO later furthered WEPSD's sustainable data mission.[112]

The drive to unite the economic, engineering, financial and bureaucratic forces of development was characteristic of other WEPSD initiatives as well. Indeed, the technological planning envisioned by the Partnership was no less dependent on "engineering donors" than it was on framing the objectives of the engineering profession worldwide.[113] Two years before its apparent ending due to lack of funding and perceived organizational inefficiency and inflexibility by founding members, the WEPSD co-organized with WFEO a meeting entitled "Recycling Waste for Agriculture: The Rural-Urban Connection," which was funded by the World Bank and took place at the Bank's offices in Washington, D.C., in September 1996.[114] The meeting was cochaired

---

[110] Motivated by the same rationale, during its Ninth General Assembly on April 26, 1996, the Federation of Engineering Institutions of Southeast Asia and the Pacific (FEISAP) endorsed the establishment of a sustainability engineering cooperative network and homepage, the "FEISEAP Net", under the leadership of Bill Rourke of the IEAust.

[111] WFEO (1997): 73 and *EFS Newsletter*, October 1997.

[112] For example, a WFEO-sponsored seminar on "Information and Sustainable Development" was held in Uruguay in November 1996, with the cooperation of UNIDO and UNESCO (WFEO, 1997). See WFEO (1997): 73 and *EFS Newsletter*, October 1997.

[113] The Partnership had proposed the formation of a "Consultative Group on Engineering Technology." This group would have constituted a team of "funding agents, engineers, scientists, economists, international financial institutions and foundations" similar to the CGIAR, formerly Consultative Group on International Agricultural Research. Sanio (1997).

[114] Interview data. To the best of my knowledge, the latest publicly available reference to WEPSD is a 1998 announcement from Don Roberts', then chairman of the WEPSD, presentation on "Sustainable Development—A Challenge for the Engineering Profession" at the Environmental Sciences Division (ESD) of the Department of Energy's (DOE) Oak Ridge National Laboratory (ORNL).

by Maurice Strong, then advisor to the president of the World Bank, and Henry Hatch, then president of the Partnership. CH2M Hill's project manager, David Burack, coauthored the 1998 report, which was conceived during that meeting: it summarizes the goals of a proposed three-phase urban waste recycling program in collaboration with UN's Development Program (UNDP), the World Bank, WEPSD and other organizations.[115]

The Partnership, in sum, smoothed the way for the mid- and late 1990s initiatives pursued by WFEO, FIDIC and ASCE, leading to subsequent adaptation of sustainability in the context of an ideology of technological change. The WEPSD also furthered the discourse on sustainable development in engineering education, by conducting the first survey of its kind on sustainability engineering curriculum in the US. The Center for Sustainable Technology (CST) was established at the Georgia Institute of Technology in cooperation with WEPSD in June 1993, after the General Electric Fund provided nearly $1 million to create a curriculum on "sustainable technology and development." An Institute-wide Sustainability Task Force was organized, and a three-course pilot sequence was developed and deployed under the coordination of professor of civil and environmental engineering Jorge Vanegas. As part of this initiative, Georgia Tech launched an Industrial Ecology Study Group, and a workshop on the metrics of sustainability engineering was convened. In addition, a significant element of Georgia Tech's sustainability curriculum involved the development of an "Environmentally Conscious Design and Manufacturing (ECDM)" program that led to the creation of three laboratories, one of which was dedicated to "technologies that support information management and use in decisions throughout a product life-cycle."[116]

The core narrative of sustainability engineering between 1985 and 1997, this section showed, was strongly consistent with an ideology of technological change. Engineering sustainability linked to technological change was the guiding spirit of the programmatic 1994 AAES volume entitled *The Role of the Engineer in Sustainable Development*: "The creation of sustainable technologies and processes is perhaps THE most practical and readily available tool to achieve the sustainable integration of the environment and technology in the foreseeable future."[117] Yet this section has

---

[115] Sanio, Michael R., David Burack, and Sadaf Siddiqui. "Reuse of Urban Waste for Agriculture: An Investment Program for Progressive Action." Alexandria, VA: World Engineering Partnership for Sustainable Development, 1998.
[116] *EFS Newsletter*, February, 1998.
[117] AAES (1994).

argued that engineering proponents of sustainability did not always give quantitative metrics like LCA palpable support. Goals converged, however, for the technological change driven major international and US engineering organizations[118] to engineer sustainability in order to sustain engineering, as was reflected in the conclusion of an ASEE-sponsored International Conference on Engineering Education and Practice held in June 1996: Engineering organizations, it was argued, need to be "proactive in responding to the need for change, change which is required because the Engineering Profession, Educators, and Practitioners are not equipped to meet the challenges of sustainable development."[119]

Meeting the Challenge to not be Left Behind (1998–2003)

By 1997, the WEPSD had begun to lose a great deal of its visibility. Two years earlier, in 1995, WFEO's US-headed technological change and liberal economics-oriented Committee on Transferring, Sharing and Assessment of Technology (ComTech) had taken over explicitly to "lead the engineering profession worldwide in the promotion and application of sustainable technology."[120] The changing organizational landscape barely disturbed the hierarchy of the ascending engineering advocacy base for sustainability. In fact, during a six-year period of intense institutional mobilization (1998–2003), the leadership of US engineering for sustainable development was systematized in lines structurally similar to those of the earlier period (1985–1997). ComTech's first president, James Poirot, was at the time chairman emeritus of CH2M Hill. AAES became the US representative to the WFEO, and WFEO sustainability-related activities were funded by a four-year grant by the NSF.[121] In 1998, Michael (Mike) Sanio of ASCE and

---

[118] EFS aside—they were the main venue for engineering society collaboration with regards to sustainability—in several instances US engineering organizations held joint "planning meetings" to coordinate and promote sustainable development activities. Such a meeting, for example, took place in October 1996, cosponsored by the Committee on Sustainable Development of AAES, chaired by Albert Grant, and the Working Group on Sustainable Development of the ASEE.

[119] "Framework for Coordinated Action Within the U.S. Engineering Community." In WFEO (1997): 71.

[120] WFEO ComTech 2002. ComTech was created in 1995 to be headed by a US president, and a 1999 NSF grant enabled the United States to continue as Secretariat for another four years.

[121] *EFS Newsletter*, October 1997.

WEPSD accepted a position with the World Bank. When Poirot retired from ComTech in 2001, he was replaced by Don Roberts.

In the early 21st century, sustainability engineering's purview remained global and development oriented. At the global level, FIDIC staffed its "Sustainable Development Task Force" (SDTF) with members of the UNEP and the WBCSD. Such mobilization notwithstanding, the image of catching up with sustainable development—first articulated in the late 1980s—remained dominant. In this vein, NAE leader David Frosch's assertion that sustainability is "challeng[ing] engineers *perhaps most of all*" is notable.[122]

WFEO's ComTech and Engineering Sustainability Through Public Private Partnerships

Dr. Norman Neureiter, then science and technology adviser to the US secretary of state, was the guest of honor at the November 1, 2002, invitation-only meeting of the Engineers Forum on Sustainability (EFS), co-sponsored by ASCE/ASEE and chaired by Hank Hatch.[123] In his talk, "Engineers for Sustainability—the view from C Street," Neureiter elaborated that governance and partnership were the two main ideas that the US delegacy had insisted upon during the World Summit on Sustainable Development (WSSD) in Johannesburg, which had taken place in the summer of the same year. In the context of empowering technological change-oriented engineering institutions and, also, in the context of neoliberal globalization Neureiter added that "[p]ublic/private partnerships are three key words" regarding global development and "here is where the American engineering community's role can be so important." He then explicated that development requires linking communities within a globalized world—"a world that today is essentially driven by technology." No longer with WEPSD, but under the banner of WFEO's ComTech, Don Roberts suggested to Neureiter that a "coalition of donors" be formed to provide millions of dollars yearly for engineering projects in the developing world.[124]

There is justification for Roberts's commitment to and pursuit of sustainability engineering through ComTech and its related activities, for it was via the WFEO that the US engineering community furthered its position in the discourse and global politics of sustainable development. Between 1995 and 2003, ComTech maintained a dual focus on contextualizing

---

[122] Frosch (1999, emphasis added).

[123] The EFS was established in 1997 by ASEE and ASCE. It was later embraced by AIChE (2002); ASME (2004); IEEE (2006); and IEEE-USA in 2009.

[124] *EFS Newsletter*, December 2002, 4; 8–9.

sustainability from the engineer's perspective and on promoting technological projects that the committee perceived as "sustainable". The three main ComTech action topics leveraging shifts in the identity of sustainable engineers were the areas of knowledge networks, international declarations, and engineering ethics.[125] Analogous to these efforts were the organization of "engineering-expert" panels at the annual meetings of the UN Commission on Sustainable Development (CSD), the implementation of "engineering considerations" in the Earth Charter—an international document on sustainability—and the reconceptualization of engineering responsibility through the integration of sustainable development in WFEO's "Model Code of Ethics" for engineers. However consonant with manipulating engineering knowledge, the rising cultural identity of "global stewardship" for sustainability left the very foundations and application contexts of such knowledge unchallenged.

James Poirot's participation as a US delegate in the Earth Summit + 5 Special Session of the UN General Assembly in the summer of 1997 in New York City was instrumental in establishing the presence of the global engineering community at the annual CSD meetings.[126] On April 27, 1998, at the request of the US EPA and jointly with AAES, ComTech presented a four-member panel on "Water Privatization Projects in the Developing World," including case studies from Latin America. The demonstration of how privatization projects could be funded, structured and introduced resonated with the US engineering community's interests in sustainable development. Thus, this was the "first time engineers had been included in the United Nations dialogue *in this way*," exclaimed the February 1999

---

[125] Under the leadership of Don Roberts, and building on WEPSD's approach to creating WENet, ComTech endorsed the idea of establishing a Virtual Environment and Sustainable Systems Engineering Library (VESSEL) within UNESCO's Framework Agreement, to enable the sharing of technological information. The Institution of Engineers, Australia (IEAust) and the IPENZ volunteered to design a prototype virtual library called the Australasian Virtual Library. ComTech also collaborated with FIDIC to facilitate virtual dialogue between engineers themselves and between public and private sector parties who shared a common interest in establishing clean technology markets and meeting the goals of UN's Multilateral Environmental Agreements. As a result, an internet-based "technology transfer" tool named SANet was created. WFEO-ComTech (2002); Rourke, William J. "A vessel for engineering education in the developing world." Paper presented at the 2nd World Congress, Sarawak, Malaysia, July 22–25, 2002. WFEO and FIDIC were instrumental partners in running a collaborative online sustainable technology communication platform developed by UNEP and the Global Environment Facility, an offshoot organization of the World Bank (WFEO-ComTech 2002).

[126] WFEO ComTech (2002).

edition of the *EFS Newsletter* with relief.[127] As shown in previous sections, it was in the context of government and corporate capitalism, undergirded by development institutions like the UN, that engineering leaders such as ComTech and CH2M Hill's Poirot approached the analysis and operationalization of engineering work. In offering a view of his company, the chairman of CH2M Hill hosted another panel on "Production/Consumption" sponsored by AAES during the 1999 CSD meeting, featuring a presentation by Otto Vydra, CH2M Hill's vice president in Europe and the Middle East.

In September 2002 came the announcement that the US would rejoin UNESCO after an absence of eighteen years. Around this time, the EFS provided institutional shelter for the North American sustainability engineering constituency whose main organizational agents took action to ensure that the emphasis remained on engineering projects for development—under the UN banner of "capacity building." Whereas other professions might have endorsed the concept, engineers saw themselves as the necessary and sufficient condition for attaining sustainability. A June 2002 Declaration by the US Engineering Community to the WSSD stated that a sustainable world would "only be realized through the application of engineering principles and a commitment to public/private partnerships…"[128]

On one level, the 2002 Declaration and a follow-up meeting held in the NSF offices in Arlington, VA—where Neureiter enacted the identity of the sustainable engineer as facilitator of global steward in public-private partnerships—reflected the assumption that sustainability and engineering would have to coevolve. What is more, engineers and financial/state institutions alike viewed technological "capacity building" and "technology transfer" as the medium for global trade liberalization and for developing countries' engagement with the market economy. On another level, engineering societies envisioned that, in light of public dissatisfaction and shrinking enrollments in engineering schools, the realization of sustainable projects and public promotion of the profession were inseparable. In summer 2003, the US engineering community told UNESCO that it should look into "Why young people around the world are turning away from engineering and how this may be understood and addressed, how best to promote the public understanding of engineering and how engineering

[127] Emphasis added.

[128] AAES, AIChE, ANS, ASME, NAE, NSPE. "A Declaration by the US Engineering Community to the World Summit on Sustainable Development." June 24, 2002.

may most effectively be applied to poverty eradication and sustainable development".[129]

In a spirit quite similar to that of ComTech's "engineering-expert" panels, US engineering leaders, manifesting support from the George W. Bush administration, submitted a $15 million proposal to the UN recommending the financing of an "Engineering and Technology Program for International Development." The proposal, written by Mike Sanio, Henry Hatch, William Kelly of ASCE and Albert Grant (then Chair of EFS), argued for a NAE and AAES led public-private partnership to initiate engineering projects in the Global South. Framed in a way that emphasized "local problems, initially environmental in nature, which... *keep true development from taking place*," the proposal was endorsed by AAES and WFEO.[130] Although it wound its way from the US government to the UN, where it was finally adopted in 2005, Hatch recently bemoaned that "sadly UNESCO has basically dropped the ball since then".[131]

This section showed that it was partly through US elite engineers' leadership in international professional organizations that the identity of the sustainable practitioner was weaved into the idea of public-private partnerships funded by development institutions and enacted through advancing engineering technology networks. The dominant ideology of technological change, in short, not only promoted the identity of "facilitator of global stewardship," but it also sought to legitimize that identity by grounding it in neoliberal economic authority and familiar notions of engineering expertise. Yet while they were in favor of the egocentric, technological change ideology of sustainable development, professional societies led by US practitioners like Roberts and Poirot were also immersed in the anthropocentric legacy of engineering codes of ethics impregnated with elements of technopolitics. At the November 1999 General Assembly meeting, WFEO's Model Code of Ethics was amended to incorporate "sustainable development."[132] The final version of the code, which was coauthored by David Thom, Bud Carroll and Donald Laplante of the Canadian Council of Professional Engineers (CCPE), was adopted in 2001. Its section

---

[129] UNESCO Secretariat. "Engineering and Technology for International Development: 'Engineering for a Better World,'" August 2003.

[130] Sanio, Michael R. "Engineering and Technology Program for International Development." (Revised Draft Proposal). April 7, 2003; UNESCO Secretariat. "Engineering and Technology for International Development: 'Engineering for a Better World.'" August 2003, emphasis added.

[131] Personal communication.

[132] *EFS Newsletter*, February 2000.

on "environmental engineering"—almost identical with the 1985 WFEO Code—retains the language of "equity," "harmony," "system aesthetics" and "eco-systemic interdependence."[133]

Claiming Technology's Territory, Stretching Sustainability's Boundaries: US Engineers and the Earth Charter

The same dissonance between sociopolitical invisibility and global environmental stewardship ideals that framed the professional identity politics in sustainability engineering would haunt much of the US technological leadership's efforts to instill an engineering view into international sustainable policy.[134]

Commissioned in 1994 and cochaired by Maurice Strong and Mikhail Gorbachev, the "Earth Charter Commission" was formed after the Earth Summit in 1992 with initial funds provided by the Dutch government. In 1996, the Commission established the Earth Chapter's "drafting committee" and appointed Middlebury College professor of religion Steven C. Rockefeller to chair it. Building on the work of the International Union for the Conservation of Nature (IUCN) and based on more than 200 civil society declarations, the Earth Charter was finalized in 2000; it was designed to provide the universal ethical framework that was missing from the Earth Summit's Declaration.[135] Besides the Earth Charter, the Rio Declaration—consisting of twenty-seven Principles and a Preamble—and Agenda 21 were the other two international documents that gave engineers a conceptual context to reimagine the profession's active involvement in global law and environmental policy realms.

By all accounts, the engineering community's response to international development institutions' calls to engage in defining sustainability embodied a paradox: on the one hand, engineers confidently laid claim to

---

[133] WFEO/FMOI (UNESCO). "Code of Ethics." 2001. <http://www.sustainable-design.ie/fire/WFEO-UNESCO_Model-Code-Ethics_2001.pdf> Accessed December 11, 2013.

[134] "Around the world, the Earth Charter process is building in intensity each month and engineering visibility is almost totally missing in the process." Jim Poirot. "Engineering Implications in the Earth Charter." Paper Presented at the World Congress on Sustainable Development: Engineering& Technological Challenges of the 21st Century, Calcutta, India, January 21, 2000.

[135] Earth Charter in Action. "A short story of the Earth Chapter." Undated. <http://www.earthcharterinaction.org/download/about_the_Initiative_history_2t.pdf> Accessed December 14, 2013.

the technological niche of development. On the other hand, the crisis of sustainable technology engendered a reaction to a professionally stressful situation. Practitioners' attempts to celebrate the engineering conditions of development sounded defensive about technology's potential, and bitter about society's failure to acknowledge how engineering has realized that potential.

In the wake of the September 1991 WFEO general assembly in Arusha, Tasmania, leaders of international engineering organizations and senior environmental engineering consultants began carving out technological territory in the global landscape of sustainable development. In 1992, the WEPSD had joined WFEO in representing the engineering profession during the Rio Earth Summit. During its 1992 annual conference in Havana, Cuba, WFEO's general assembly had approved the "Havana Resolution," inviting national WFEO members to incorporate the recommendations of Agenda 21. And soon after the UN Conference on Environment and Development, members of WEPSD had initiated a systematic analysis of the Agenda 21 text. For instance, during the Partnership's March 1–3, 1992, workshop, members of WEPSD examined the fifth chapter of Agenda 21, which advocated how engineers can use information technologies to enable "capacity-building." In part, established to "present some unifying engineering thinking on Agenda 21," WEPSD's analysis "identified 1700 action items of 2500 that had engineering/technology implications" and recognized that "at least 241 of them [had] main engineering/technology implications."[136]

Like Agenda 21, the very existence of the Earth Charter, conceptualized as an international declaration for a global sustainable future, substantially challenged engineering proponents of sustainability. In fact, elite engineers like Jim Poirot and Don Roberts saw these documents as opportunities to define the boundaries of sustainable engineering projects and to amplify engineering's social and political legitimacy.

"Too often politicians and other professions set the policies engineers must work within, and, in the case of the Earth Charter, the engineering profession has an opportunity to help shape the principles of sustainable development," recalled Jim Poirot.[137] Engineers, he argued, should provide input for international policy declarations. They should do so, Poirot indicated, because without technology development and transfer

---

[136] Leonard (1992); Carroll (1993); WFEO (1997).

[137] Poirot, James W. "The Earth Charter, Engineering Roles and Responsibilities." The Earth Charter Initiative, Virtual Library, 1999.

of technological expertise the Earth Charter's principles could never be applied. Moreover, engineers must acknowledge the "prideful" element in the drafting process of the Earth Charter. Such "Principles of Sustainable Development"—as defined by the Charter's initiators—would set a standard, Poirot believed, regarding "the engineer's compliance" with his/her professional codes of ethics.[138] Similarly, recognizing the value-laden nature of the Earth Charter, WFEO encouraged engineers to consider that organizational intervention was vital, since it could entail "some exceptions to certain principles."[139] WFEO members' concern with possible prevention of genetic engineering is a case in point, as well as other engineers' uneasiness that technological regulation is "bothersome" and that certification and labeling seemed "to imply a huge amount of bureaucracy."[140]

Following the Rio+5 Forum in 1997, the "Earth Charter Benchmark Draft" was issued and the US National Earth Charter Committee was appointed. Among the most prominent of American practitioners, Jim Poirot represented the WFEO on the Earth Charter International Resource Committee from 1997 to 2001. In doing so, he offered his US colleagues the chance to review and provide input to the Earth Charter. In draft after draft, Poirot highlighted the principles he thought had "direct engineering implications," urging members of the EFS to review and comment upon them.[141]

The challenge of sustainable development has since made engineers reconsider key dimensions of their professional identity, and the Earth Charter principles that Poirot did not judge to be engineering-relevant—no less than the ones he did—continue to shape this identity. Most engineers in

---

[138] Poirot, (2000).

[139] During its November (15–20) 1999 General Assembly meeting, which took place in Madrid, Spain, WFEO adopted four resolutions addressing the Earth Charter. According to the 3rd Resolution: "The WFEO continue to be involved in the future by bringing a full Earth Charter including the sub-principles as finalized by the United Nations before the WFEO General Assembly in 2003 recognizing that action by WFEO could include some *exceptions to certain principles*" (emphasis added). WFEO Executive Council. "Earth Charter Resolution." September 20, 2000.

[140] Poirot, (2000).

[141] A "Benchmark Draft no. II" of the Earth Charter was distributed in October 1998 while the WFEO was invited by the Earth Council to contribute to the Earth Charter Continental Conference held in Cuiaba, Brazil in December 1998. Another "benchmark draft no. II" of the Earth Charter was published in April/May 1999 and yet another one on November 15, 1999. Poirot requested that the EFS members provide comment on the current draft of the Earth Charter highlighting seven principles (3, 5, 6, 10, 14, 15, and 16) that "need review and comment" (*EFS Newsletter*, February 1999).

the 1990s and early 2000s viewed the sustainability challenge as being primarily technological in nature. In other words, with respect to building "...participatory...societies," treating "all living beings with compassion...," promoting "gender equality..." and—most controversially—ensuring that engineering projects promote development "in an equitable...manner," the boundaries of sustainability engineering were significantly less flexible.[142]

This ambiguity of the socially and politically charged measures of sustainability revealed itself in the comments that WFEO, and Poirot personally, received about the Earth Charter. "Many engineers," Poirot admitted, "are prone to reject efforts of this nature [drafting the Earth Charter] as extremist, radical and detrimental to financial progress." In addition to this, "rigid" language such as "stop" or "ensure" were deemed inappropriate because "there would be little support for the document." And more importantly, some US engineers' reactions to the Earth Charter prescribed that "sharing the wealth" or promotion of its "equitable distribution" either supported "an extreme socialized government," or "will not be supported in free democratic nations." These preoccupations reflected the heavy influence of the perspective dictating a realistic context for engineering verbiage: As Poirot himself reported, "it is vital to engineers that the Earth Charter becomes a practical and realistic document."[143]

The same ambiguity was true of WFEO's reactions to the Rio Principles. The WFEO commentary on the inseparability of environmental protection and the development process (Principle 4), as well as on the participatory elements of sustainable technology (Principle 10) took a step back from the argument that promoting sustainability is central to revitalizing enthusiasm for technology. Engineers, the commentary detailed, are either "constrained" or "in a very difficult situation" regarding the application of such principles, due to their dependence on client organization policies.[144] And as he sought to answer similar questions in the context of engineering

---

[142] In quotations are the principles of Earth Charter Benchmark Draft No. II that Poirot did not emphasize as having "especially" engineering implications. Some of the principles he did emphasize include: "Advance worldwide the cooperative study of ecological systems, the dissemination and application of knowledge, and the development, adoption, and transfer of clean technologies," "Protect and restore the integrity of Earth's ecological systems, with special concern for biological diversity and the natural processes that sustain and renew life." Poirot, James W. "The Earth Charter, Engineering Roles and Responsibilities." The Earth Charter Initiative, Virtual Library, 1999.

[143] Poirot, James W. "The Earth Charter, Engineering Roles and Responsibilities, The Earth Charter Initiative." Virtual Library, 1999 and Poirot (2000).

[144] WFEO 1997.

implications of the Earth Charter, Poirot wrote discontentedly: "We have yet to effectively explain [to the public] that the decisions to construct... [environmentally harmful] facilities were made by elected officials and corporate officers..."[145] Regarding the precautionary approach (Principle 15), which the WFEO still "recommended," it was argued that "the level case of information available does not permit comment on engineering experience."[146]

In short, Poirot's vision for engineering involvement in the drafting process of the Earth Charter did not take root in the world of US engineering professional societies, despite his conviction that international policy would give engineers a chance to reassert their identity as decision makers in sustainable development. As indicated in the recommendation by an *ad hoc* AAES Task Force created in 1999, the society decided to "neither endorse nor discard the Earth Chapter as it does not purport to become a treaty..."[147] Yet Poirot's legacy, like Hatch's, lies at the very heart of contemporary efforts by societies like the ASCE to create direct links between professional leadership and engineer perspectives on the one hand with sustainability *and* social justice on the other. In that sense, Poirot was a sustainability engineering visionary: he not only attempted to reconfigure the cultural status of engineers as "respected partners in the future built environment," but he also implicitly stirred a process, still ongoing, where engineering contextualizations of sustainability continually stretch the boundaries of engineering reality, technological methodologies, culture and professional identity[148] (Figurc 2.3).

It is essential to the formation of a sustainable engineering identity, this chapter has shown, that debates occurred between those who advocated for greater sustainability but differed in how they approached it. Another sustainability engineering visionary, Don Roberts, advanced Poirot's ideas.

---

[145] Poirot (2000). Along similar lines, a decade earlier, the US Corps of Engineers James (Jim) Waddell warned: "This [guiding the decision making process towards sustainable development] will be inherently difficult if the engineer's perspective is broader than that of the client or local community. At times it may be impossible, and decisions will be made contrary to what the engineer recommends as sustainable" Waddell (1991).

[146] Less ambivalent on the role of precaution in engineering, FIDIC's commentary argues that the "precautionary principle could provide a basis for policies relating to complex systems that are not yet fully understood and whose consequences of disturbances cannot yet be predicted." FIDIC (1994).

[147] A taskforce formed by AAES' International Committee (IntAC) with invited participation from the AAES Engineers Public Policy Council reviewed Draft II of the Earth Charter (with notes by Jim Poirot) on March 29, 1999.

[148] Poirot (2000).

Figure 2.3 "'Mr. Cool' himself…a leader, planner and 'deep thinker.'" Jim Poirot was described by his CH2M Hill colleagues as having "dedicated himself to advancement of the profession." Adapted from *Engineering News Record*, February 18, (1988): 30–31.

Roberts' perspective clearly echoed the notion of a unified—virtually connected—profession under the mandate of sustainable development. Around the time that Poirot was seeking allies among the US engineering societies, Roberts designed a ten-year plan to establish an "effective communications network" for sustainability engineers and to evaluate national progress in sustainable technology exchange and capacity building. Among other points, Roberts' plan included the development of guidelines to conduct sustainable engineering in different types of projects, engineering mentorship programs, and the collaboration of retired engineers with a "Youth Corps" that would educate engineers in developing countries.[149]

---

[149] Roberts, Don V. "Improving Engineers' Response to Agenda 21, Chapter 34." 2001, manuscript. This was the same project for which, a year later, Roberts would suggest to

Conclusion

No *a priori* relationship exists between sustainability and the engineering profession. Such a view takes both "engineering"—along with notions of professional identity, practices and understanding—and "sustainability" for granted, whereas this book aims to address their coevolution. A contemporary engineer, for example, may trace the roots of sustainability to the protoplasts, or quote Thomas Jefferson, purporting that the latter "summed up sustainability eloquently."[150]

But history is not teleological: that is, there is no predetermined script connecting the past with the present; nor is it linear. In other words, there is no *necessary* nor unidirectional relationship between Adam and Eve, Thomas Jefferson, Jay Forrester, the World Engineering Partnership for Sustainable Development and current debates around the requirement specifications for attaining sustainability in an engineering context.[151] The engineering world, no less than the world itself, is constantly undergoing a process of profound change as a result of friction between conflicting views of technology. Moreover, the cultural and technological grammar that informs the integration of competing technology ideologies also shifts over time and across place. The contextualization of sustainability in American engineering is an example of such an ideological integration.

Chapter 2 has recounted the story of how—in order to catch up with the "sharp end of changing social values" in the late 1980s and 1990s—engineers and their organizations adapted their technological methodologies, practices, rhetoric and professional identities to meet the challenge of a degrading environment.[152] The standard language of engineering proponents of sustainability was that of "systemic interdependence." For example, the WEPSD, the epitome of organizational rationality of technological change, was described as an "integrated partnership."[153] One of the most

Norman Neureiter, science adviser to the US secretary of state, that a "coalition of donors" be formed to support it.

[150] Allen, Don, M. Carla, and D. Mark Webster. "Introduction." In *Sustainability Guidelines for the Structural Engineer*, edited by Dirk Kestner M., Jennifer Goupil, and Emily Lorenz, 1–7. Reston: ASCE, 2010.

[151] "The challenges related to stewardship of resources and sustainability" wrote one engineer in 2009, "have been with us since the Lord told Adam and Eve that he was putting all of his creation under their care, while at the same time telling them to have many children. I doubt that they will be solved in the near future. Bazan-Arias (2009).

[152] WFEO (1997): 28.

[153] Ellis (1994): 6.

profound ways, then, by which engineering ideologues of sustainability in the 1980s and 1990s contributed to the contemporary definition of sustainable development was not through discursive competition, but by *sharing* the same grammar—that of ecosystems theory—to provide accounts of natural and social reality. As a result, conflicting elements between the ideologies of technological change and technopolitics often faded when "sustainable technology" was described in the context of systemic interdependence. The very existence of a common sustainability engineering grammar blurred the ends and the criteria for the application of sustainable technology.

During the historical period (1985–2003) examined in this chapter, sustainability in an engineering context meant, in the words of Joe Herkert, "strik[ing] a balance between growth and [environmental] protection."[154] At the same time, and primarily as a response to criticisms such as Herkert's, a trend in progressive engineering began recognizing a social dimension of sustainability. In fact, technological change ideologues have been referring to a three-pillared conceptualization of development at least since 1995: more recently, in 2002 the US engineering community's Declaration to the World Summit on Sustainable Development recognized that "engineers must deliver solutions that are technologically viable, commercially feasible and, environmentally and socially sustainable."[155] The formulations and techniques of engineers, like LCA, were influenced by cultural boundary shifts within the profession.

The integration of the two ideologies of sustainability engineering examined in this chapter manifested itself in the progressive incorporation into engineering input-output techniques of what were initially viewed by most practitioners as extra-engineering dimensions of "the system." The engineering contextualization of social and political dimensions of sustainability was operationalized with the boundaries of LCA to include qualitative impacts of technologies and engineering projects.

The relationship between sustainability and engineering was debated extensively in the articles and declarations of academics and spokesmen of the profession, as the next chapter explains, but it was also significantly mediated by activities in the industrial and engineering consulting sectors

---

[154] Herkert (1996).

[155] Vanegas, Jorge A., R. Jennifer DuBose, and R. Annie Pearce. "Sustainable Technologies for the Building Construction Industry." In *Designing for the global environment: a symposium*, edited by Saeid Lohbani Sadri [16 pages]. Atlanta: College of Architecture, Georgia Institute of Technology, 1995; AAES, AIChE, ANS, ASME, NAE, NSPE (2002).

(e.g., FIDIC). In fact, Wayne Tusa, the consulting engineer and chair of ASCE's task committee on global resource planning at the time, suggested in 1993 that "some companies may be more aware [of sustainability] than engineers." "Civil engineers," he asserted, are often ahead of the curve on issues, but not on this one."[156] "Efficient enterprises," concurred Maurice Strong, "are at the head of the movement to sustainable development."[157] And they were rightly noticed, for it was consultants like Roy F. Weston Inc.'s Jim Fava, and corporate engineering managers like DuPont's Charles Holliday who operationalized sustainability in an engineering technique such as LCA and solidified its status in cooperation with international development institutions such as the WBCSD and WRI. "At the end of the day," urged Dow Chemical's David T. Buzzelli, "you must have something you can measure." In other words, "[w]e break it down into areas that people are familiar with," recalled Dow's director of environment, health and safety, and business performance Scott Noesen. "We know how to get all those [economic and environmental] numbers," reassured DuPont's Dawn Rittenhouse.[158]

Chapter 2 illustrates that an ideology of technological change provided practitioners an interpretive framework in which sustainability became meaningful to engineers while softening the toll on traditional professional values. Technological change ideologues conceptualized sustainability as compatibility between economic development and environmental protection. By emphasizing their managerial qualities—particularly on the assumption that industrial inputs and outputs have measurability for practitioners—engineers cast themselves as facilitators of sustainable development. Their legacy is well established in the popular worldview, according to which society is understood as a vision articulated by uncontrolled, yet manageable and therefore sustainable, technological forces. Seen as the way to navigate the unintended consequences of technological change, sustainability made ideological space for the persistence of the notion of engineering progress in society.

---

[156] Prendergast (1993).

[157] Quoted in Manion, Mark. "Ethics, Engineering, and Sustainable Development." *IEEE Technology and Society* 21, no. 3 (2002): 39–48.

[158] Morse, Paige Marie. "Taking a measure of sustainability: Chemical companies are using a variety of resources to determine effective ways to chart progress in sustainable business." *Chemical and Engineering News* 77, no. 30 (1999): 19–22.

3

Resisting Sustainability

In the 1990s, the same affirmations of rationalizing technological leadership in the context of global eco-change that promised engineers an escape from cultural and political invisibility, substantially challenged dominant professional values. Despite Don Roberts' "exhortations" about sustainability, reported *Engineering News Record*, the likelihood of heightened "politicism" by FIDIC members became highly controversial during the first-ever engineering conference on sustainable development held in Oslo, Norway, in June 1990. "It is a new aspect for us to have a double loyalty—one to the client, and one to society," articulated an Oslo conference attendee.[1] That is, "suppose a client asks your firm to build a highway in a country where there are a few cars," wondered another engineer six years later at a "sustainable construction" symposium. "If you really think this is not the best use of the world's resources, do you tell him so and turn down the job?"[2] In order to address these questions, professionals had to revisit their values of

---

[1] Rubin (1990).

[2] Johnson, Darcie B. "What are the ethical responsibilities towards sustainable development?" *Journal of Professional Issues in Engineering Education and Practice* 125, no. 2 (1998): 35–37.

engineering servitude. So while General Hatch's calls for sustainability were frequently met with resistance from fellow practitioners, he nevertheless insisted that only those "few professional organizations [which] are holding back" are maintaining the idea that the engineer's duty is to conform with clients' requirements. Hank Hatch, the embodiment of the technological change ideology of sustainability, "respectfully reject[ed]" that idea.[3]

What I suggest in this chapter is that contemporary notions of sustainability engineering took shape along with engineering identity politics of professionalism. Engineering professionalism and sustainability have been both interconnected and, often times, contradictory. As discussed earlier, the FIDIC's 1990 Policy Statement calling engineers to accept the challenge of a threatened planet ignited heated debate in Oslo. The statement, inviting consultants to exert influence over policy making, also argued that the treatment of environmental problems be reinstated or adjusted to include all phases (from planning to decommissioning) of a project.[4]

Nor was the Policy Statement the only instance showing that for engineers, as the June 1990 *Engineering News Record* editorial put it, "advocacy will find you no matter how hard you try to avoid it." It was consulting engineering leader Knut Ekenberg himself who substituted for Norwegian prime minister Gro Harlem Brundtland to give the FIDIC 1990 keynote address, when—one day prior to the conference—Brundtland withdrew, boycotting the presence of delegates from South Africa. Thus, for consulting professionals entangled in controversy over the role of the engineer in the LCA of technological project environmental impacts, the Oslo conference substantiated a reality that engineering advocates of sustainability became quite familiar with in the course of the 1990s: "even apolitical technicians are not immune to political events."[5]

Jumping on the Bandwagon

Given that sustainability may be "the only salvation" for a world threatened by environmental decline, wrote James Waddell of the US Army Corps of Engineers in 1991, "one might ask why engineers as a group have been relatively slow to 'jump on the bandwagon' and embrace this philosophy."[6] Seventeen years later, a *Civil Engineering News* column remarked, "few subjects have been more debated" amongst practitioners than sustainable

---

[3] Atkisson (1992).
[4] FIDIC (1990).
[5] Unknown. "*Engineering News Record* (1990).
[6] Waddell (1991).

development. Just as the FIDIC 1990 Policy did not find universal appeal in the engineering profession, so too have many individual practitioners and engineering constituencies been reluctant to make the sustainability commitment. Some practitioners, continued *Civil Engineering News*, "opine that adding the adjective 'sustainable' to a project lowers the probability of the work being conducted." In this context, some engineers also believed that sustainability is extra-engineering subject matter.[7]

Together these features illustrate, for example, why it was not until 2006 that the "First Fundamental Principle" of ASCE's code of ethics required engineers to use "their knowledge and skill for the enhancement of human welfare *and the environment*".[8] And even as some of its most influential members sought to encourage fellow engineers to accept the concept, the ASCE Board of Direction adopted the first formal engineering definition of "sustainable development"—echoing the 1987 Brundtland articulation—in 1996.[9] Likewise, the environmental, economic *and* social pillars of sustainability were first officially linked together by ASCE's Board of Direction in October 2009.[10] The efforts of elite engineering proponents of sustainability to contextualize

---

[7] Bazan-Arias, Cathy. "Guidance on ethical dilemmas and the history of sustainable development." *CENews*, December 30, 2008.
<http://www.cenews.com/print-magazinearticle guidance_on_ethical_dilemmas_a-6236.html>
Accessed October 25, 2013.

[8] Emphasis added. The first time that the ASCE code included the word "environment" was in paragraph (f) of the 1976 version of the *guidelines to practice* for Canon 1: Paragraph (f) read that "Engineers should be committed to improving the environment to enhance the quality of life." Fourteen years later, in August 1990, IEEE revised Canon 1 of the society's Code of Ethics which currently reads: "[IEEE members agree] to accept responsibility in making engineering decisions consistent with the safety, health and welfare of the public, and to disclose promptly factors that might endanger the public or the environment."
IEEE. "Code of Ethics." Undated.
< http://www.ieee.org/about/corporate/governance/p7-8.html>
Accessed May 9, 2015.

[9] The definition read: "Sustainable development is the challenge of meeting human needs for natural resources, industrial products, energy, food, transportation, shelter, and effective waste management while conserving and protecting environmental quality and the natural resource base essential for future development."

[10] "Sustainable Development is the process of applying natural, human, and economic resources to enhance the safety, welfare, and quality of life for all of the society while maintaining the availability of the remaining natural resources."
ASCE. "Code of Ethics." Undated.
<http://www.asce.org/Ethics/Code-of-Ethics/>
Accessed December 25, 2013.

the concept were met with resistance by individual engineers, professional organizations and industry. Why did some engineers view sustainability not as a force of professional recognition and technological supremacy but as a site of contesting values threatening professional engineering normalcy?

"Engineers have tended to be wary of involvement in politically and emotionally charged issues, full of ambiguity," proclaimed Hank Hatch in his address to the 1989 ASCE annual convention in New Orleans, an event that marked the entrance of sustainability in US engineering discourse. "These issues," he further conceded, "are not compatible with our pragmatic style. Moreover, social and environmental questions pose *uncomfortable ethical issues*".[11] For most engineers in the 1990s and early 2000s, the path to professional change was technology—not politics, nor philosophy. Or as one engineer lamented, sustainability is to be blamed for the dependence of policy on emotiveness and deliberation than on factual and rational judgment.[12] Though central to the study of engineering interest in contextualizing sustainability, "philosophical statements will not carry the day," observed former DuPont chairman Edgar Woolard.[13]

Yet the concurrent anti-sustainability arguments made by some within the engineering profession during the 1990s indicate that resistance to sustainability was primarily on philosophical and ideological grounds—and not in terms of "measurable…target[s]," as Woolard proclaimed.[14] A British engineer working for an international engineering company in Hong Kong remarked how many engineers still associated environmentalism with extremism in an article that traced the principles of sustainable development in Taoism and Confucianism.[15] That the grounds for acceptance or rejection of sustainability were fundamentally social and political was also the focus of one sustainability critic who maintained that the question "to be openly addressed is whether civil engineers would remain on the environmental bandwagon or would… resist extending the extremist (socialist even) agenda into virtually every corner of human lives." Would engineers endorse a "crisis mentality" and accept a status of dependence on "government funding," he questioned, or would they begin to contradict and debunk every sustainability claim?[16]

---

[11] Hatch (1989), emphasis added.

[12] Chadderton (1995).

[13] Woolard (1999).

[14] Ibid.

[15] Kennard, Robert M. "Environmental Challenge for Engineers: Hong Kong Perspective." *Journal of Professional Issues in Engineering Education and Practice* 121, no. 2 (1995): 140–142.

[16] Chadderton (1995).

A good deal of the engineering counterargument to sustainability focused on the ideological ramifications of the concept, particularly in the implication of a "socialist" mode of governance or reconsideration of the dominant patterns of resource distribution. A "worthy vision" indeed, but as the IEEE Electronics and the Environment Committee observed in 1995, sustainability is "inescapably expressed in value-laden terms...implying for some people redistribution of wealth or a need to restrict current consumption." Sustainability was thus considered by IEEE's committee to be inappropriate guide for policy making.[17] The same counter-perspective is clearly discernable in one practitioner's response to an article that Joe Herkert published in 1997 in IEEE's journal *The Institute*. "Was the Herkert article intended to say that engineers... should support some kind of socialist malarkey on 'sustainable development' and one-world government type 'globalization,'" the respondent wondered sarcastically—ignoring, however, the standard use at the time of "globalization" in IEEE discourse.[18]

The sustainability engineer—as portrayed by Hatch, Roberts, Weston and other technological-change ideologues—would be an innovator and Earth steward, a local engineer and renowned facilitator of global policy, a leader of other professions in the technological change race, a planter of the seeds of eco-transformation who shares the fruits of development with "third world" countries. But as one engineering student pointed out in the late 1990s, "that does not mean there [would] be no opposition to the new ways" of being an engineer and doing engineering work.[19] At the same time, and utilizing the social and political discourse they felt was taking over the "engineering dimensions" of development, proponents of sustainability engineering gave their own interpretation of the resistance that was forming against the emerging professional paradigm. Sustainability "faces political resistance," wrote WRI's Robert Repetto in 1990, "from some groups that would face either short-term adjustment costs, or the long-term loss of policy entitlements."[20]

The intellectual foundations of sustainability were also debated on philosophical grounds. While endorsed by technological change and

[17] IEEE (1995).

[18] Herkert. Joseph. "ABET's Engineering Criteria 2000 and Engineering Ethics: Where do we go from Here?" Paper Presented at the OEC International Conference on Ethics in Engineering and Computer Science, March 1999.

[19] McWhorter, Ron. "What are the ethical responsibilities towards sustainable development?" *Journal of Professional Issues in Engineering Education and Practice* 125, no. 2 (1998): 37–39.

[20] Repetto, Robert. *Promoting Environmentally Sound Economic Progress: What the North Can Do*. Washington, DC: World Resources Institute, 1990. [Repetto (1990) (b)].

technopolitics ideologues alike, the standard sustainability engineering metaphor of natural systems that Roberts first described in detail in 1990 did not go fully unchallenged. The idea of a system "is flawed" if adopted as the conceptual foundation of sustainability reflected one practitioner in a 1997 issue of the *Journal of Professional Issues in Engineering*, because "in the long perspective, no natural system can be sustained to fulfill *the same function* throughout time."[21]

To professionals who thought that extra-engineering discourse was counter-productive to the proliferation of the principles of sustainable development, the merging of "rational" and "ideological" or "political" elements of sustainability may have seemed oxymoronic. Yet, as this chapter details, this was not the case. The civil engineer and chair of the Oregon Tech Sustainability Committee, Charles Riley, nicely captured this often dismissed dimension in the history of sustainability in the engineering profession: "The students are at times inclined to incorporate sustainability concepts and at times not depending on their personal background and their view of its profitability."[22] Even if very practical concerns like unfamiliarity with certain techniques certainly hampered practitioners' willingness to comply with the emerging paradigm of engineering project development, downplaying the ideological and philosophical underpinnings of sustainability engineering leaves half the story untold.

The Never-to-be ASCE Eighth Canon of Ethics

"I consider... [sustainability] a fad and a farce of enormous waste to our people," an engineer observed critically in 2009. "It's the reason I quit ASCE," he wrote; because of "that 'code' of ethics bit."[23] Such a reaction against sustainability from practicing engineers was nothing new to the profession. In fact, precisely the story of "that code of ethics"—and its intersection with the concept of sustainable development—is typical of the resistance that engineering proponents of sustainability faced from their fellow professionals in the 1990s.

Long before, in 1976 the Environmental Impact Analysis Research Council (EIARC) of ASCE had drafted an Eighth Fundamental Canon for the society's Code of Ethics: the suggested Canon's purpose was "to expand

---

[21] DeMers, Gerald L. "Discussion: Evaluating Technological Alternatives on Basis of Sustainability." *Journal of Professional Issues in Engineering Education and Practice* 123, no. 1 (1997): 43.

[22] Personal communication.

[23] Personal communication.

ASCE ethical commitments to include environmental values."[24] It seems that environmental considerations played little role in engineering codes of ethics at the time, for the 1976 Canon was never enacted. Six years after its first articulation, and two years before the development of WFEO's remarkably progressive "Code of Environmental Ethics for Engineers," a second version of the 8th Canon was proposed by EIARC, to be rejected unanimously in January 1984 by ASCE's Professional Activities committee.[25] Legal concerns and the assumption that its provisions were covered by ASCE's 1980 Policy 120 were, according to two engineering practitioners Vesilind and Gunn, the main reasons which led to rejection of the Canon that read: "Engineers shall perform service in such a manner as to husband the world's resources and the natural and cultured environment for the benefit of present and future generations."[26] Another attempt to enact the 8th Canon was put forth almost immediately within the ASCE organization, through the creation of the Committee on Engineering Responsibility in 1985. Nevertheless, the committee's vision was not endorsed by the leaders of the society and ASCE therefore soon disbanded that committee.[27] Like sustainability itself four years later, the 1985 8th Canon of the Committee on Engineering Responsibility was introduced to engineers by a leading member of the US Corps—in this case, by Gerald Galloway, a water management specialist and leader of the short-lived Engineering Responsibility committee.[28]

By the early 1990s, consideration of "sustainable development" in their codes of ethics had become a key self-positioning strategy for engineering professional societies. This was especially true within ASCE but occurred in other societies as well. For example, in 1993, ASCE's newly formed task committee on sustainable development began, once again, to assess the feasibility of adding an 8th Canon to its Code. By 1994 both ASCE and Australia's IEAust had drafted very similar canons, not only mirroring

---

[24] EIARC. "A Proposed Eighth Fundamental Canon for the ASCE Code of Ethics." *Journal of Professional Issues in Engineering* 110, no. 3 (1984): 118–122, 118.

[25] Vesilind, Aarne P., and S. Alastair Gunn. "Sustainable Development and the ASCE Code of Ethics." *Journal of Professional Issues in Engineering Education and Practice* 124, no. 3 (1998): 72–74.

[26] Vesilind and Gunn also note that a "former chairman of the EIARC later admitted that...[the legal concern objection] seemed implausible." ASCE. "Assuring a Desirable Quality of Life." ASCE Policy Statement no. 120. New York: ASCE, 1980. The statement recommended that "[c]ivil engineers must recognize the effect their efforts will have on the environment, by increasing knowledge and competence in incorporating ecological considerations in design."

[27] Ibid.

[28] Dr. Gerald Galloway curriculum vitae, manuscript.

the Brundtland commission's emphasis on natural resource preservation for meeting long term societal needs, but recognizing explicitly non-engineering participation in project design as a vital element of sustainability.[29] The third version of a proposed 8th Canon was disseminated to the ASCE membership for comment in February 1995, but was again met with opposition until it was finally disapproved during an ASCE Board of Directors meeting in October 1995.[30] This suggested Canon—drafted under the leadership of *EFS Newsletter* editor and technological change ideologue of sustainability, Al Grant—would have added two elemental dimensions to the ways engineers perceived themselves and the role of non-engineers in sustainable project development.[31] According to Don Roberts, one of the five principles mentioned in the code's 1995 draft stipulated that engineers "include[d] environmental costs and benefits for the life-cycle of the product."[32] Importantly, the draft considered provisions on "Public Involvement" and on "Multidisciplinary Systems Approach," the language of which is glaringly absent from current versions of ASCE's Code of Ethics:

> Public Involvement. Engineers should encourage full involvement of all project participants in the planning, execution, and management of civil engineering projects. This includes early disclosure of information appropriate to project planning. *Engineers should encourage public decision-making processes*, which include project participants, designed to reach acceptable solutions without unnecessary conflict or litigation.

---

[29] Canon 3 [Engineers should practice Engineering in Accord with a Sustainability Ethic that Leads to Sustainable Development] of the 1994 IEAust Code of Ethics had many similarities with the 1994 ASCE suggested eighth Canon. Ellis (1994): 20.

[30] Unknown. "A question of Ethics: Development of Sustainability Provisions in ASCE Code of Ethics." Undated.
<http://www.asce.org/Publications/ASCE-News/2008/11_November/A-Question-of-Ethics/> Accessed January 3, 2014.

[31] The Renewable Resources Journal in 1995 wrote of Grant: "His exploration of the environmental codes of ethics of various international engineering societies helped shape the debate over the additions of such a code for ASCE… He was responsible for the preparation of the first ASCE Strategic Plan [on sustainable development]." Grant. Albert A. "The Ethics of Sustainability: An Engineering Perspective." *Renewable Resources Journal*, Spring (1995): 23–25, 23.

[32] Quoted in Duffell, Roger J. "Toward the Environment and Sustainability Ethic in Engineering Education and Practice." *Journal of Professional Issues in Engineering Education and Practice* 124, no. 3 (1998): 78–90.

> Multidisciplinary Systems Approach. Engineers should *adopt and apply an integrated, multidisciplinary systems approach* where the relationship of the parts to the whole is considered. Engineers should be aware of the architectural, engineering, historic, cultural, archaeological, geological, and biological worth of sites and the existing structures and living systems within them. Further, engineers should strive to include the preservation of high quality, unique, and rare natural resources and systems in project formulations.[33]

The 8th Canon polemic was carried out in ASCE meetings and vote was taken by voice of show of hands.[34] Referring to the 1995 version of the Canon, which posited that "[e]ngineers shall perform services that help sustain the world's resources, while protecting the natural and cultural environment," the emeritus professor in civil engineering at the University of Hertfordshire, Roger Duffell, opined that it "generated the most fierce national debate ever within ASCE, with the membership split almost 50/50." Younger engineers, commented Duffell, were more likely to embrace the Code as opposed to older professionals who "feared for job security and litigation."[35] Rather than identify sustainability with a measurable goal—a circumstance considered lamentable by most engineers—the controversy over ASCE's 8th Canon attests to the fact that advocacy for sustainability, and the search for methods to operationalize it, were practically inseparable.

"It's the swallow-without-eating directive of the proposed sustainability canon that makes no discernible sense..." complained one engineer from California, in a letter to *Civil Engineering*. "[The] canon is ethically naïve," he claimed, for it "would be unenforceable, partly because most breaches would be unrecognizable and partly because thousands of engineers simply do not routinely perform the services the canon purports to require..."[36] In keeping with visions of technological change, and echoing the 1980 ASCE Policy 120 that warned against "overregulation," as an economic inhibitor another opponent expressed his concern that a possible endorsement of the canon would actually stifle technological progress.[37] Even so, in both cases

---

[33] Emphasis added.
[34] Interview data.
[35] Duffell (1998).
[36] Sonnen, Michael B. "Sustainability Canon Is Not Justified." *Civil Engineering* 65, no. 9 (1995): 35–36.
[37] ASCE (1980); Weldon, David G. "One Environmental Canon Is Enough." *Civil Engineering* 65, no. 9 (1995): 35–36.

sustainability adversaries left the "environmental" aspects of engineering outside the scope of their criticism.

The 8th Canon debate in the pages of *Civil Engineering* reveals how deeply entangled in philosophical and ideological premises the engineering contextualization of sustainability has been. It also shows that any resolution on sustainability in an engineering context depended—as it does now—on forging ideological compromise between technological change and technopolitics. In their postulations about sustainability, professionals sometimes combined definitions of development at the intersection of economic growth and environmental assessment with arguments for equitable distribution of resources.

A student engineer's letter of support to the Canon featured two familiar elements: increasing engineering visibility through sustainability and fostering closed-loop ecosystem function as the gauge-metaphor of development. "I find sustainability to be an opportunity for engineers to recognize the overwhelming, unmatched importance of the profession on society..." he wrote, while "respectfully" disagreeing with critics. Echoing Roberts's and many other sustainability enthusiasts in the profession who later followed in his steps, he backed up his claim, arguing that practitioners "should appreciate the significance of nature as a closed system in which the waste of one is always the food of another..."[38]

Perhaps the most balanced description of the stakes raised by the 8th canon for ASCE members was provided by Lawrence Quinn, an industrial and systems engineer who after a long career in water project management completed his PhD work in the development of an LCA (ISO14001) methodology to assess the sustainability of international water development projects. "[T]he proposed changes to the Code... are simultaneously too nebulous (when using terms such as 'sustainable development principles')," Quinn remarked in 1996, "and too weak when one considers the overriding nature of the issue." And to justify his assertion that he "had hoped for more substance" to the Code, Quinn went on to suggest an alternative formulation of the 8th Canon, assuring that its passage would help engineers "regain their positions of influence regarding economic-cum-environmental planning..." Yet Quinn's version of the 8th Canon was equally vague: it premised engineering leadership, and "multidisciplinary

---

[38] Johnson, Richard R. "Sustainability is key to engineering relevance." *Civil Engineering* 65, no. 12 (1995): 28.

techniques that recognize the inherent interrelationships of people and the natural and cultural environment."[39]

By late 1996, the controversy over the 8th Canon of ASCE's ethics code had reached closure with a modification of the society's "First Fundamental Canon" to include sustainable development for the first time: "Engineers shall hold paramount the safety, health and welfare of the public and shall strive to comply with the principles of sustainable development in the performance of their professional duties." This development may have put sustainability "on par with public safety in the ASCE Canon," as Hank Hatch boldly announced to his World Bank audience in 1998, or it may not have. The guideline (If) of the Canon read that "engineers *should* be committed to improving the environment...." Indeed, the use of "should" blunts the effectiveness of the Canon, a fact that made critics wonder whether it was "unkind" or not to assume that the Canon's modification reflected not "a meaningful change in the actions of civil engineers," but an effort to enhance the society's public image.[40]

In sum, the struggle over the 8th Canon may, however, be one window to understanding the intersection of sustainability and professional engineering identity politics in the US. Like the engineering for "social responsibility" groups that were first founded outside the US indicating the dominance of technocratic ideas in American engineering, the passage of the Australian Engineering Code of Ethics in 1994 whose complementary "Environmental Principles" further recognized "the rights of the community to be involved in project formulation and development and actively encourage such involvement"[41] implied that what happened in the US reflects a particularly American engineering culture with stronger liberal economy, market and corporate ties.

Easing Resistance Through Education and LCA Operationalization

Engineering education is a key locus for professionals to recontextualize themselves and the type of work they perform.[42] To proponents of the

[39] Quinn, Larry. "Sustainability: Another New Paradigm." *Civil Engineering* 66, no. 10 (1996): 6.
[40] Vesilind and Gunn (1998).
[41] For example, the 3.11 provision of the IEAust Code stated that engineers need to "recognize the rights of the community to be involved in project formulation and development and actively encourage such involvement" (Ellis (1994): 20).
[42] Conlon, Edward. "The new engineer: Between employability and social responsibility," *European Journal of Engineering Education* 3, no. 2 (2008): 151–159.

technological change ideology of sustainability, engineering education presented both a site of opportunity and a set of cultural resources for envisioning the "sustainability engineer." Like his General in Chief in the US Corps, James Waddell identified as early as 1991 the " 'feelers' and 'visionary intuitors' who learn and thrive by interpersonal relationship and interaction, enjoying the sense of being a member of a team;...who analyze information in a more circular iterative manner" as "precisely the type of person[s] who can more easily accept and carry out the concepts of sustainable development, *adjusting their vision of problem-solving to include a more holistic perspective*."[43] By 2009 the belief that education is key to implementing sustainability—namely, that a broader understanding of the options and impacts of engineering work was needed if the profession was to move toward a more viable trajectory—was commonplace within both ideological camps of sustainability engineering.[44]

As visionaries of sustainability engineering filled journals with their speeches and programmatic articles, and as sustainable development conferences from an engineering perspective became more frequent and attracted more practitioners from professional societies and academia, industry engineers were still "struggling" to understand how, "in practical terms," sustainability would provide a guide for future action.[45] Industrial news venues like *Chemical & Engineering News* often reported that major US chemical companies needed help from the government in translating fuzzy goals and concepts into engineering services.[46] Over time, to address the pressing issue of operationalizing sustainability, more and more engineers suggested that a modified environmental impact analysis including effects on future societies and ecosystems be considered.[47] More and more, engineers and their professional organizations reformulated their advocacy and policy statements to mirror an identification of sustainability, with a holistic life cycle approach to the design of engineering projects. For example, in 2001 a "subcommittee on sustainability" was organized by ASCE's Technological Activities Committee (TAC) to identify, pull together and publicize the role of the Society in calling attention to the new paradigm; and the Society revised its 418 Policy Statement on the "Role of

[43] Ibid. emphasis added.
[44] Bazan-Arias (2009).
[45] (CERF) 1996, emphasis added.
[46] Johnson, Jeff. "Environment and the bottom line: Rand study of four leading companies shows environmental concerns are being joined to overall corporate planning." *Chemical and Engineering News* 77, no. 25 (1999): 25–26.
[47] Vesilind and Gunn (1998).

the Civil Engineer in Sustainable Development." Reflecting the attitudes of the subcommittee's members—Al Grant (chair), Hank Hatch, Mike Sanio and others—the Policy statement required that engineers "[a]dopt and apply an integrated systems approach for project decisions in which costs, benefits and effects on sustainability are considered for the whole lifetime and enduring effects of the project."[48]

Engineering identity is strongly connected with quantitative balancing of "costs" and "benefits". In that context engineers "should think about the ultimate fate of manufactured goods in a total life-cycle analysis," explained Virginia Tech's Jon Novak, voicing—according to a 1999 edition of *ASEE Prism*—one of the "two approaches to incorporating sustainability perspectives into the engineering curriculum."[49] For all the difficulty in translating sustainable development in practical terms that industry engineers could understand and apply in commercial operations, engineering educators viewed the introduction of "sustainability" into the engineer's lexicon as a natural extension of the systems approach.

"Design constraints relating to sustainability concerns in the project life-cycle are introduced during the problem definition phase of the systems engineering design phase," was one response in a 2002 survey of the Engineering Deans Council of ASEE when asked about ways to introduce sustainability into the engineering curriculum. Specifically, "[t]he concepts of designing unit operation treatment processes for water and wastewater treatment are discussed in terms of their long-term integration into the natural environment," said another engineering educator.[50] Like his colleagues in ASEE, William Kelly of the Catholic University of America viewed LCA not only as the exemplar application of sustainability engineering, but also as the metaphor to describe the idea itself. Kelly—with a history of involvement in engineering education with both ASEE and the Accreditation Board for Engineering and Technology (ABET)—looked with longing toward the full implementation of LCA in sustainability engineering. In 2003, while chair of ABET's Engineering Accreditation Committee, he assured fellow engineers that "Life-cycle design"—although "far from normal practice"—is one potential way

---

[48] *EFS Newsletter*, May 2001.

[49] The other approach was Georgia Tech's. Gibney, Kate. "Cleaning up on cleaning up: sustainable development is a new way of doing business." *ASEE Prism* 8, no. 5 (1999): 20–25.

[50] *EFS Newsletter*, August 2002.

through which engineering professionals could make a difference in sustainable development.[51]

A connection thus may exist between the operationalization of sustainable development through LCA and the normalization of sustainability in engineering education. A 2002 *European Journal of Engineering Education* article, for example, had it that—according to engineering students themselves—the ability to perform calculations with LCA software was the most popular response indicating "expert understanding" of sustainability "content knowledge" (Figure 3.1).[52] By 2010, LCA would be the main denominator of sustainability engineering metrics and sustainability engineering education. The most comprehensive survey to date of sustainability in US engineering education was published in 2009 by David Allen and coauthors: they acknowledged that sustainability-relevant courses "concentrate primarily on...the firm (gate-to-gate or design for environment) or product (cradle to grave or environmental life-cycle analysis)." And in concluding that LCA was the cognitive vehicle of sustainability engineering operationalization, it was also understood that LCA defined the field's "system boundaries"—along with "Design for Environment," "Industrial Ecology" and "Cultural and Social Dimensions." Finally, although this operative definition of sustainability engineering, according to the report's authors, "can be viewed as a return to systems approaches to engineering design," they also concluded that while "the analysis tools needed to evaluate economic metrics are generally covered in current engineering programs...," and "the tools needed to assess environmental metrics are covered in some engineering education programs..., [t]he tools needed to evaluate social metrics are largely absent from engineering curricula."[53]

Sustainability engineering was defined against the background of the dialectic between technological change and technopolitics and by the criteria set by the politics of engineering professionalism. This section showed that the fact that sustainability was debated in extra-engineering terms made it difficult for the majority of US professionals in the 1990s to see

---

[51] Kelly, William E. "Sustainability in Engineering Education." *EFS Newsletter* March (2003): 1–3.

[52] Carew, Anna L., and A. Cynthia Mitchell. "Characterizing Undergraduate Engineering Students Understanding of Sustainability." *European Journal of Engineering Education* 27, no. 4 (2002): 349–361.

[53] Allen, David, *et al.* "Benchmarking sustainable engineering education: Final report." Austin: University of Texas at Austin, Carnegie Mellon University, Arizona State University, 2009. Pages iii, 2, 8 and 12.

Expert ways of knowing	Elements of expertise (table 2)	Expert sustainability knowledge (statement 1)	Example of sustainability-relevant learning outcome (some from table 1)
Content knowledge	Declarative knowledge	'To have factual and theoretical knowledge of sustainability . . .'	Use computer program for life cycle analysis calculations (e.g. Sima-Pro®)
	Theoretical knowledge		Understand the principles of life cycle thinking
Structure of knowledge	Procedural knowledge	'. . . to appropriately apply sustainability knowledge to contextualized decision-making . . .'	Principle 2: Recognize the interdependence and intradependence of ecosystem, socio-system and economy (table 1)
	Conditional knowledge		Principle 8: Recognize the unique contextual factors in each decision-making situation (table 1)
	Critical thinking	'. . . judging the ethics and sustainability of one's decisions and decision outcomes'	Principle 9: Take responsibility for the impacts resulting from decisions (table 1)

Table 3. Examples of the elements which may contribute to development of an expert-like understanding of sustainability.

Figure 3.1 "Examples of the elements which may contribute to development of an expert-like understanding of sustainability." Adapted from Carew and Mitchell (2002): 353.

its practicality and accept it as part of their engineering identity further occluding efforts—like the 8th Canon—to broaden the boundaries of the concept. In the next section I discuss AESR, a parallel, but intersecting story of more radical, politically oriented engineers. The story of AESR is part of an internal professional politics between those who were advocating for sustainability in engineering. Instead of settling on LCA, however, this group shows a different trajectory that sustainability proponents have taken.

Negotiating Ideas about Self and Nature: the Case of American Engineers for Social Responsibility (AESR), 1988–1994

American Engineers for Social Responsibility (AESR), 1988–1994

In 1987, Jim Evans, a professor of structural engineering at the University of Washington, and Tom Munsey, a hydraulic engineer working for the US Army Corps of Engineers, met in Portland, Oregon, to follow up on a letter Munsey had submitted to *Civil Engineering*, the journal of ASCE. Evans contacted Munsey and the two then coauthored an article about the need for the engineering practitioner to reconsider the social and ethical implications of her work.[54] Their intentionally controversial essay, entitled "Engineers, Ethics and Nuclear Weapons," was published in the ASCE journal *Professional Issues in Engineering*.[55] The paper provoked reflection among academics and engineering practitioners, and fiery debates appeared in the journal's July 1988 volume.[56] AESR was then formed following a meeting that took place at Munsey's home in Manassas, Virginia; "[a] dignified and professional approach to issues of social responsibility among engineers was the goal desired..." by those who attended that first gathering.[57] AESR's founding members admittedly aspired to, but were not directly associated with, Physicians for Social Responsibility (PSR).[58] However, the striking coupling of "human welfare" with "life on earth" in AESR's original "Fundamental Principles" echoed PSR's mission statement, which was published in PSR's annual report in the summer of 1988.[59] Although ESR members were espousing environmental issues from

---

[54] *AESR Newsletter* 1, no. 1 (1988): 4–5.

[55] Evans, Roger J., and E. Thomas Munsey. "Engineers, Ethics, and Nuclear Weapons. *Journal of Professional Issues in Engineering* 113, no. 3 (1987): 268–275.

[56] See for example: Britzius, Charles W. "Discussion of 'Engineers, Ethics, and Nuclear Weapons.'" *Journal of Professional Issues in Engineering* 115, no. 1 (1989): 100.

[57] *AESR Newsletter* 1, no.1 (1988).

[58] Before he met with Tom Munsey and co-founded AESR, Jim Evans was involved with a Noam Chomsky reading group organized by the Seattle Chapter of PSR (Jim Evans, personal communication).

[59] In the decades after World War II, members of the Federation of Atomic Scientists (FAS) adopted and modified the conceptual framework of *social responsibility*, initially conceived by elite American engineers in the first decades of the 20th century. In the post-WWII context of FAS, the Bulletin of the Atomic Scientists, and the Pugwash Conferences on Science and World Affairs, *social responsibility* became synonymous with scientists' duty to educate the "peoples by spreading among them a wide understanding of the dangers and potentialities offered by the unprecedented growth of science" (Third Pugwash Conference, 1958; cited in Mitcham, Carl. "Responsibility and technology: the expanding relationship." In *Technology and Responsibility*, edited by Carl Mitcham, 3–39. The Netherlands: Springer, 1987). PSR was formed in Boston, MA, two years after the Third Pugwash Conference and twenty-four years later was awarded the Nobel Prize for its work on educating the public about the medical implications of the use of nuclear weaponry. AESR's first Newsletter is dated Spring 1988 and includes the following mission statement: "AESR is an organization dedicated to encouraging the adherence of engineers to an ethical standard which *holds the long term enhancement of human welfare and life on earth above all other concerns*" (emphasis added). PSR's mission statement was published in the organization's "Summer 1988" report: "Physicians for Social Responsibility is a professional organization whose primary objective is to prevent nuclear war. Members believe on the basis of known fact and deliberation that nuclear war constitutes an overwhelming threat *to human life and the global biosphere*" (emphasis added). Finally,

their organization's foundation in 1983, AESR's "Fundamental Principles" comprised the first time ever that an American engineering group included a phrase in its *Principles* that connoted the natural environment.[60]

Evans and Munsey argued in their essay that engineers must "shed the image of uncritical accommodation of authority and... begin to adhere to their code of ethics:" Engineers' blindness to social concerns is paradoxical because the work and products of engineering shape every aspect of human affairs. Evans and Munsey stated that the ASCE Code of Ethics must be respected—meaning that "the welfare of the public should be held paramount"; furthermore, by honoring this code the engineering community would benefit in the long term. AESR founders were inspired by PSR's mission statement, which assumed that one of the principal implications of enacting *social responsibility* at the organizational level was to prevent nuclear war. Evans and Munsey were thus concerned with the consequences of "Star Wars" and the question of whether or not a "Strategic Defense Initiative" or SDI—a ground and space-based defensive system proposed by President Ronald Reagan in 1983—could be an acceptable alternative to nuclear deterrence on technological, strategic and moral grounds. They wrote that the threat of nuclear holocaust was so significant that American engineers were obligated to address the issue as a top priority. Their call was based on their acceptance of the First Fundamental Principle of the ASCE ethics code that engineers "should use their skill and knowledge to enhance human welfare."[61] Strategic weapons production should cease, the authors asserted, and adherence to all treaties could be

according to AESR's Newsletter (Volume II, no. 1) "the similarities [between AESR and PSR] are such that the two organizations should be able to cooperate on a number of mutual objectives related to *the future health and safety of the human race and life on earth*" (emphasis added).

[60] ESR's membership flier (1983) says: "The selection and achievement of long term goals for humanity underlies present day problems. The provision of adequate food, water and shelter for an increasing world population, in a world with finite resources required more of our technological efforts than ever before." Also Gerry Coates' article "The Responsibility of Engineers" (*New Zealand Engineering*, September 1983) mentioned: "Some of the problems we need to address ourselves to are: Environmental issues – health and safety, pollution, depletion of resources, unemployment, energy policy, communications policy, privacy." According to AESR's "second fundamental principle," the organization "recognizes that human wellbeing is inextricably linked with the wellbeing of all life" *AESR Newsletter* 1, no. 1 (1988): 3.

[61] Sharon Beder, already in 1995, argued that engineering codes of ethics leave it up "to the individual engineer to decide whether environmental protection is an essential ingredient of community welfare" (Beder, 1995).

a reasonable and appropriate strategy towards sustaining a professional engineering code of ethics.[62]

Specifically, Evans and Munsey argued against the SDI "not on grounds of cost, not on grounds that it will not work, but on grounds that it would eventually lead to an escalation of the arms race".[63] Finally, they suggested that engineers should more actively address social issues in technology. Consequently, they put forth the concept of an organization like AESR as the means to promote ethical behavior in engineering. Contrary to other areas of professional employment such as law and healthcare, they concluded, "there is no active organization [in the US], e.g., 'Engineers for Social Responsibility in engineering'".

Having coauthored "Engineers, Ethics and Nuclear Weapons," Munsey traveled to New Zealand, where he met with John Peet and Gerry Coates, two of the leading members of ESR.[64] AESR's founding members—and the Newsletter's "interim editors" Jim Evans, Tom Munsey, and Orson Smith, who at the time was doing a post-doc in North Carolina State University—wrote to organizations with vaguely similar aims to announce the formation of AESR. AESR, nevertheless, was not well received by most US professional engineering societies; only five of the forty-five societies contacted published AESR's announcement.[65] Following its formation, Evans, Munsey and Smith envisioned AESR based on a membership with chapters (regional, state) each with its own funds, leadership, and activities.

---

[62] Evans and Munsey (1987).

[63] *AESR Newsletter* 1, no.1 (1988).

[64] Evans had already been introduced to ESR during his Sabbatical at the University of Canterbury, Great Britain, from January until the summer of 1987. During that time Evans also traveled to Sydney, Australia, where he met with members of the Society of Social Responsibility in Engineering, Australia (founded in 1983). "We had a long meeting with those people [in Australia] and that was very rewarding… Terrific people. All men. And then I met with them in England, several times I met with Architects and Engineers for Social Responsibility [formed in 1991]" (Jim Evans, Tom Munsey and Gerry Coates, personal communication).

[65] The five professional societies that published AESR's announcement were: The American Ceramic Society; the American Society of Civil Engineers through their *Civil Engineering Magazine*, *Engineering Education News*, the American Society of Safety Engineers (*Safety Engineering Magazine*), and the National Association of Corrosion Engineers. "One could speculate," Munsey wrote in fall 1988, "on why AESR has not had a warm welcome among established engineering organizations…" (AESR Newsletter 1, no. 2 (1988): 1). Evans said that AESR tried to initiate a conversation on *social responsibility* at the professional society level, "but they were not interested" (personal communication).

AESR was a very small engineering society, with a membership that never exceeded 300, representing 0.05% of the 600,000 licensed engineers in the United States in 1990.[66] Its founding was based on two pillars: On the one hand was the aversion against the global arms race and the use of highly sophisticated, otherwise "good engineering," and, in particular, the use of nuclear weapons; on the other hand, instead of just being another organization against nuclear war, AESR defined an engineering-specific perspective to address environmental problems at the US and international levels. Engineers are manipulators of the environment, Evans wrote; yet, "[t]here is abundant evidence that much environmental manipulation is actually diminishing human welfare."[67] (Figure 3.2). Although initial AESR positioning "anticipated…getting requests for expert advice on issues of AESR membership" *including* environmental issues, interviews with members and the Newsletter records indicate that such requests never came in.[68] Targeted, according to its members, by US government counterintelligence programs, and perceived by fellow engineers as "possibly subversive organization" the AESR case study is indicative of how daunting the challenges of restructuring engineering with a technopolitics mindset may be.

AESR and Engineering Conceptualizations of Sustainability: Negotiating Ideas about Self and Nature

> The mechanisms of [economic] conversion are important, but so is the broader question of what we must convert to. Ultimately we must convert to a way of life, to an economy which is sustainable. Surely, we should be looking at the meaning of sustainability itself.[69]

In personal communication, AESR's founding members claimed that they "knew about sustainability twenty years ago."[70] Moreover, "many…in AESR…[came] to believe that, if [they] could find one task that would be the long-term focus of [their] organizational energy into the foreseeable future,

---

[66] National Council of Examiners for Engineering and Surveying, 2011. <http://www.ncees.org/Licensure/Number_of_licensees_by_state.php> Accessed April 9, 2012.

[67] Evans, Jim. "Engineers, Ethics and Nuclear Weapons." *AESR Newsletter* 1, no. 1 (1988): 4–5. [1988 (b)]

[68] *AESR Newsletter* 1, no.1 (1988).

[69] Evans, Jim. *AESR Newsletter* 3, no. 4 (1990): 4.

[70] Jim Evans and Tom Munsey, personal communication.

Figure 3.2 Manipulators of the environment. ESR's original logo as it appeared in their first newsletter in July 1983. ESR (New Zealand) had also conceptualized engineering work as contributing not only to human but also to the planet's welfare. Image by, and courtesy of, Gerry Te Kapa Coates.

that task would be planning for a sustainable society."[71] When the Toronto, Canada Chapter of the Institute of Electrical and Electronics Engineers (IEEE), and the Society on Social Implications of Technology cosponsored a 1991 conference entitled "Preparing for a Sustainable Society", AESR was actively represented by cofounder Tom Munsey, engineer-consultant Richard Duncan, and Richard Devon from Penn State University. For the first time at a conference in North America the engineering community effectively engaged in debating the meaning of sustainability.

The development of AESR's group identity went hand-in-hand with members' attempts to comprehend and define sustainability. In general, sustainability included but was not limited to transitioning toward a military-free economy, sustainable energy, "a unified theory of history," and indigenous spirituality. AESR as a case study attests to the co-construction of professional identity and requisite conceptualizations of nature and sustainability based on the adoption of a sixty-year-old vision of social responsibility in American engineering. In that sense, the inception of AESR

[71] *AESR Newsletter* 3, no. 2 (1990): 1. Along these lines, in the September 1991 issue of the Newsletter Evans was arguing that "AESR is focusing on sustainability as our long term goal..."

in the US is a revival of an older social responsibility idea.[72] "Now, however, there seems to be more social responsibility, as concerns are expressed about encroachment of human culture on the natural environment."[73]

AESR's commitment to define sustainability from an engineering perspective ran parallel to mainstream, US-based or international, engineering societies' action on the topic. The April 1992 Newsletter, for example, called on members to "look closely" at a draft of what later became the famous Agenda 21 (UN 1992), in order to inform AESR's "organizational position". And like other engineering societies at the time, AESR members deliberately positioned themselves in terms of engineering for sustainability.[74]

AESR members viewed the idea of "economic conversion" (EC) as a suitable activity for the organization.[75] EC was AESR's major project, for which they received a grant from the NSF.[76] Three reasons drove interest in EC: 1) First, most engineers self-identify as problem solvers; since professionals with engineering degrees are heavily represented in military-related industries, AESR members viewed EC as a problem set

---

[72] For examples of the types of considerations that have spurred engineers to reflect on their social responsibility throughout the 20th century see Sakellariou, Nicholas. "A framework for social justice in Renewable Energy Engineering." In *Just Engineering: How Can Engineering Education Educate Engineers Who are Socially Just?*, edited by Juan C. Lucena, 243–267. New York: Springer, 2013. In that book chapter I argued that the discourse on social responsibility has—until recently—monopolized engineering reflexivity.

[73] Evans, Jim. "Seattle Happenings." *AESR Newsletter* 3, no. 1 (1990): 13.

[74] Echoing AESR's idea about the ability of EC to make engineering practices sustainable, members of the Finish group "High Technology Professionals for Life" argued that "[a] necessary condition for a sustainable development is that the limits posed by nature are not surpassed by maintaining production of unnecessary items, laying an extra burden on the environment. Conversion must be seen as inevitable, freeing resources to satisfying basic needs all over the world." *AESR Newsletter* 4 no. 1 (1991): 2.

[75] Economic conversion refers to the changeover from military to civilian work for defense-dependent firms, employees and communities.

[76] AESR's interest on EC, which culminated in the publication of MacCorquodale, Patricia L., W. Martha Gilliland, P. Jeffrey Kash, and Andrew Jameton, eds. *Engineers and Economic Conversion: From the Military to the Marketplace*: Springer-Verlag, 1993, grew out of discussions between Martha Gilliland, Tom Munsey and Jim Evans during the inaugurating AESR meeting which took place in March 1989 in Munsey's home in Manassas, Virginia. "As the first drafts of the proposal for support to the National Science Foundation (NSF) emerged, all involved felt NSF reviewers would, at best, view economic conversion as less than serious research and at worst reject it on the basis that we were radicals wasting time studying a future that had no realistic possibility of occurring" (MacCorquodale *et al.* 1993: vii).

with mechanisms they understood better than any other occupational group.[77] The "mechanisms of economic conversion" were considered part of legitimate engineering work; 2) Second, within the conceptual framework of social responsibility members maintained that EC was beyond controversy. EC contributed logically to human welfare, AESR proclaimed, and was thus consistent with the society's mission statement and its commitment to environmental sustainability. For example, some AESR members' conceptualizations of sustainability could only be properly understood if one looked into the context of EC; 3) Third, EC provided a rare opportunity for a small and controversial engineering society like AESR to serve its progressive mission statement and simultaneously enter the mainstream of American engineering.

"Hunter-Gatherer Days" and Economic Conversion

> AESR was the outcome of our previous work on Economic Conversion! In other words, sustainability was seen as what EC would hopefully be all about.[78]

Jim Evans served as AESR's president from the organization's inception until a few months before its dissolution. He was a proponent of both a philosophical and a more practical attitude toward engineering sustainability. Several factors influenced Evans' philosophy of sustainability: his father's Marxist legacy was a determining factor; as was his broad interest in things intellectual; combined with a continuous involvement in grassroots activism spanning his graduate years at Berkeley to his career as an engineering professor and ethicist at the University of Washington. For example, while writing his "President's columns" on sustainability for AESR's Newsletter, he was also reviewing the activity of Sustainable Seattle, a pioneering organization that later became a world leader in constructing indexes to quantify and measure progress toward urban sustainability.

Just a few months before Sustainable Seattle published its first set of urban and regional "Indicators of Sustainable Community," in the AESR's

---

[77] According to labor statistics from the National Science Foundation (NSF), in 1987, the year AESR was founded, "approximately 16 percent of the engineers and 11 percent of the (natural, computer, and math) scientists working in the United States were involved in defense work." See:
NSF. "The Impact of Defense Downsizing on Technical Employment." Undated. <http://www.nsf.gov/statistics/seind93/chap3/doc/3s293.htm>
Accessed October 6, 2012.

[78] AESR member, personal communication.

Newsletter, Jim Evans developed the idea of a "hunter-gatherer day" as a way to look at energy consumption.[79] The concept of a hunter-gatherer day "was intended to be a quantitative measure of the amount of energy a human can produce per day: Evans suggested 1500 K-Cal., based on the average food intake per person. "It seems to me," he wrote, "that the important measure of a species' environmental impact is the rate of total energy intake[;]....for us it would be the daily human energy intake (HEI), the number of people times the average energy intake per day."[80] In personal communication, however, his views were more strident:

> So, when people talk about sustainability, I think they should talk about the seventh generation, you know biblical. Let's make sure that in seven generations we still have resources available. In which case we'd be—basically, we wouldn't be using oil for transportation. So, that to me, sustainability is an awful word. I hate the thought of it because nobody knows what it means. You should have to pass an examination of thermal-dynamics before you're allowed to talk about sustainability. That's why the idea of a hunter-gatherer struck me as a good way to look at how much energy you were using... I mean, how many hunter-gatherer days would it take to mow a lawn, for Christ's sake. You know, a big lawn would be...five or six hunter-gatherers all day. The only thing I know that's sustainable is hunting and gathering. And when there's not enough to gather, and not enough to hunt anymore the population goes down. That's sustainable.[81]

Evans formulated his views regarding sustainability in concordance with AESR's Fundamental Principles—namely, using engineers' skill and knowledge to increase human welfare, and acknowledging that human welfare is inextricably linked with the welfare of all life. But what is human welfare, he pondered?[82] And how can human welfare be reconceptualized to account for resource scarcity and the accumulation of industrial waste? He concurred with the idea that human welfare is inherently qualitative, not subject to quantification, and argued that the qualitative features of human welfare could be the basis for a "more realistic" definition of

---

[79] "I still [November 2012] think it is a useful way of looking at energy consumption particularly in relation to a sustainable population density" (Jim Evans, personal communication).

[80] Evans, Jim. "Further Thoughts on Sustainability." *AESR Newsletter* 5, no. 1 (1992): 1–2.

[81] Evans, Jim, personal communication.

[82] Evans, Jim. "Sustainability and the Enhancement of Human Welfare." *AESR Newsletter* 3, no. 4 (1990): 4–5.

sustainability than the one proposed by the Brundtland Commission in 1987.[83] A dialectic between technopolitics and sustainability thus implies intellectual and professional space for participatory engineering design. Which in turn implies that, beginning with the 1990s, whatever the dominant engineering ideology of sustainability there is also tacit within it some space for challenging traditional notions of engineering identity and expertise.

AESR members came to view EC as an opportunity to operationalize society's moral vision of engineering; and Evans' outlook, that sustainable engineering should improve human welfare, comes full circle with his more practical perspective of AESR's potential contribution to the goal of EC. Making sustainability engineering a reality through implementing EC, gradually became the founding members' vision for the future of AESR. Clearly, Evans hoped that AESR could make a difference in American engineering, for "the very idea of economic conversion is in its infancy… [and] at present, *very few engineers are involved…*"[84] His philosophical interests in human welfare led him to question the standard definition of sustainability and "EC was a step along the way."[85] In fact, if AESR was to become a viable organization able to seek its own proper place and sphere of influence in the geography of US engineering societies, this prospect seemed not only "natural" but also quite pressing.

Overall, during AESR's five years in existence, members expressed concerns regarding the Newsletter's philosophico-cultural orientation and/or the board of directors' "left-wing politics." And after 1990 the task of convincingly addressing such criticism became vital for AESR's future survival. In short, sustainability was given priority status in EC, and vice versa—this principle was chosen by the AESR founders specifically to affect the (controversial) path the organization was seen to be taking.

Backward to the Future: System Dynamics and the Olduvai Theory

Richard Duncan holds undergraduate and postgraduate degrees in electrical engineering from the University of Oregon. He received his

---

[83] Evans, Jim. "Why are we doing what we are doing?" *AESR Newsletter* 4, no. 2 (1991): 1–2. According to the Brundtland Commission, sustainable development is "development that meets the needs of the present without compromising the ability of future generations to meet their own needs." WCED (1987):8.

[84] Evans, Jim. "Editorial: Economic Conversion." *AESR Newsletter* 1, no. 2 (1988): 14–16, emphasis added.

[85] Evans, Jim. "Why are we doing what we are doing?" *AESR Newsletter* 4, no. 2 (1991): 1–2.

doctorate in systems engineering from the University of Washington in 1973. Duncan founded the Institute on Energy and Man in 1992 in Seattle, and is perhaps the most publicly recognized member of AESR. He made noteworthy contributions to the organization, including introducing his "Unified Theory of Human History" to fellow members and his role in representing AESR at the 1991 IEEE Toronto "Preparing for a Sustainable Society" conference.[86]

As a graduate student at the University of Washington, Duncan attended a talk by cosmologist Sir Fred Hoyle (1915–2001) who argued that "[w]ith coal gone, oil gone, high-grade metallic ores gone, no species however competent can make the long climb from primitive conditions to high-level technology. This is a one-shot affair".[87] Hoyle's hypothesis, which assumes that human civilization suggests a "one-shot affair" not subject to cyclical repetition, has haunted Duncan's professional career ever since.

Duncan's intellectual trajectory forms an effort to verify the accuracy and integrity of Hoyle's claims—to estimate the quantifiable life expectancy of modern, industrial civilization. His view on sustainability is similar to Jim Evans' concept of a "hunter-gatherer day." Duncan conceptualized social change in terms of global average energy-use per individual or world energy production per capita (e)[88]. The very name "Olduvai Theory" (OT) was taken from the Olduvai Gorge in Tanzania, a place that Duncan visited shortly after he began assembling his argument and used metaphorically to connote industrial civilization's forward shift in the Stone Age.[89] OT is based on a four level (environment, production, governance, human genetics) explanatory scheme that builds on the processes of biological and cultural evolution over the course of the last four billion years. OT argues that human history can be divided into three phases: Pre-industrial, Industrial, and De-industrial.[90] The study of each of these phases, and more specifically the transitions between them, are key to Duncan's views of

---

[86] *AESR Newsletter* 3, no. 1 (1990): 2–5.

[87] Hoyle, 1964 quoted in Duncan, Richard C. "The Olduvai Theory: Sliding Towards a Post-Industrial Stone Age." Institute on Energy and Man, June 27, 1996, manuscript.

[88] Duncan, Richard C. "Evolution, technology and the natural environment." Paper presented at the Annual ASEE Conference, Binghamton, NY, United States, 1989; Duncan, Richard C. "World Energy Production, Population Growth and the Road to the Olduvai Gorge." May 2001, manuscript.

[89] Duncan (1989).

[90] Duncan, Richard C. "A unified theory of human history: Summary presentation." *AESR Newsletter* 3, no. 3 (1990).

sustainability. In its most recent version, the OT forecasts that "the terminal decline of industrial civilization will begin circa 2008–2012."[91]

In a series of articles written between 1990 and 1996 Duncan used energy and population data from various authors and sources, including British Petroleum (BP) and the UN, to back up his earlier (1989) assertions that industrial civilization will end sometime during the middle of the 21st century.[92] Further on, he explored some of OT's implications for public policy. He focused on "the mounting problems with the high voltage electric power networks worldwide" and he indicated their declining reliability.[93] More recently he compared US and OT intervals (e) to contend that the former anticipate the latter.[94]

Duncan places himself in a lineage of thinkers that sowed "the seeds of the Olduvai Theory."[95] The OT genealogy includes early predecessors such as the Greek lyric poet Pindar (ca. 522-443) and the American historian Henry Adams (1838–1918)—but also scholars such as architect Frederick Lee Ackerman (1878–1950), anthropologist Leslie A. White (1900–1975) and geophysicist King M. Hubbert. Ackerman, White and Hubbert also perceived social evolution in terms of an energy/population ratio. Perhaps the most accurate description of OT's intellectual background is Jay Forrester's "system dynamics."

Duncan's assumptions are premised on the idea that the "human life support system" is a closed network which tends towards a state of balance via negative feedback-loop structures. This is the foundational supposition of system dynamics and of the intellectual tradition associated with the "Limits to Growth" framework.[96] Duncan himself thought that not only his theory succeeds where mainstream and steady-state economics fail

---

[91] Duncan, Richard C. "The Olduvai Theory. Toward Re-Equalizing the World Standard of Living." *The Social Contract*, (2009): 67–80.

[92] E.g., Duncan, Richard C. "Sustainability—Is there a middle road?" Moses Greeley Parker Lecture Series, Lowell, MA.

[93] Duncan (2001).

[94] Duncan, Richard C. "The Olduvai Theory. Terminal Decline Imminent." *The Social Contract* (2007): 141–151.

[95] Duncan, Richard C. "The Olduvai Theory. Energy, Population, and Industrial Civilization." *The Social Contract* (2005–2006): 1–12.

[96] Both Duncan and Evans were influenced by the ideology of "limits to growth" associated with The Club of Rome. Duncan (1990), for example argued that "...driven beyond its limits, the world system is highly unstable." Similarly, Evans lamented society's "inability to reduce the rapacious depletion of finite natural resources" (AESR Newsletter 1, no. 1 (1988): 5). Such inability, he thought, was evidence of greed—"taking what we can rather than what we need" (ibid).

(i.e., they do not include energy production), but that OT and the Limits to Growth computer model are "consistent and complementary" as well.[97]

What technological-change ideologues of sustainability view as "sustainable" is "greener," "better" or "smarter" technology. Duncan, however, defined sustainability in terms of social/cultural evolution. This explains Duncan's thesis[98] that "industrialization isn't evolving toward sustainability." In his *AESR Newsletter* article, for example, Duncan argued against two "myths" of industrial society: the naturalness of continued growth and the ability of social institutions to exert control upon the world system. Duncan thought the only viable form of control was through the manipulation of the world system's cultural variables. However, due to certain qualities "innate in human beings" such as "self-centeredness" and "entropy blindness", Duncan concluded that "we have been programmed by Mother Nature for a world that no longer exists".[99]

Eduard Naudascher, AESR member and professor of hydraulic engineering at the University of Karlsruhe at the time, critiqued Richard Duncan's views of sustainability from the perspective of "a definitive interconnectedness of man's inner and outer worlds." Despite OT's gloomy predictions, Naudascher proposed that what could matter for fellow AESR members in dealing with their frustration about the social and environmental ramifications of their work is "the priority given to the spiritual dimension of man and environment which, as a rule, is disregarded in science and technology".[100] Duncan, influenced by Sir Fred Hoyle, saw humans' obsession for wealth and power as an explanation for industrial civilization's "one-shot affair"; Naudascher, who had studied the writings of Mahatma Gandhi (1869–1948), maintained hope for qualities not considered such as faith and love. More references to environmental spirituality are found in the pages of the Newsletter: for example, Naudascher warned that "technology bereft of religion is a death trap" and Gerry Coates of ESR invited fellow engineers to appreciate the importance of "emotional tools" such as the Council of All Beings.[101]

---

[97] Duncan (1990); (1996); (2001); (2005).
[98] Duncan (1996).
[99] Duncan (1990).
[100] Naudascher, Eduard. "Discussion of Richard C. Duncan's 'A Unified Theory of Human History: Summary.'" *AESR Newsletter* 3, no. 2 (1990): 10–11.
[101] Based on ongoing work by Joanna Macy and John Seed.
Rainforestinfo.org. "The Council of All Beings." Undated
<http://www.rainforestinfo.org.au/deep-eco/coab.htm>
Accessed November 12, 2012.

Indeed, up until recently, fellow engineers have been critical of OT.[102] This is due mostly to Duncan's lack of faith in technology as *the* appropriate force to restore the world's equilibrium; practitioners have been less critical of Duncan's postulations that technological growth, moving too fast for "the world governing system" (1990) to track, has contributed to the destruction of the earth's biosphere.[103] It is therefore OT's *premise* that the "controllability of the life support system" (1991) cannot be defined in terms of technological management, and not its gloomy *conclusion*, which makes it inconsistent with the ideology of technological change.

Social Responsibility and the Broadening of Engineering Sustainability

AESR's programmatic statements mirrored the philosophical outlook shared by PSR in the late 1980s. On the one hand, this was part of the post-WWII scientists' movement formed explicitly to address concerns of nuclear energy and which conceived of social responsibility on the basis of providing education to the democratic public.[104] On the other hand, AESR should be seen as part of another movement that began in the 1970s, labeled by Funtowicz and Ravetz as "post-normal" science.[105] AESR's short life and limited impact on engineering affairs makes it impossible to test Funtowicz and Ravetz's assumptions. However, AESR did not shy away from "new responsibilities" that implied a radical restructuring of the engineering profession, such as EC. Moreover, with their emphasis on the political and cultural aspects of sustainable technology, the society's founding members aspired to transform the very nature of what it means to do engineering work.

New appropriations of intellectual contexts challenge and also expand the old. The AESR is part of an intellectual history that began in the late 19th century—with the invention of social responsibility by elite technocrats and

---

[102] Multiple authors. "Letters Responding to Richard Duncan's 'Olduvai Theory' Essay." Social Contract, 2006.

[103] In 1990 Duncan wrote that "human behavior is the driving force in [the OT] model."

[104] Mitcham's (1987): 11 description of the scientists' movement fits well with PSR, AESR and INES' philosophies: "...the aim of the Post-World II scientists' movement was simply to take science away from the military and place it under civilian democratic control and, ultimately, under a world government, for which the international community of scientists itself is proposed as a forerunner or model."

[105] Funtowicz, Silvio O. and R. Jerome Ravetz. "A New Scientific Methodology for Global Environmental Issues." In *Ecological Economics: The Science and Management of Sustainability*, ed. Robert Costanza. New York: Columbia University Press, 1991, 137–152.

engineering spokesmen, who crafted a vision of engineering professionalism as a prerequisite towards achieving national prosperity and social progress. Although AESR members were aware of that history, it is unclear whether this historical knowledge—or the lack thereof—was considered potentially relevant for the future of the organization.[106] Yet AESR's vision differed from that which informed the 1920s idea of *social responsibility*, and that which engineering historian Edwin Layton Jr. has famously termed "engineering ideology".[107] AESR founders and members enthusiastically embraced the ideal of the engineer as an agent of technological change and as an active force for human welfare—the first component of "engineering ideology." Within their particular appropriation of social responsibility, AESR reflected upon their identity as professional engineers, but also as humans. The remainder of that engineering ideology's constitutive elements did not fit AESR's perspective, however: namely, the engineer as the optimal social leader due to his ability to think free of bias; and the engineer with a special social responsibility to protect progress:

> [Defining human welfare] will be an ongoing task which will take up a lot of our collective energy and which will, presumably, bring us into contact with new or emerging paradigms where regionalism and nationalism are replaced by internationalism and where anthropocentrism itself is challenged and either replaced or accepted as human nature.[108]

> We as engineers, cannot claim to have a special social responsibility in the sense that it is ours alone. If we think in terms of welfare of the planet, we are faced with ideas which were alien to many engineers of the past when nature posed frontiers rather than limits and when national welfare was paramount....Similarly, the Taylorites committed a fallacy...i.e. believing that every system behaves like a mechanical system and hence follows laws which, when discovered, can be used to manage that system.[109]

AESR was ahead of its time in reconsidering the interaction of cognitive and cultural elements of engineering identity in terms of sustainability. It

---

[106] In his review of Layton's Revolt in the Newsletter, Jim Evans mentioned that the book "contains information pertinent to AESR, not so much as to what we might do, as to what has been done in the past" (*AESR Newsletter* 2, no. 1 (1989): 8–10). His reading of Layton's book was that "[i]t is a story of unattained goals and of overriding conflicts between public welfare and the interests of business" (ibid).

[107] Layton (1971).

[108] [38]Evans [1988, (a)].

[109] Evans, Jim. "A brief (and incomplete) history of social responsibility in engineering." *AESR Newsletter* 2, no. 1 (1989): 8–10.

was one of the first engineering organizations to explicitly link the concept of human welfare and planet Earth's welfare. AESR contributed to the debut engineering conference on sustainability in 1991, and the society asked its members to consider the assumptions of the UN's draft statement for the 1992 Rio de Janeiro Conference on Environment and Development. AESR members sought to inspire an organizational identity characterized by engineering sustainability and economic conversion—"a political, rather than technological issue."[110] In addition, AESR dedicated a fair amount of its intellectual and organizational efforts to include environmental ethics in the engineering curriculum.[111] Between April and May 1990, for example, AESR cosponsored a series of four lectures and a follow-up workshop entitled "Environmental Ethics in a Modern Technological Setting."[112] One of the last volumes of the Newsletter (December 1992) ends with an announcement of the society's decision to author a position paper on sustainability.[113] And as a whole, in exploring the parameters of sustainability engineering, AESR members debated human nature, cultural norms and contemporary politics, not technics.

Conclusion

If the technological change ideology of sustainability resonated with both the dominant engineering cultures and professionals' grounding in

---

[110] Evans (1991).

[111] Vivian Weil, who served as director of the Ethics and Values Studies Program of the NSF between 1990 and 1991, wrote to the Newsletter (Volume 4, no. 1: 5) to congratulate AESR for their discussion of engineering ethics. The Newsletter also featured a review by Greg McIsaac of Weil, Vivian. *Engineering Ethics in Engineering Education: Report on a Conference, June 12–13, 1990*. Chicago: Center for the Study of Ethics in the Professions, Illinois Institute of Technology, 1992. In the April 1992 Newsletter McIsaac wrote: "Engineers for social responsibility may be delightfully surprised by a collection of papers presented at the 1990 meeting of the National Academy of Engineering and published in Engineering as a Social Enterprise. According to its authors, many people, including engineers, need to revise their understanding of the processes of technological innovation."

[112] The workshop, which took place in Seattle, hosted renowned experts in the field of environmental ethics such as the American philosopher Michael E. Zimmerman and the New Zealand ethicist of engineering Alastair S. Gunn.

[113] The last issue of the *AESR Newsletter* was published in October 1993. According to Greg McIsaac, editor of the Newsletter at the time, "I don't recall getting much if any content with which to produce another newsletter after the Oct. 1993 edition," personal communication.

materials balance analyses such as LCA, so too the technopolitics ideology of sustainability profoundly questioned engineering cultures and the very boundaries of LCA. It was not "engineering leadership" *per se* that challenged technopolitics ideologues of sustainability, but the politics of engineering artifacts and engineering values. Echoing recent developments in humanities courses for engineers, AESR's Gregory (Greg) F. McIsaac wrote in 1998 that the "globalization of the economy suggest[s] that engineers will need to understand and work with people who do not share their cultural assumptions..."[114] Practitioners of engineering, McIsaac and coauthor Nancy C. Morey suggested, "may be better equipped to provide leadership toward sustainable development and may be more effective in the global marketplace by moving toward a more 'loose' culture."[115]

Although active in defining sustainability from conceptual, institutional and operational perspectives, engineering leaders across sustainability's ideological spectrum also felt a profound ambiguity about the visibility of their interventions and of the profession more broadly. In 1996, for example, Roberts was expressing his disappointment that the WEPSD has had "little impact to date at the grassroots level of individual organizations..."[116] Similarly, Chantal Toporow of Altruistic Engineering Consultancy lamented over the "slow incorporation" of environmental issues in engineering societies, pointing to the fact that IEEE's Society for the Social Implications of Technology (SSIT) and ASME's Technology and Society Division (T&S) have the smallest proportion of engineering membership as compared to other groups within the respective societies.[117]

Engineers' contributions to defining the conflicts of sustainable technology in the postmodern world come into view, ironically, when, as in the integration of technological change and technopolitics ideologies of sustainability, traditional cultures and identities are being challenged and engineers revisit the legitimacy of their operating principles (Table 3.1),

---

[114] McIsaac and Morey (1998): 118. In 2006 Downey *et al.* proposed that a conceptualization of "the global competency of engineers... [must begin] by showing that the often-stated goal of working effectively with different cultures is fundamentally about learning to work effectively with people who define problems differently." Gary Lee Downey, Juan C. Lucena, Barbara M. Moskal, Rosamond Parkhurst, Thomas Bigely, Chris Hays, Brent K. Jesiek, Liam Kelly, Jonson Miller, Sharon Ruff, Jane L. Lehr, and Amy Nichols-Belo. "The Globally Competent Engineer: Working Effectively with People Who Define Problems Differently." *Journal of Engineering Education* 95 no. 2 (2006): 1–17.

[115] McIsaac and Morey (1998): 118.

[116] Roberts (1996).

[117] Toporow (1991).

Table 3.1 Operating principles of engineering ideologies of sustainability.

Engineering ideologies of sustainability	Premises	Core assumption	Operating principle
Technological Change	Engineering creates prosperity—yet engineers have unintentionally contributed to environmental problems. "Sustainable" means "environmentally sustainable." Environmental problems can be eliminated by technological means without a sacrifice of prosperity.	The reorientation of environmental technology is autonomous, thus the exploitation of natural resources should neither be decreased nor increased, but ought to be effectively managed.	Engineering activity must account for its own environmental costs in the context of a free-market economy. Sustainability is an engineering problem.
Technopolitics	The design and social integration of engineered systems reflects normative—yet not readily recognizable—assumptions and values; therefore, development has equivalent effects on the social order as any other form of political act. Engineering and science play a supportive—not central role—in the quest for sustainability. The logic of competitive productivism must be reconsidered.	Provided that technology is autonomous, the pressing question in sustainability engineering is reconceptualizing and redirecting the democratic control of technological means and technological expertise.	Engineering activity must continually challenge the adequacy of sustainability metrics as well as the very processes by which their objectivity is established. Sustainability is a political problem: assessing social and environmental impacts of a project or a product requires equal involvement on the part of technological experts and non-engineering "stakeholders," such as local communities or users.

methodologies of engineering knowledge and practices. The technological change ideologue of sustainability presumes that "...the ethos of [sustainable engineering] practice will be different...; it's not that what we learn of engineering science will be different."[118] The technological change ideologue of sustainability presumes that engineering knowledge has a nature of its own. The technological change ideologue of sustainability presumes that sustainability is, by and large, a function of technological expertise. For as Hatch announced in 2002 to "Listen to Those You Serve," "cooperation and coalitions replace confrontation" is one thing, but it is quite another thing to expect —through non-technological expert involvement in sustainable project development—that what engineers learn and do in engineering science *will* be different.[119]

As this chapter explains, it is only a partial interpretation to argue that the technological change ideology of sustainability was monolithic. The expansion of the technological change ideology of sustainability becomes more visible in juxtaposition with a technopolitics view of sustainable development. The continuous integration of these two ideologies is consistent with the idea of calculating impacts across an engineered system's life cycle and its most recent variation: SLCA. For example, it is through integration of the two ideologies explored in this chapter that the interpretation of SLCA, as a technique that encompasses social, economic and environmental dimensions of sustainability, becomes tangible and—to recall Hatch—"realistic" in an engineering context.

In sum, the differences between, and integration of, sustainability ideologies of technological change and technopolitics have produced some of the most significant tensions of modern engineering methodology and practice. It is the next chapter's aim—through a case study of SLCA—to further describe how technological change and technopolitics ideologies of sustainability overlap and coexist, rather than occupy separate engineering discourses or discrete areas of meaning.

[118] David Thom in Ellis (1994).
[119] Hatch (2000).

Part II

LIFE CYCLE ANALYSIS

4

When Social Values Make an Entry[1]

Introduction

Many engineers are increasingly acquiring an identity as "sustainable" practitioners. Nonetheless, engineering for sustainability is still quite instrumental and the more radical technopolitics perspective remains on the margins, implying a fundamental power imbalance in the engineering world. Still, starting around 2000, a small group of engineers and, vitally, companies and consulting firms with an eye to addressing technology's "social impacts" have laid a basis for developing an interesting sustainability tool called Social Life Cycle Analysis (SLCA). Technology is at the heart

[1] Primary data for this chapter is based on thirty in-depth, semi-structured interviews with North American and European LCA and SLCA experts. Other primary data were collected from participant observation in SLCA webinars and workshops.

of SLCA—it is a shared faith in technology as the solution. At the same time there is growing appreciation amongst SLCA proponents that such technology must be construed more critically. Although it remains a subaltern current within LCA, SLCA is evidence of how technological change and technopolitics are starting to converge and influence each other—a probe toward a more reflective form of engineering discourse and toward the formation of a new hybrid sustainability ideology.[2] SLCA, I argue in this chapter, is an ideological hybrid where there are many spots of dissent and disagreement but also some surprising fundamental alignments between those who see engineering as technics and those who believe that engineering needs to be socially and politically contextualized. Besides looking outside the profession to examine the viability of drawing conceptual links between sustainable development and LCA, this chapter details how contemporary engineers can also rely on a model of professional identity advancement that interweaves the technological change and technopolitics ideologies of sustainability.

SLCA began emerging when sustainable businesses—companies that take into account their own "corporate responsibility"—perceived a lack of transparency in their supply chains, preventing awareness of and reactions to corporate exposure to risks such as forced or child labor, prohibition on freedom of association, etc. SLCA, then, is based on the combined support of two of sustainability engineering's most fundamental conceptual roots: First, the ontological assumption that the world is a system comprised of interlocking processes which produce global unintended impacts or "footprints"; second, the methodological supposition that footprints become meaningful—namely, their awareness can help researchers inform decisions—when measured in life cycles, after these global processes have been modeled by input-output data.

In 2002, the Life Cycle Thinking Initiative, an international partnership between the United Nations Environment Program (UNEP) and the

[2] As one European engineer and SLCA practitioner put it, "I know... [that in 2014] several companies...are working on SLCA; trying to define suitable indicators to assess social aspects along...[product] life cycle[s]...to apply the SLCA guidelines...[and] to integrate [SLCA] into existing sustainability reporting." Groupe AGÉCO, a Canadian consulting group, reported in 2013 that it has conducted fifteen SLCAs, while LCA pioneer PRé Consultants—serving 80% private companies and 20% government services in sixty countries around the world—explained that more and more clients express interest in SLCAs. Interview data; webinar on "Combined Environmental and Social Life Cycle Assessment in the Food, Beverage and Agricultural Products Sector." Webinar offered by New Earth on June 26, 2013.

Society of Environmental Toxicology and Chemistry (SETAC), was formed to propagate the application of LCA.[3] During the first three to four years of the Initiative, LCAs continued to measure, almost exclusively, material, energy and economic flows without considering context-dependent impacts. [4] That changed in 2009, when a UNEP document, prepared by a team of predominantly European engineers led by a Canadian anthropologist with a background in social impact assessment in the mining industry, Catherine Benoît-Norris, argued that LCAs are "also made of stories about production and consumption impacts on the workers, the local communities, the consumers, the society and all value chain actors."[5] Engineers thus ascribed to the LCAs the objective of measuring and protecting "human wellbeing" across industrial product life cycles.[6] In other words, one of the emerging sustainability engineering fields where extra engineering values and engineering practice do not appear as opposing factors is in linking life cycle thinking to social impact assessment in global value chains. And as the sustainability engineering perspective expands, surpassing ideological, cultural, professional, and methodological boundaries, it spreads normative assumptions about technology, economy and society, as well as about the role of engineers in facilitating development.

LCA practitioners tend to be comfortable with their practices within the realm of technics, whereas in SLCA this no longer holds true. Yet, regardless of the fact that the translation of quantitative data into engineering decisions is no less problematic in LCA than in SLCA, most engineers view LCAs as more robust and technically sound. SLCA is then an attempt by sustainability practitioners to make *social impacts* more "sound" in engineering cultures.

The construction of a life cycle social impact assessment method calls for moving away from the non-contextual, yet tangible—e.g., CO2 emissions—to measure contextual intangibles, e.g., human dignity. In the mid-2000s, a small international engineering community argued that the

[3] Life Cycle Initiative. "Why Take a Life Cycle Approach." Paris, France: UNEP, 2004.

[4] Environmental LCA only considers damage on people, which occurs as a consequence of impacts on the environment.

[5] Hereafter "UNEP Guideliness." Life Cycle Initiative. "Guidelines for Social Life Cycle Assessment of Products." Paris, France: UNEP, 2009. Elsewhere the report writes that SLCA is meant to "account for [such] stories and inform systematically on [social] impacts that otherwise would be lost in the vast and fast moving sea of our modern world," (2009): 5.

[6] Similarly, the UNEP Guidelines state: "…the ultimate goal of S-LCA technique is to promote improvement of social conditions throughout the life cycle of a product, human wellbeing is a central concept," (2009): 22.

collision of value systems inevitable in the development of sustainable technology requires a new technological remedy, one that incorporates traditional engineering techniques with non-traditional engineering data.[7] In developing such a remedy, SLCA champions extended the standardized LCA concept of "Areas of Protection" (AoP)—which includes "human health," "natural environment," "natural resources," and "man-made environment"—to "human wellbeing", thus allowing extra-engineering values to enter into engineering discourse and practice. By explicitly accepting that value systems are inherently contained within social impact assessment—"not a deficiency of SLCA but...[a characteristic of] its very nature... necessitat[ing] an honest...approach"—SLCA advocates offered sustainability engineering a new venue to envision technology as a medium of social transformation, and an opportunity to reflect openly upon engineering assumptions, cultures, and professional identities.[8]

There were ramifications for the identity of sustainability engineers with the opening of LCA boundaries; they were captured with clarity in a dialogue between mechanical engineer, Harvard School of Public Health professor and SLCA champion Greg Norris and the Dalai Lama during a "Mind and Life—Ecology, Ethics and Interdependence" series that took place in His Holiness's office in Dharamsala, India, on October 18, 2011, moderated by renowned psychologist and *New York Times* best-selling author Daniel Goleman. The dialogue captured the context within which the conception and application of SLCAs has been taking shape: "[t]here is now [in the West] more movement from the cost alone toward the social responsibility view, toward a...broader view of ethics," explained Goleman to the Dalai Lama. Engineering ideas of navigating development's unintended consequences are recognizable in Buddhist maxims of "existence itself...[inherently] causing problems" and the need to "reduce suffering [and to] bring happiness to all beings."[9]

---

[7] "Generally," argued the UNEP Guidelines, "practitioners of S-LCA will need to incorporate a large share of qualitative data, since numeric information will be less capable of addressing the issues at hand..." (2009): 9. For example, "bypassing data on worker impressions in favor of more "objective" data (such as variability in observed worker arrival times, or other attempted proxies for perceived degree of control) would introduce greater uncertainty in the results, not less" (2009): 40.

[8] Reitinger, Claudia, Mattias Dumke, Mario Barosevcic, and Rafaela Hillerbrand. "A Conceptual Framework for Impact Assessment within SLCA." *International Journal of Life Cycle Assessment* 16, no. 4 (2011): 380-388.

[9] Unless otherwise noted, quotations in this introduction are adapted from the author's transcript of the October 18, 2011, "Mind and Life—Ecology, Ethics and Interdependence" series.

Figure 4.1 "Handprints: A Remedy for Reputation Risk." Image courtesy of the Social Hotspots Database, SHDB.

A "footprint" is, in Norris' words, "an idea in a way,...[;] a portrait of the supply chain of a product...which almost looks like a mandala"—the ancient Hindu concept that was meant to describe, but also to account for, a world model whose structure was that of concentric circles.[10] From this perspective, SLCA "lessons" are seen by Norris as instances when systemic interdependence manifests itself in social and political life: "When we buy one product, we are really touching the planet...and millions of people with our purchases... in many different ways." At the same time, in Norris' view, "footprints are only part of the story...[;] so we could actually introduce the idea of *handprints* to quantify the positive impacts of our [engineering] creativity."[11] And in fact, "if the benefits of our creativity are bigger than our footprints... then we are healing the earth" (Figure 4.1).

The high-profile dialogue between the Dalai Lama and the engineer Greg Norris is an example of how the verbalization of LCA can provide one of the socially accepted frameworks that undergird the contemporary sustainability discourse in engineering: indeed, with such a framework we can imagine extending this discourse to include technology's "social impacts". What is more, this is a rare instance where the sustainable engineer-philosopher of the postmodern world becomes abundantly visible. In concluding the October 18, 2011, panel discussion, the Dalai Lama recognized that the awareness provided by engineering techniques like SLCA is a first

---

[10] The UNEP Guidelines agree that a "product life cycle is an idea" too (2009): 55. "While the boundaries can unambiguously be defined...this is not true for a 'product life cycle [because]...it's hard to see where the effects of social interaction...diminish" (2009): 9.

[11] Emphasis added.

step to shaping sustainable social change. "I think…this *realistic approach* [life cycle thinking]," said His Holiness, turning to Greg Norris at the end of their two-hour conversation, should be "led…not by [the] politician… but…by the scientist—[the] real Guruji."[12]

Optimism about SLCA is fueled by faith in engineering progress, ideologically rooted as it is in the integration of engineering ideologies of sustainability and oftentimes depicted as an "instrument of renewal".[13] Yet the ideology of technological change, while important, is only one of the drivers of SLCA. As SLCA researchers are celebrating sustainability through fair trade and smart corporate management, they also show greater willingness to open the "black box" of their technique. Like the technopolitics ideologues of sustainability, SLCA pioneers claim that their engineering knowledge's justification lies in its explicitly value-laden nature—"not a deficiency of SLCA but…[a characteristic of] its very nature… necessitat[ing] an honest…approach."[14]

In an elaborate feasibility study on the possibilities of developing SLCAs, the Danish management engineer Andreas Jørgensen described the development of life cycle frameworks to assess technology's social impacts as an exercise in engineering reflexivity.[15] Correspondingly, during the 2nd International Seminar on SLCA held in May 2011 in Montpelier, France, practitioners were prompted to "question their own methodological choices" (i.e., what social impacts to assess and how to demarcate SLCA boundaries).[16] "Framing the [SLCA] question is…an important task that

---

[12] Emphasis added.

[13] Spillemaeckers, Sophie. "The Belgian social label: The first governmental social label based on social life cycle analysis." Paper presented at the CALCAS Workshop Governance and Life-Cycle Analysis. Opportunities for Going Beyond ISO-LCA, Brussels, Belgium, 2007.

[14] Reitinger *et al.* (2011).

[15] In his 2010 dissertation, Jørgensen admitted that the SLCA field's methodological 'openness' may potentially be used to consciously select indicators or data in favour of one alternative" (page iii). "To mitigate this possibility for manipulation," he argued, "a more comprehensive demand…[must be] considered which is to always include an assessment of the completeness and uncertainties in SLCAs accessible to the public" (ibid). Jørgensen, Andreas. "Developing the Social Life Cycle Assessment—addressing issues of validity and usability." PhD diss., DTU Management Engineering, 2010.

[16] Macombe, Catherine, Pauline Feschet, Michel Garrabé, and Denis Loeillet. "2nd International Seminar in Social Life Cycle Assessment—Recent Developments in Assessing the Social Impacts of Product Life Cycles." *International Journal of Life Cycle Assessment* 16, no. 9 (2011): 940-943. Similarly, Dreyer *et al.* (2006) had noted that "[i] t is crucial for the legitimacy of the Social LCA that…value-judgements [are made] transparent…" Dreyer, Louise Camilla, Z. Michael Hauschild, and Jens Schierbeck. "A

should come before any modeling choice," proclaimed Alessandra Zamagni, of the Italian National Agency of New Technologies.[17] Interestingly, while contextualizing their assumptions, however, practitioners communicate their developing perspectives on SLCA by *decontextualizing* sustainability engineering discourse—by taking metaphors like "holism" or "feedback" for granted—hence furthering the discursive integration of technological change and technopolitics. Thus blurring the ideological boundaries of sustainability engineering was manifested, for example, in a 2013 article published in the *International Journal of Life Cycle Assessment,* which surmised that because life cycle methods are inherently value-laden, they render technological egalitarianism credible.[18]

At the core of SLCA technique—embodied particularly in the social impact assessment step of the LCA method—is researchers' specifications for weighing the different social risk impact categories (i.e., "child labor" or "excessive work time") and the application of such conjectures in aggregating different social risk issues.[19] "I am an academic and I am looking for assumptions," recalled an engineer and SLCA practitioner in personal communication: "That's what I'm looking for. But if you want to just go to General Motors and ask them about the assumptions built into SimaPro [LCA software developed by PRé Consultants]...the guy doing that work isn't going to know."[20] In other words, once an engineering practice like LCA is in place, its ideological politics are seldom discussed, and they are sometimes literally encoded in software or input-output databases. Presently vice president for sustainable development at KTH Royal Institute of Technology and active in the field of SLCA, Swedish chemical engineer Göran Finnveden wrote in 1997: "A prerequisite for an increased agreement on valuation methods [how different 'footprints' or unintended consequences are valued against each other in LCA methodology]...may

Framework for Social Life Cycle Impact Assessment." *International Journal of Life Cycle Assessment* 11, no. 2 (2006): 88-97.

[17] Zamagni, Alessandra. "Inclusion of Economic Mechanisms into Life Cycle Analysis: Start with 'Framing the Question.'" *Integrated Environmental Assessment and Management* 6, no. 4 (2010): 777-785.

[18] Sala, Serenella, Francesca Farioli, and Alessandra Zamagni. "Life Cycle Sustainability Assessment in the Context of Sustainability Science Progress (Part 1)." *International Journal of Life Cycle Assessment* 18, no. 9 (2013): 1653-1672.

[19] This is the so-called aggregation problem, which is part of every LCA technique (environmental, social or economic LCA). The nature of the data used in SLCAs which can be qualitative, quantitative or semi-quantitative, makes aggregation particularly challenging in this case.

[20] Interview data.

be that ethical and ideological values of relevance are discussed more explicitly."[21] In this chapter, I propose that sustainability's operationalization via SLCA be subject to heightened scrutiny through the lens of the ideological visions of technological change and technopolitics.

Encoded in SLCA studies is the question of the technique's legitimacy, which is related to political and ideological assumptions—and the issue of whether or not SLCAs emerge from an opaque political perspective. The fact that each manufacturer may produce an LCA "that places its products in a favourable light..." wrote one higher European Union official in the mid-1990s, means that LCA "obviously no longer helps in decision-making."[22] In essence, the period of sustainability's intense operationalization in engineering contexts through life cycle thinking metrics—beginning sometime around 2003-2004—is no less ideological than the "philosophical" period examined in chapter 2 (Figure 4.2). "Technologies can contribute to sustainable development...and at the same time cause sustainability problems...[,]" asserts TU Berlin's Annekatrin Lehmann, one of the first engineers to utilize the SLCA framework in assessing the sustainability of technological systems—as opposed to studying particular products or processes in corporate supply chains.[23] There is agreement in the Centre for Good Governance Guide for Social Impact Assessment which argues that "the impacts of development interventions take different forms... [and that such social] impacts not only need to be identified and measured but also need to be managed in such a way that the positive externalities are maximized and the negative externalities are minimized."[24]

The student of engineering ideologies of sustainability will not be surprised that the interpretations of and debates around SLCA are fraught with assumptions regarding the social value of both what the technique helps to assess, and SLCA itself. "From an SLCA point of view, you might have...put the boundary somewhere around the mangrove swamp [in Myanmar] at one end and...the consumption of the prawn [in the West]

---

[21] Finnveden, Göran. "Valuation Methods Within LCA—Where are the Values?" *International Journal of Life Cycle Assessment* 2, no. 3 (1997): 163-169.

[22] Schleicher, Ursula. "The Uses of Life Cycle Assessment for European Legislation." *International Journal of Life Cycle Assessment* 1, no. 1 (1996): 42-44.

[23] Lehmann, Annekatrin, Eva Zschieschang, Marzia Traverso, Matthias Finkbeiner Matthias and Liselotte Schebek. "Social Aspects for Sustainability Assessment of Technologies—Challenges for Social Life Cycle Assessment (SLCA)." *International Journal of Life Cycle Assessment* 18, no. 8 (2013): 1581-1592.

[24] Centre for Good Governance. "A Comprehensive Guide for Social Impact Assessment." Hyderabad, India: Centre for Good Governance, 2006.

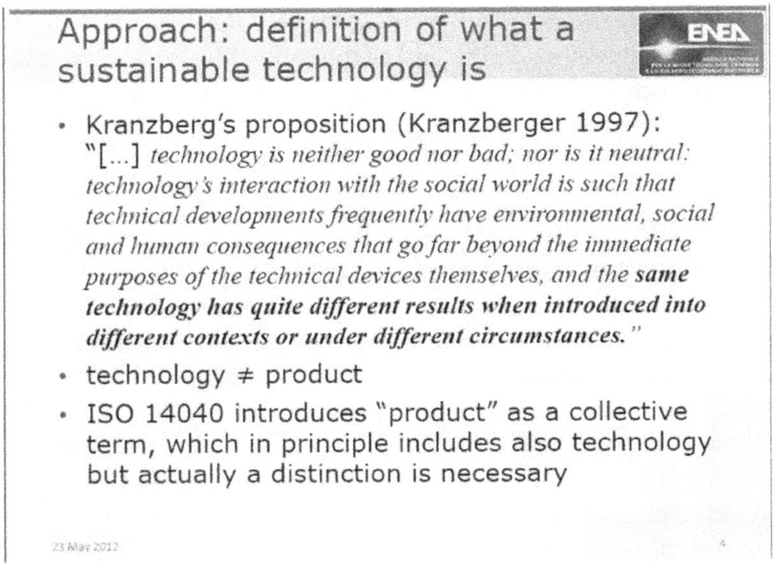

Figure 4.2 Kranzberg's proposition in the context of sustainable technology and SLCA. Image courtesy of Alessandra Zamagni.

at the other end, and measure all of those impacts that prawn production has," surmised Martin O'Brien—himself a sociologist and social justice activist, who collaborated with UK sustainability engineer Roland Clift to coin the term SLCA in a 1996 seminal paper.[25] "But of course what I wanted to know was why the World Bank gave them all this money in the first place," he recalls. "What were the consequences of funneling… something like 250 million dollars into Burma for the production of prawns?"[26] Thus it seems that investment strategies, for example, do not fit well into SLCA because they are designed before any limiting factors have been identified—not only for the prawns, but also for the chemicals that support prawn fishing, and for their demand in the marketplace. Investment strategies are not captured by SLCA boundaries because they fall under the realm of global politics. The most recent sustainability engineering vocalizations, this chapter illustrates, have shifted over time along the continuum of technological change and technopolitics. Merging these two ideologies in SLCA mobilizes engineering cultures in support of

---

[25] O'Brien, Martin, Alison Doig, and Roland Clift. "Social and Environmental Life Cycle Assessment (SELCA): Approach and Methodological Development." *International Journal of Life Cycle Assessment* 1, no. 4 (1996): 231-237.
[26] Personal communication.

qualitative social impact assessment; as these issues become a legitimate part of engineering knowledge, and as they are addressed by professionals in everyday practice, convergence gradually expands the boundaries of engineering cultures themselves.

Targeting the Social Risks: Fitting the Life Cycle Perspective into Corporate Social Responsibility

Reflecting on the purpose of SLCA studies in 2011, German engineer Andreas Ciroth wondered, "How high is the probability for [corporate] scandals? How much can you lose due to an image damage?"[27] His thought was echoed by an abstract on SLCA that was submitted to the 2012 Symposium on "Sustainability Assessment in the 21st Century: Tools, Trends & Applications", which declared: "Social issues are becoming among the most relevant factors for competitiveness and profitability of enterprises".[28]

Violations of company social and environmental responsibility frequently occur in distant, and often obscure, nodes within complex corporate supply chains. The life cycle of photovoltaic (PV) technologies, for instance, involves "blind spots" which may engender social and environmental sustainability problems. The polysilicon manufacturing processes—key to producing the majority of PV systems, located mostly in Chinese industrial facilities—have been linked to the dumping of silicon tetrachloride on agricultural lands. This situation raises serious ethical and environmental considerations with regard to the sustainability of PV's life cycle, for silicon tetrachloride is a toxin causing skin burns, lung cancer and crop infertility.[29] Similarly, a series of studies show a connection between illnesses among workers in factories that are producing semiconductors out of cadmium for

---

[27] Ciroth, Andreas and Juliane Franze. "Conducting a Social LCA." Presented at the Workshop: Social Aspects of Products Over the Whole Life Cycle, Berlin, May 30, 2011.

[28] Petti, Luigia, K. Paola Sanchez Ramirez, L. Cássia Maria Ugaya, R.V. Vicoli, and S. Vaira. " A tomato case study using subcategory assessment method for social life cycle assessment." Presented at the SETAC 18th LCA Case Study Symposium, Sustainability Assessment in the 21st Century Tools, Trends & Applications, Copenhagen, 26-28 November, 2012.

[29] Cha, Ariana Eunjung. "Solar Firms Leave Waste Behind in China." *The Washington Post* March 9 (2008): A1; Nath, Ishan. "Cleaning-up After Clean Energy: Hazardous Waste in the Solar Industry." *Stanford Journal of International Relations* 11 no. 2 (2010): 6-15.

use in manufacturing PV cells.[30] At the beginning of this decade, a January 2011 *Daily Mail* article reported that the Chinese state-owned Baogang processing plant—marketing neodymium magnets for wind turbines—was creating an "immense lake" of carcinogenic and radioactive waste.[31]

Exposure of such violations can have a considerable impact on a company's profit and reputation. Many stocks in the solar sector, for example, dropped by over 10% the day after a polysilicon manufacturer violation in Gaolong, China was reported by the *Washington Post*.[32] Furthermore, according to a 2010 report, such responsibility violations may lead to costs reaching 0.7% of a firm's revenue.[33] In addition, contemporary anti-trafficking and anti-forcible labor regulations in the US and elsewhere align the profit/reputation-related concerns of socially responsible businesses with the goals of state and national governments.[34] Companies thus see social auditing tools as techniques that use life cycle thinking to enable not only improved communication of social risks along supply chains, but also to transform business into an actor for social change.

Over the 2000s, sustainability engineering continued undergoing two important historical transformations. First was the concept of the

---

[30] See, for example, Kan Usuda, Koichi Kono, Keiko Ohnishi, Shin Nakayama, Yumiko Sugiura, Yasuhiro Kitamura, Akihiro Kurita, Yuko Tsuda, Motoshi Kimura and Yasuhisa Yoshida. "Toxicological aspects of cadmium and occupational health activities to prevent workplace exposure in Japan: A narrative review." *Toxicology and Industrial Health* 27, no. 3 (2011): 225-233.

[31] Parry, Simon and Ed Douglas. "In China, the true cost of Britain's clean, green wind power experiment: Pollution on a disastrous scale." *Daily Mail*, January 26, 2011. <http://www.dailymail.co.uk/home/moslive/article-1350811/In-China-true-cost-Britains-clean-green-wind-power-experiment-Pollution-disastrous-scale.html>
Accessed January 10, 2014.

[32] Derbyshire, Katherine. "Public policy for engineers: solar industry depends on policymakers' goodwill." ELECTROIQ, August 2008.
<http://www.gslb.cleanrooms.com/index/display/semiconductors-article-display/335746/articles/solid-state-technology/volume-51/issue-8/features/cover-article/public-policy-for-engineers-solar-industry-depends-on-policymakersrsquo-goodwill.html>
Accessed January 10, 2014.
See also Cha (2008).

[33] Lefevre, Clement, Damien Pellé, Shabnam Abedi, Raul Martinez, and Pierre-Francois Thaler. "Value of Sustainable Procurement Practices." Pricewaterhouse Coopers, 2010.

[34] Prelogar, Brittany, Laura Ardito, and Michael Navarre. "New Human Trafficking Laws and US Government Initiatives Make Anti-Trafficking a Compliance Priority for Businesses in 2013." Steptoe & Johnson LLP. February 2013.
<http://www.steptoe.com/publications-8618.html>
Accessed January 12, 2014.

so-called "triple bottom line," coined in 1994 by sustainability leader John Elkington and adopted in the title of the 1997 business report put out by the Anglo-Dutch oil company Shell, which had become a key element of the engineering discourse on sustainable development (Figure 4.3).[35] In 2002, *Walking the Talk* was cowritten by DuPont's Charles (Chad) Holliday and Swiss corporate titan Stephan Schmidheiny, with the authors maintaining that sustainable development was "partly about social justice."[36] Representing a "small but growing number [of corporations that] are moving towards sustainability reports…" Holliday challenged fellow engineers to embrace the "social side of sustainable development."[37] In response to such challenges, the SLCA community—heavily represented by individuals with engineering backgrounds—was officially launched in 2004 to expand the LCA methodology to a triple bottom line tool.[38]

---

[35] Elkington, John. "Enter the Triple Bottom Line." In *The Triple Bottom Line, Does it All Add up? Assessing the Sustainability of Business and CSR*, edited by Adrian Henriques, Julie Richardson, 1-16. Sterling: VA, 2004. A 2002 WFEO report read: The role of science and engineering in the current policy making process at national and global levels regarding the three pillars of sustainable development—social, economic and ecological is insufficient…[;] while progress has been steady in the economic area, the scientific underpinnings of the social pillar is less developed and requires increasing attention." ICSU and WFEO. "Addendum No. 8: Dialogue Paper by Scientific and Technological Communities. Role and Contributions of the Scientific and Technological Community (S&TC) to Sustainable Development." United Nations Economic and Social Council E/CN.17/2002/PC.2/6.Add.8. Advance Copy, 28 January 2002.

[36] Holliday, Charles O., Stephan Schmidheiny, and Philip Watts. *Walking the talk: The business case for sustainable development*. San Francisco: Berrett-Koehler, 2002, 13, 163, and 106.

[37] Ibid., 163, and 106.

[38] See for example, Benoît, Catherine, A. Gregory Norris, Sonia Valdivia, Andreas Ciroth, Asa Moberg, Ulrike Bos, Siddarth Prakash, Cassia Ugaya, and Tabea Beck. "The Guidelines for Social Life Cycle Assessment of Products: Just in Time." *International Journal of Life Cycle Assessment* 15, no. 2 (2010): 156-163. As a result of discussion within Working Groups of the December 2003 meeting of UNEP/SETAC members in Lausanne, Switzerland it was decided that a new "Task Force" be created to focus on the inclusion of social issues in LCA. In 2003 an LCA leader wrote that "[t]here seems to be a consensus about these three pillars, but not about the relative weights of these aspects." Klöpffer, Walter. "Life-Cycle Based Methods for Sustainable Product Development." *International Journal of Life Cycle Assessment* 8, no. 3 (2003): 157-159. Another early instance where SLCA and sustainability were cast in the context of the triple bottom line is Udo de Haes, Helias A., Reinout Heijungs, Sangwon Suh, and Gjalt Huppes. "Three Strategies to Overcome the Limitations of Life-Cycle Assessment." *Journal of Industrial Ecology* 8, no. 3 (2004): 19–32.

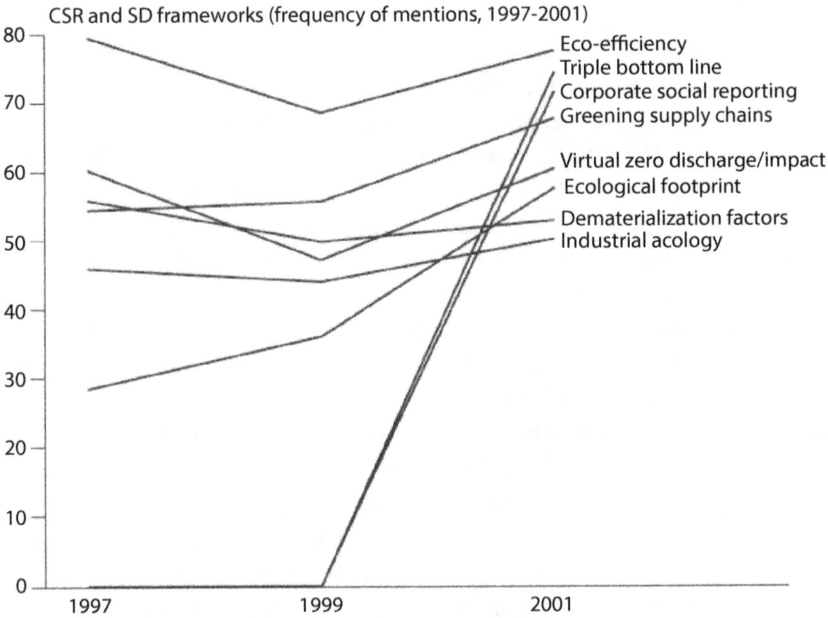

Figure 4.3 Frequency of mentions of "Triple Bottom Line," 1997-2001. Adapted from Elkington (2004): 2.

The second historical transformation to promote sustainability in industry was the development of "corporate social responsibility" (CSR). Adhering to the central idea that external values challenge the engineering profession, Technical University of Denmark's management engineer Louise Camila Dreyer, in the first PhD to cover SLCA in 2009, contextualized her subject matter as helping companies meet "society's expectations…[as regards]…a wider responsibility for the social impacts of their business activities."[39] These two closely related transformations grew in large part out of development-oriented institutions, corporate engineers, and LCA practitioners' growing engagement with elements of a technopolitics vision of sustainability.

Social Sustainability Tinkerers

According to the International Organization for Standardization (ISO) "the business sectors most interested in implementing the eventual standards

[39] Dreyer, Louise C. "Inclusion of Social Aspects in Life Cycle Assessment of Products: Development of a Methodology for Social Life Cycle Assessment." PhD diss., Technical University of Denmark, 2009.

[of LCA] are the ones who provide experts to develop the standards."[40] Or as Wolfgang Gawrisch, vice president for research and technology at chemical multinational Henkel, stated in a 1999 issue of *Chemical and Engineering News*, the reason behind corporations' involvement with institutions like the WBCSD is "to contribute new ideas to point the way forward, rather than seek their support in the implementation of sustainable development."[41] To contribute to the development of life cycle tools was not only to provide technical expertise; it was also to help society make progress toward sustainable futures.

From an engineering expert's perspective, the first effort to tinker with social, environmental, and economic dimensions of sustainability was a method developed in 1987 by Germany's Öko-Institut. In succeeding years the method was modified by German chemical company Hoechst AG and became known as Product Sustainability Assessment (PROSA).[42] Starting in the late 1990s, another German chemical company, BASF, would collaborate with Öko-Institut and Karlsruhe University to articulate SEEbalance®, the first SLCA method developed explicitly for industrial product comparative assessment and application. BASF offered a corporate engineering vision based on the World Bank's "four capital approach," according to which companies were supposed to strike a balance between providing for their social networks, caring for their workers' productive capacities and for the natural environment, and being efficient regarding their produced capital.[43]

In parallel to but independent of Öko-Institut's sustainability assessment, SETAC in 1993 published its "Conceptual Framework for Life Cycle Impact Assessment," which proposed the consideration of a "social welfare impact category."[44] The same perspective was built into the 1992 Nordic guidelines for product LCA and embraced in the scholarship of a small group of Swedish and Danish researchers in the 1990s.[45] Most

---

[40] Stellar.in. "Frequently Asked Questions (FAQ's). Undated. <http://www.stellar.in/faqs/> Accessed January 13, 2014.

[41] Morse (1999).

[42] Öko-Institut. *Produktlinienanalyse*. Cologne, Germany: Kölner Volksblatt Verlag, 1987.

[43] Schmidt, Isabell, Manfred Meurer, Peter Saling, Andreas Kicherer, Wolfgang Reuter, and Carl-Otto Gensch. "Managing Sustainability of Products and Processes with the Socio-Eco-Efficiency Analysis by BASF." *Greener Management International* 45 (2004): 79-94.

[44] SETAC (1993).

[45] Nordic Council of Ministers. *Product Life-Cycle Assessment-Principles and Methodology*. Copenhagen, Denmark: Nord 1992, 9th Nordic Council of Ministers, 1992. See also: Andersson, Karin, Merete Høgaas Eide, Ulrika Lundqvist, Berit Mattsson. "The feasibility

such independent projects appeared in engineering settings, like Patrick Hoffsteter's 1998 dissertation, which built on Mary Douglas's cultural theory to integrate the "value sphere" in LCA—but they were not followed up by other researchers until very recently.[46] For example, a 1995 study in the *Journal of Cleaner Production* proposed a combination of quantitative and qualitative impact categories for assessing occupational hazards and argued that "the best solution for the external environment is not always the best solution for the work environment and vice versa..."[47] That same tension would characterize much of the SLCA literature and the social sustainability movement as a whole during the succeeding decades.

From the mid-1990s on, the importance of social issues in consideration of sustainable development grew substantially, as did the range of constituencies involved in defining social sustainability metrics. At the corporate level, "chemical companies have actually been involved in this social [metrics] area for many years," remarked a corporate litigation attorney in 1999 in *Chemical and Engineering News*.[48] Companies sought guidance from fellow corporate efforts and legitimacy from the same global institutions that contemporary SLCA studies rely upon to define categories of social indicators. "We have spent a lot of time looking at others' efforts on social metrics; [w]e are finding ways to measure the social aspect [of sustainability] through global community perception [the UN] and employee opinion surveys, but the metrics are not as sophisticated, standardized, or developed," recalled two Dow Company corporate executives at the end of the 1990s.[49] If the challenge of a shrinking environment that put the engineering profession on the mission to sustainable development in the first place were the result of a short-term, profit-focused industry,

of including sustainability in LCA for product development" *Journal of Cleaner Production* 6, (1998): 289-298—the method outlined fills a gap between traditional LCA and so-called "Socio-Ecological Principles." See also: Wenzel, Henrik, Michael Hauschild, and Leo Alting. *Environmental Assessment of Products. Vol. 1 – Methodology, tools and case studies in product development.* United Kingdom: Kluwer Academic Publishers, 1997. Hauschild, Michael, and Henrik Wenzel. *Environmental assessment of products. Vols. 1 and 2.* London: Chapman and Hall, 1998.

[46] Hofstetter, Patrick. Perspectives in Life Cycle Impact Assessment. A Structured Approach to Combine Models of the Technosphere, Ecosphere and Valuesphere." PhD diss., Swiss Federal Institute of Technology Zurich, 1998.

[47] Antonsson, Ann-Beth, and Helene Carlsson. "The basis for a method to integrate work environment in life cycle assessments." *Journal of Cleaner Production* 3, no. 4 (1995): 215–220.

[48] Morse (1999).

[49] Ibid.

then business performance standards that incorporated environmental and social metrics would achieve industrial sustainability integration. To accomplish this goal, companies would need the support of engineering practitioners and their professional organizations.

In November 2002, the American Institute of Chemical Engineers (AIChE) issued a statement which read: "[The Sustainable Engineering Forum] SEF will do its part to add some scientific rigor to analyzing sustainability and to use appropriate metrics to determine comparative merits of [industrial product] alternatives."[50] AIChE's SEF was not chartered until 2003, so the first engineering "sustainability metrics," intended for the chemical processing industry, were coauthored in 2002 by the UK's Institution of Chemical Engineers (IChemE) Roland Clift, who had also coauthored the first article ever on "Social and Environmental Life Cycle Assessment" six years earlier.[51] The IChemE "Sustainable Development Progress Metrics" were conceived as an amalgam of engineering practicality and supply chain systems thinking. IChemE's sustainability leadership cast the Metrics in accordance with the triple bottom line—i.e., "the impact of industry in sustainability...summarized"—and saw themselves as initiating a relationship between companies developing performance standards for internal benchmarking on the one hand, and engineering practitioners implementing and interpreting those standards on the other. IChemE was also interested in developing "social indicators" by looking at two major social themes—namely "employment situation" and "health and safety at work"—while also discussing the broad category "Society" for which inventory data such as "Number of meetings [per year] with external stakeholders concerning company operations" were suggested. The introduction to the Metrics, echoing AIChE's 2002 statement, explained that the concept of sustainability required practitioners to reflect on their identity as technical workers.[52]

The same year Roland Clift collaborated with his colleagues at IChemE's Sustainable Development Working Group to develop the Metrics, he published an *Engineering Management Journal* article entitled "Engineering

---

[50] AIChE. "Statement on Sustainable Engineering Forum of AIChE Ideas and Scope," November 2002.
<http://webpages.eng.wayne.edu/~as8971/AIChE_SEF/sef_aiche_statement.htm>
Accessed January 20, 2014.

[51] O'Brien *et al.* (1996).

[52] IChemE. "The Sustainability Metrics: Sustainable Development Progress Metrics Recommended for Use in the Process Industries." Warwickshire: Institution of Chemical Engineers, 2002.

with a human face." As the title suggests, Clift's piece focused on engineers facilitating a more humane transition to sustainable development. Similar to their predecessors the "heroic materialists" of Watt and Brunel's caliber, sustainability engineers drew on the insights gleaned from tinkering with the interface of natural and technological environments, taking pride in carving social progress while having "fun of the impossible." But unlike their predecessors, the sustainability engineer "has emotions," and was likely to be female. In "Engineering with a human face" Clift expressed the view that the addition of social concepts into engineering methodology was a means to simultaneously enhance engineering practice's technical sophistication and to mirror the holistic principles of the profession. The "new role" that the sustainability engineer faced, according to Clift, was "more challenging and satisfying than merely acting as a technician." Clift recorded not only the changes in engineering science ("does my project recognize that science is uncertain?"), but the transformation that sustainability brings into the relationship between the engineer and her public audiences. He went on to describe the shift in engineering thinking presently manifested in the development of SLCA: "...management of the product started by eliminating risks to the workers who manufactured it, and to the customers. Over time this has grown to address risks to suppliers and distributors, and to encompass 'design for the environment.'"[53]

Clift's 2002 article might have appealed to technological change proponents of sustainability, in large part because it echoed the twin focus on engineering leadership and on managing material equilibrium flows through market exchanges, or because of his professional concern that a lack of an applicable definition of sustainability was contributing to the low enrollment figures in engineering schools. Yet, whereas early technological change theorists of sustainability held an environment-economy view of development, Clift—like most of his fellow engineering ideologues of sustainability at the time—rejected the idea that "Socio-centric" issues are extra-engineering material. Simultaneously, Clift's piece echoed the technopolitics-inspired engineers who in the past decade or so had been arguing for new, participatory, "deliberative processes" through which users could become part of technological design.

In the context of industrial transformations, like the development of CSR, this section has showed, engineers began to tinker with social sustainability. Insofar as it had already been dealing with voluntary efforts since

---

[53] Clift. Ronald, and Neil Morris. "Sustainable Development: Engineering with a human face." *Engineering Management Journal* 12, no. 5 (2002): 226-230.

the early 1980s in the US at least, the chemical industry was leading the way in defining social sustainability indicators. Importantly, as the work of engineers like Clift illustrates, whatever their proponents' self-understanding, the life cycle thinking and social impact improvement corporate discourses *combined* helped SLCA appeal to both technological change and technopolitics proponents of sustainability. In combination, the next section will discuss, the historical development of these discourses—influenced by European cultures of environmental responsibility—reverberated and promoted a new engineering identity, one that embraced a social element in engineering knowledge *per se*, and an identity operationalized along product life cycles.

Sustainable Development, Corporate Social Responsibility, and Life Cycle Assessment

SLCA stands at the crossroads between sustainable development, CSR and LCA. As discussed in chapter 2, the official birth of the tripartite relationship between sustainable development, LCA and CSR can be attributed to the creation of the Business Council for Sustainable Development (BCSD), founded by Stephan Schmidheiny in 1991. The expanding action of partnerships between international agencies, governments and private companies to develop metrics and methods to assess resource consumption and environmental impacts "throughout the full life cycle of products and processes" was one of the main mandates of Agenda 21, and it prompted Schmidheiny to found BCSD in preparation for the 1992 Rio Summit.[54]

"After Rio, there was a sense of OK, what's next?" Schmidheiny is quoted as having asked. The answer was corporate efforts to define extension of his ideas around the concept of "eco-efficiency". Another answer was BCSD's initiative—under the leadership of Swiss company Anova Holding AG's Frank W. Bosshardt—to reach out to ISO regarding the development of international environment standard ISO 14000. ISO worked closely with the International Electrotechnical Commission (IEC) through an ISO/IEC Strategic Advisory Group on Environment (SAGE), which culminated in the creation of the TC 207 technical committee that started to develop the standard for LCA in February 1993. In 1995, the BCSD merged with the World Industry Council on the Environment (WICE) to open the WBCSD secretariat in Geneva, Switzerland. Sustainability, in the lexicon of the

[54] United Nations Conference on Environment and Development (UNCED). *Agenda 21: Programme of Action for Sustainable Development*. United Nations, 1992, page 60.

corporate engineer, was translated into taking responsibility for industrial products throughout their life cycles.[55]

These sustainability initiatives, and the standardized knowledge they propagated, facilitated the application of LCA across various business contexts. In the early 1990s, 150 companies, including some major US corporations, adopted the "Business Charter for Sustainable Development," which was based on sixteen principles developed by the International Chamber of Commerce (ICC). The Charter was launched during the Second World Industry Conference on Environmental Management (April 10-12, 1991, Rotterdam, The Netherlands).[56] In the mid- to late 1990s the ICC—an organization which engineer-founders of the WEPSD considered essential for the Partnership's future—worked together with UNEP and FIDIC to develop the so-called Environmental Management System (EMS) Training Resource Kits based on the ISO 14001 and the EMAS (European Environmental Management and Audit Scheme).[57]

The same year UNEP, the ICC and FIDIC published a revised, 1997 Version 1.0 of their "Environmental Management Systems Training Resource Kit" (EMSTRKit) for industry, two female engineers from the US claimed it was time for corporate environmentalism to move from "rhetoric to results." Ann Rappaport, assistant professor of civil and environmental engineering at Tufts University, and Margaret Flaherty, then with WBCSD and previously at the ICC, traced the origin of corporate responsibility to globalization and the liberalization of material, human and capital flows. Under globalization, they wrote, corporate environmental responsibility "has taken on new meaning both in operational and philosophical terms."[58] The two engineers, who in 1990 had surveyed US-based multinationals to find that three out of four companies supported the development of envi-

---

[55] Timberlake Lloyd. *Catalyzing Change: A short history of the WBCSD*. Geneva: WBCSD, 2006, 21; Piper, Lennart, Sven-Olof Ryding, and Curt Henricson. *Continual Improvement with ISO 14000*. Stockholm, Sweden: SINF miljö&kvalitet AB, 2003; Garcia-Johnson, Ronie. *Exporting Environmentalism. U.S. Multinational Chemical Corporations in Brazil and Mexico*. Cambridge: The MIT Press, 2000.

[56] Lois (1991).

[57] Coates (1993).

[58] UNEP/ICC/FIDIC. *Environmental Management System Training Resource Kit*. UNEP, 1997. Flaherty, Margaret and Ann Rappaport. "Essential Matter. Corporate Environmentalism: From Rhetoric to Results." United Nations Research Institute for Social Development, Sept. 1997. <http://unrisd.org/unrisd/website/newsview.nsf/(httpNews)/33A02E752C5E6C9980256B7500539602?OpenDocument>
Accessed February 15, 2014.

ronmental management standards, championed companies' "accountability and standing among stakeholders" as well as the operationalization of sustainability through new methods of engineering practice that consider the "full range" of industrial processes and decisions.[59]

This is how—European-based—organizations like the WBCSD and the Society for Promotion of Life Cycle Assessment Development (SPOLD) became sources of rhetorical, but also data and software, resources for government officials and corporate engineers in Europe and the US, who saw in LCA an opportunity to address practical issues of environmental policy and business responsibility, respectively. In 1996, the inaugural issue of the *International Journal of Life Cycle Assessment* included then vice president of the European Parliament Ursula Schleicher's comments in regard to activities promoted by SPOLD:

> Industry and policymakers need reliable information in order to operate conscientiously. This applies to both data and scientific knowledge. The demand for so-called "Life Cycle Assessments" as aid to decision making belongs to this area. There is a definite tendency to demand more accountability from industry in the field of environmental policy—"product responsibility" being the term used.[60]

SPOLD, created in 1992 by a variety of multinationals interested in the promotion of LCA as a corporate management tool for sustainability, offered a vision of CSR as exemplified in the social value of LCA and linked by inventory data, methods to exchange such data (SPOLD format), and other environmental assessment techniques such as Material Flow Analysis. After SPOLD's dissolution in 2001, its mission was continued by the, again European-based—UNEP/SETAC Life Cycle Initiative, which was created a year later.[61]

The fact that many SLCA practitioners depict CSR as the normative and institutional backbone of their professional identity is not surprising. According to Catherine Benoît-Norris, first author of the UNEP Guidelines and cofounder—with Greg Norris and Deana Aulisio—of the Social

---

[59] Rappaport, Ann B. "Global Industry and Hazardous Waste Practice: Heading Toward Sustainability with Risk Management." *Practice Periodical of Hazardous, Toxic, and Radioactive Waste Management* 1, no. 4 (1997): 144-147. Flaherty and Rappaport (1997).

[60] Schleicher (1996).

[61] 2.0 LCA Consultants. "What is (was) SPOLD." Undated.
<http://lca-net.com/spold/whatis.html>
Accessed February 20, 2014.

Hotspots Database (SHDB), "there are many different streams of research that we call normative social responsibility."[62] At the level of social groups directly related to a corporation, engineering work in the SLCA discipline must accommodate the interests and concerns of various "stakeholders." At the broader societal level, SLCA engineering identity must address issues that subsequently feed back into corporate behavior itself, informing its social mission. Conceptually, CSR's most recent articulations, as those of SLCA, are based on "stakeholder theory," which, according to its developer Edward Freeman, "begin[s] with the assumption that values are necessarily and explicitly a part of doing business…"[63] In practice, however, CSR meant that business pursued questions of accountability and transparency through participation in voluntary standards.[64]

By the late 1990s theoretical or practical approaches to integrating environmental, economic and social indexes of sustainability had been utilized by actors as diverse as the World Bank, the WRI the Body Shop and Sustainable Seattle.[65] The period 1995-2000 was marked by the proliferation of voluntary sustainability standards for corporate reporting—the WRI, for example, reported in 1998 the existence of

---

62 Participant observation, "SLCA Short-Course," June 27, 2013.

63 Freeman, Edward R., C. Andrew Wicks, and Bidhan Parmar. "Stakeholder Theory and 'The Corporate Objective Revisited.'" *Organizational Science* 15, no. 3 (2004): 364–369.

64 These standards, which include examples such as the Organization on Economic Co-operation and Development, OECD's Guidelines for Multinational Enterprises, the Dow Jones Sustainability Index, the International Labor Organization ILO's Tripartite Declaration of Principles Concerning Multinational Enterprises and Social Policy, the UN's Global Compact or the same organization's Norms on the Responsibilities of Transnational Corporations with Regard to Human Rights, are covering—primarily—human and worker rights.

65 Serageldin, Ismail. 1996. *Sustainability and the wealth of nations: first steps in an ongoing journey*. Environmentally sustainable development studies and monographs series; no. 5*ESSD Environmentally & Socially Sustainable Development Work in Progress. Washington, DC: The World Bank, 1996. <http://documents.worldbank.org/curated/en/1996/07/696375/sustainability-wealth-nations-first-steps-ongoing-journey> Accessed January 30, 2014.
Ranganathan, Janet. "Sustainability Rulers: Measuring Corporate Environmental and Social Performance." Washington, DC: World Resources Institute, 1998. Nicholson-Lord, David. "Stake-out at the Body Shop. Social auditing: the stakeholder is alive and living in South America as Anita Roddick's empire undergoes an ethical inspection." *The Independent*, Sunday 21 January 1996.
<http://www.independent.co.uk/news/business/stakeout-at-the-body-shop-1324992.html> Accessed February 2, 2014.
See also See chapter 3—section on American Engineers for Social Responsibility (AESR).

"over fifty initiatives...developing measures of business sustainability."[66] The development of life- cycle-based applications for business, metrics, standards, and volitional reporting processes illustrates the convergence of sustainability engineering ideologies underlying the political purpose of SLCAs: to balance environmental LCA aimed at measuring the negative impacts of development with a parallel technique focused on the *positive* impacts—minimizing technology's unintended consequences and maximizing its positive capacities is the core principle of technological change as mentioned in chapter 1.

This section explained that whereas the US had fostered technocratic LCA, European engineers and practitioners seem much more politically attuned and willing to tackle social complications reflecting different engineering traditions and broader cultural concerns arising from European country contexts.[67] Under the aegis of SPOLD, European scientists and engineers asserted in a technopolitics fashion in 1996 that the barriers between LCA practitioners and society at large needed to be broken down so that both "LCA results be presented in a [socially] meaningful way... [and] societal priorities...be expressed in terms that enable LCA 'technological experts' to produce an answer."[68] In other words, as sustainability engineers become more diverse in who they are and where they come from, they bring different values compared to the economistic Americans. The integration of technological change and technopolitics as SLCA may thus also be a story about how American-centered engineering cultures are facing challenges.

The creation of corporate life-cycle-based tools and metrics, this section highlighted, also helped weave a sustainability engineering identity centered on practitioners with a goal to investigate social impacts of industrial processes who, when asked to justify the nature of their work, grounded it in the engineering technique of LCA. In LCA pioneer Bo Weidema's

---

[66] Ranganathan (1998).

[67] Latin American engineers may also be more socially and politically minded: "Social and economic benefits are not normally considered in a life cycle assessment, but these are significant aspects in the context of developing countries toward sustainable development," members of the Chilean Research Center for Mining and Metallurgy wrote in a 2011 SLCA study. Rada, Ramón J., Armando Caldeira-Pires, Catherine Benoît, and A. Claudia Pena. "Environmental Indicators Align with Integral Sustainability: The Chilean Copper Ongoing Process." *International Journal for Sustainable Innovations* 1, no. 1 (2011): 16-21.

[68] Unknown. "Synthesis Report: The Social Value of LCA." *International Journal of Life Cycle Assessment* 1, no. 2 (1996): 65.

words, "[w]hen these concepts [CSR or Ethical Trade] are extended to the entire value chain, the relationship to Life Cycle Management...becomes obvious."[69]

Weidema, who coauthored the UNEP Guidelines and teaches SLCA seminars in the recently launched International Life Cycle Academy based in Barcelona, Spain, argues that the notion of following stories of social impacts in industrial production is not simply CSR rhetoric; rather, LCA *is* the quantification of CSR in the value chain. However, a closer look at the historical roots of SLCA in the 1990s formulation of social auditing standards suggests that there is no such thing as a natural or "obvious" conceptual classification of social sustainability. Rather, many of the currently unsettled arguments about the nature and purpose of SLCA can be traced to the ideas and practices of those individuals behind the ground-breaking social auditing methods created in the mid- to late 1990s.

Social Auditing Pioneers and the Global Reporting Initiative (GRI)

One of the first "full range" reporting standards for industry that "walked the talk" of the triple bottom line was AA1000, developed in 1999 primarily by Simon Zadek, the AccountAbility founder, self-proclaimed "Buddhist economist," and until recently non-resident senior fellow at Harvard University. The purpose of AA1000, according to Zadek, was to fuse elements of ethical and socially responsible behavior in 21st century corporate management and business operations.[70] Zadek had brought to AccountAbility six years of experience in corporate social auditing, as Development Director at the New Economics Institute. The Institute was cofounded in 1986 by leaders of "The Other Economic Summit" (TOES), namely economist James Robertson, and Schumacher Society Council member Alison Pritchard, who protested the G7 Economic Summit meeting in London the same year. A prolific writer and activist-scholar, Zadek was, in 2014, collaborating with the Chinese government to implement green public procurement policies, while his books are standard references in the CSR literature.

In his early articles (1993-1999), Zadek extended the accountability, transparency and sustainable production notions that pervade current SLCA discourse, linked them to Schumpeterian economics and appropriate technology, and employed them as analytical tools to operationalize social

---

[69] Weidema, Bo. "Quantifying Corporate Social Responsibility." Undated manuscript.

[70] Institute of Social and Ethical AccountAbility. *AA1000 Framework*. London: ISEA, 1999.

change at the structural and individual levels. A 1997 article concluded: "This [Buddhist Economics] vision was (and is) largely consistent with a number of historical and contemporary calls for an approach to economics that embodies principles of self-realization, social justice, harmony and creativity, and an appreciation in practice of one's role in the cycles of nature."[71]

Zadek made explicit connections between appreciating the nature of being human and overcoming the "deep lethargy of the democratic project."[72] Such a link also resonated with technopolitics ideologues of sustainability, who postulated that market capitalism's hegemony had to be challenged, if not overthrown, if environmental sustainability and social justice were to inform both the means and the ends of engineering knowledge and practice. Read in the context of SLA's contemporary development as an engineering technique that is seeking transparency in the valuation of social impacts, Zadek's assumptions represent another level of integrated technopolitics ideology—particularly the idea of professional and methodological reflexivity—in mainstream sustainability discourse.[73]

The same year Zadek advanced his version of Buddhist economics, individuals from the Coalition for Environmentally Responsible Economies (CERES) and the Tellus Institute began cofounding the Global Reporting Initiative (GRI) with the support of a $3.75 million grant from UNEP—the largest donation of its kind in the history of Non Governmental Organizations (NGOs).[74] Considered by management scholars and SLCA researchers alike as the cutting edge of sustainability reporting, the GRI has exerted substantial influence on business sustainability as a multistakeholder procedure for designing and communicating reporting guidelines. In 1997, for example, the GRI began collaborating with the WBCSD to develop the reporting standard for greenhouse gas (GHG) emissions that was first published in 2001. In the absence of legitimate social metrics of sustainability, SLCA studies would have been even harder to imagine, let alone execute. Thanks in large part to the vision of GRI founding members Robert Kinloch (Bob) Massie IV, Allen L. White, and Paul Raskin, the SLCA community has found an institutional reference point and a methodological standard in the face of constantly shifting expectations about

---

[71] Zadek, Simon. "Towards a Progressive Buddhist Economics." 1997.
<http://www.inebnetwork.org/thinksangha/tsangha/zadekdsbook.html>
Accessed February 15, 2014.

[72] Zadek (1997).

[73] Ibid.

[74] The GRI was founded between 1997 and 1999.

the purpose of sustainability engineering practice.[75] Insofar as a fairly small set of people who moved between NGOs, consulting firms and companies entailed the development of the institutional structure of social sustainability, we see a similar pattern emerging between sustainability engineering, detailed in chapter 2, and SLCA.

A risk analysis and Total Cost Assessment (TCA) expert, Allen L. White, proposed new premises for socially responsible companies to reposition their role in society. A pioneer in designing methods for social and environmental assessment—such as the PoleStar software, which simulates socio-ecological systems—Paul Raskin helped bridge the relationship (often conceptualized as a gap) between sustainability rhetoric and situated, meaningful, practice. And an early champion of the CSR movement—he served as executive director of CERES from 1996 to 2003—Bob Massie promoted integrated reporting in places where it was resisted most: multinational engineering corporations.

Allen White is what Halina Szejnwald Brown and coauthors have termed an "institutional entrepreneur."[76] Such individuals are promoting ideas and tools of corporate transparency and stakeholder participation to affect sustainable social change. Along with Bob Massie, White was involved with CERES and the development of its environmental reporting standard since the early years of the organization. After cofounding GRI in 1997 and acting as its CEO until 2002, in 2004 he also cofounded Corporation 2020, created to facilitate integrating social mission into corporate purpose. He currently serves at the board of directors of Greg Norris and Catherine Benoît-Norris' New Earth, which in 2009 launched the SHDB, the only currently available database explicitly designed to make the UNEP Guidelines SLCA framework operational.

The complementary interests in life cycle, risk analysis approaches to pollution prevention and TCA that White developed during his tenure at the Tellus Institute in the early 1990s would later become the backbone of his strategies of institutional entrepreneurship. In 1993, White was directing Tellus Institute's risk analysis group and pioneering Total Cost

---

[75] Szejnwald Brown *et al.* (2007); Levy, David L., Halina Szejnald Brown, and Martin de Jong. "The Contested Politics of Corporate Governance: The Case of the Global Reporting Initiative." *Business & Society* 49, no. 1 (2010): 88-115; Labuschagne, Carin, C. Alan Brent, and Ron P.G. van Erck. "Assessing the Sustainability Performances of Industries." *Journal of Cleaner Production* 13 (2005): 373-385.

[76] Szejnwald Brown *et al.* (2007).

Assessment.[77] Six years later, he published an IEEE article entitled "Life Cycle Design Practices at Three Multi-National Companies" with Tellus Institute colleague Karen G. Shapiro, in which they argued that LCAs should not be identified with measuring environmental performance only. Their article proclaimed that corporations were entering an era of global competition that mandated transparency at the product life cycle level. With its ability to benchmark corporate practices and make plain the intricacies of organizational complexity and technological innovation, life cycle design, they claimed, was the appropriate tool to "operationalize stewardship." In other words, older approaches such as Product Environmental Performance (PEP) were inadequate to meet the need for efficient translation in the practice of corporate production units. In contrast, Shapiro and White wrote that LCA was evolving into a pragmatic methodology that struck an engineering balance between societal expectations and corporate reality.[78] More than a decade later, in the wake of the Occupy Movement, White linked business accountability with the question of a "Version 2.0" social contract: "[W]hich version of the social contract will prevail?" he questioned. Will it be that of "unfettered markets…[and] gradual undoing of the concept of collective good…or [that which] rewards business leaders who respect…the historical citizen-government covenant."[79] In White's view, the sustainable corporation and society at large are not separate entities. By identifying positive social impacts in industrial supply chains, SLCA boundaries are also shifting the boundaries of the social contract between business and society.

Paul Raskin, White's colleague and founder of the 1976 Tellus Institute of Boston, also exemplifies the concept of institutional entrepreneurship. A theoretical physicist, a philosopher—he studied philosophy at Berkeley under Paul K. Feyerabend, the proponent of epistemological anarchism—and a pioneer in socio-ecological modeling, Raskin has contributed to legitimizing the notion that socially oriented corporate auditing could transform the identity of business and, with it, the future of the economy.

---

[77] White, Allen L., Monica Becker, and E. Deborah Savage. "Environmentally Smart Accounting: Using Total Cost Assessment To Advance Pollution Prevention." *Pollution Prevention Review* (1993): 247-259. The authors contend that, regardless of its contribution to enhancing a company's profitability, TCAs do "not incorporate the social costs of a proposed project…"

[78] Shapiro, Karen G., and L. Allen White. "Life-Cycle Design Practices at Three Multi-National Companies." Paper presented at the 1999 International Symposium on Electronics and the Environment, Danvers, MA, United States, 1999.

[79] White, Allen L. "Time for Social Contract 2.0?" Undated manuscript.

The models he designed—the Long-range Energy Alternatives Planning (LEAP) system, the Water Evaluation and Planning (WEAP) system, and the PoleStar System—are all examples of integrated approaches to managing resources and to exploring future socio-economic and environmental scenarios.

Like Feyerabend, Raskin had no doubts about the limitations of scientific materialism—namely the idea that a materialistic cosmology could provide an accurate account of human nature. But unlike his mentor at Berkeley, Raskin associated self-understanding, the key in mediating between human and non-human nature, with the world of business. The year he founded the Tellus Institute, he published an essay on the "Ecology of Scientific Consciousness" that appeared in *Telos*, a critical theory journal, following a piece by Theodore Adorno that debated the historical process of understanding the meaning of human existence in the context of Beethoven's Missa Solemnis.[80] In 2002, Raskin had found an appropriate name for that historical process: Great Transition. His 2002 book featuring that title is an excursion into future global socio-technological scenarios. "New values" have taken root, wrote Raskin, leaving their mark on the global business landscape. Unlike the "unfettered market" (a phrase used also by White) which can only deliver the goods of efficiency, "[e] nvironment, equity and development goals are supra-market issues..." Instead of collapsing social sustainability into the corporate bottom line, the accountability movement is characterized by "forward-looking corporations [that] seize the new reality as a business opportunity and a matter of social responsibility." In addition, corporate operationalization of transparency would be marked by the "establish[ment] [of] tough standards for sustainable businesses and innovative practices to meet them."[81] The next challenge that institutional entrepreneurs would have to take up was to promote the ideal in multinational corporations that awareness of how social impacts "move" across products' life cycles is embedded in company visions of financial and environmental sustainability.

In late September 1997, a year before launching the GRI with fellow institutional entrepreneurs White and Raskin, Bob Massie, then executive director of CERES, delivered a speech in front of 1,200 engineers and

---

[80] Bernow, Stephen S., and D. Paul Raskin. "Ecology of Scientific Consciousness." *Telos* 28 (1976): 55-80.

[81] Raskin, Paul, Tariq Banuri, Gilberto Gallopín, Pablo Gutman, Al Hammond, Robert Kates, and Rob Swart. "Great Transition. The Promise and Lure and the Times Ahead." Boston, Massachusetts: Stockholm Environment Institute-Boston, 2002.

environmental officers at the General Motors Global Plant Engineering and Environmental Conference plenary session on "Reputation and Social Responsibility in the Twenty-First Century." A historian, theologian, business scholar, politician and survivor of both classical hemophilia and HIV/AIDS, Massie was not to be awed either by "a large crowd in a large room… [nor] the size and the power of the collection of human beings known as General Motors." He instead fused social responsibility rhetoric and practice to help engineers and corporate executives reimagine their identities and work in the context of business sustainability, transparency and engineering measurement of socio-environmental performance. Only a year out of Harvard's Divinity School, where he served as director of the Project on Business Values and the Economy, Massie was, and still is, well versed in engineers' identity politics—particularly those involving GM.[82] "Twenty-five years ago, General Motors' annual meetings were battlegrounds in which activist shareholder raised questions about board diversity, South African investment, toxic waste, and many other issues," he told his audience in his introductory remarks. But as attested to by the reason behind his visit to Detroit's Renaissance Center—GM's 1994 decision to endorse the CERES Principles—a new era of public accountability was thought to be on the rise. This impression was reinforced by the fact that Massie had just married Ane Tate, sustainable architecture pioneer, professor, and present director of the project "Urban Eden" at Rhodes Island School of Design; hence, his following depiction of the socially responsible engineer: "I deeply respect the work of designers and engineers," he admitted. "Design and engineering are, to my mind, a kind of poetry of the possible, an endless application of intelligence to the solution of problems in the physical world."[83]

Much of that poetry of the possible, whose verses are currently written by sustainability engineers and champions of SLCA, had first blossomed in the minds of Massie and his fellow institutional entrepreneurs. "Will we find a way to build a Global Reporting Initiative, so that international trade does not lead to an increase in secrecy and a decrease of environmental and labor standards?" Massie asked the GM engineers at the climax of the CSR movement in the late 1990s. Like Norris's briefing with the Dalai Lama, Massie's 1997 talk to GM employees was permeated with pragmatism and urgency, but also with a sense of optimism stemming from a faith

[82] In personal communication Massie added that "just because I said these words to the engineers doesn't mean they did it!"

[83] Massie, Robert K. "Poetry of the Possible." Manuscript, 1997.

that engineers must honor their responsibility to society. At the same time, in front of 1,200 practitioners, Massie's words supported the technopolitics ideologues of sustainability:

> I want you to imagine that in that [hypothetical stadium at the center of which we are all sitting now]… every section, every row, every seat is filled by those who are silently watching us, watching what we say, what we decide, and what we do. There are many different kinds of watchers, and many different reasons they are silent. There are the poor of this world…*They have no voice in our [technological] legislatures* and no ownership in our firms. They cannot dispute our deeds and decisions; *they must simply live with them or die by them.*[84]

Just one year after Massie's speech to GM engineers, Zadek, whose career encapsulates the evolution of the field of SLCA as a whole, had joined the GRI. The 1998 GRI Sustainability Working Group, chaired by Zadek, remarked that "…[sustainability reporting] is complementary to the existing financial reporting, i.e., by addressing the non-financial aspects of business and the intangible business assets and values…" To that end, "sustainability reporting," the Working Group continued "*fills the gaps* left by the traditional financial reporting and *brings to completion* the picture of a company's performance and balance sheet."[85] The most current (2013) version of GRI's guidelines divides social performance indicators into labor practices and decent work, human rights, society and product responsibility. And like SLCA studies, GRI's G4-PR2 guideline focuses on reporting "the total number of incidents…concerning the health and safety impacts of products and services during their life cycle…"[86] Due to the convergence of institutional, ideological and technological developments, the assessment of social impacts of corporate behavior and industrial operations in life cycles soon became a central feature of the definition of both sustainability and CSR.

---

[84] Ibid., emphasis added.

[85] Szejnwald Brown, Halina, Martin de Jong, and Teodorina Lessidrenska. "The Rise of the Global Reporting Initiative (GRI) as a Case of Institutional Entrepreneurship." Working Paper No. 36 of the Corporate Social Responsibility Initiative, May 2007, emphasis added.

[86] Global Reporting Initiative. "G4 Sustainability Reporting Guidelines: Reporting Principles and Standard Disclosures." Amsterdam, The Netherlands: GRI, 2013, 80. By 2007, there was already many sources of social performance indicators includind company CSR rating schemes. Good Guide, for example, founded in 2007 by UC Berkeley professor Dara O'Rourke had initiated a process of aggregating between different data sources.

A year after the formation of the GRI, Joachim Spangenberg, biologist by training and then advisor to the German federal government's board on sustainability indicators as well as secretary of the Social Democratic group in the federal parliament, published a white paper that considered sustainability metrics from the perspective of societal change. A member of the Executive Committee of the International Network of Engineers and Scientists for Global Responsibility (INES) while exploring the status of social auditing with the "Compass on the Road Towards Sustainability," Spangenberg was heading Wuppertal Institute's Division for Material Flows and Structural Change. As his title implied, Spangenberg was primarily interested in developing "interlinkage indicators" that would enable the connection between micro-level (i.e., corporate) sustainability standards to enable stakeholder communication and business efforts with macro-level, normative, and political questions. Spangenberg's report moves smoothly through a literature review of environmental and social indicators—the World Bank's work, the UN Human Development Index (HDI) and OECD's Pressure State Response (PSR) among others. When it comes to life cycle metrics, however, Spangenberg assumed a more critical tone: The "ISO 14000," he wrote, is not "suitable for the development of meaningful sustainability indicators as a communicative management tool." Like the driving political forces of structural change, sustainability indicators "must be *simple*…[namely,] the methodology of calculating them [must be] transparent…[and] *directionally safe* [meaning] that it should be obvious what they indicate is [politically] *relevant*…"[87] In a 2012 interview, Spangenberg echoed the SLCA trend toward measuring "the livelihoods we create and sustain" but was less enthusiastic about development agencies' ability to tailor their practices to the degree of change that is relevant to social justice.[88]

This section detailed how SLCA grew out of CSR—which a good number of companies in Europe and the US were backing already by the late 1990s demanding more nuanced LCAs. As a result, in the 2000s definitions of "sustainability" embodied in SLCA began to find more fertile ground. Part of this activity was driven by "institutional entrepreneurs" like Zadek,

87 Spangenberg, Joachim H. and Odile Bonniot. "Sustainability Indicators—A Compass on the Road Towards Sustainability." Wuppertal Paper No 81, February 1998.
88 Romanik, Clare. "A conversation with Joachim Spangenberg on sustainable development and social change," October 2012.
<http://europeandcis.undp.org/blog/2012/10/25/a-conversation-with-joachim-spangenberg-on-sustainable-development-and-social-change/>
Accessed February 16, 2014.

White, and Norris building institutional partnerships as well as designing and selling SLCA tools to industry. Yet another part is corporate power and coalitions like WBCSD and SPOLD playing a key role in advancing SLCA alongside engineering actors. That is, sustainability engineering identity being increasingly formed and normalized from within the corporate world. This contrasts with the resistance that sustainability faced by corporate managers in the 1990s as described in chapter 1. Also, this growing attentiveness toward social sustainability interestingly contrasts with the marginality of AESR between 1988 and 1994—although members of social responsibility engineering groups exerted some influence in the creation of SLCA. Timing, at any rate, is part of the explanation as the strategies and connections of social sustainability entrepreneurs that proved much more fruitful than AESR's radical politics. The creation of SLCA via CSR, however, also reflects the shift in larger politics and the role of government—particularly in Europe—promoting LCA and relevant policies.

Conclusion

What is "Natural"?

Enough evidence exists to argue that the historical development of sustainability engineering overlaps considerably with "sustainable development," largely defined by global institutions; "corporate social responsibility," largely run by multinational corporations; and "Life Cycle Analysis," increasingly integrated within both development and CSR discourses. Yet, even as the concepts of sustainable development, CSR, and LCA provide the intellectual and institutional mold within which SLCA becomes conceivable, these concepts may also obscure the historicity of sustainability engineering.

A common formulation amongst SLCA champions is that sustainable development is an overarching concept at the policy level; social responsibility is the business application or framework of a broader policy mandate; and SLCA is a technique that "stakeholders" may use to assess social conditions in product supply chains throughout their life cycles[89] (Figure 4.4). The same narrative also recognizes the influence of institutions such as the GRI in initiating within the CSR movement a focus

[89] Interview data.

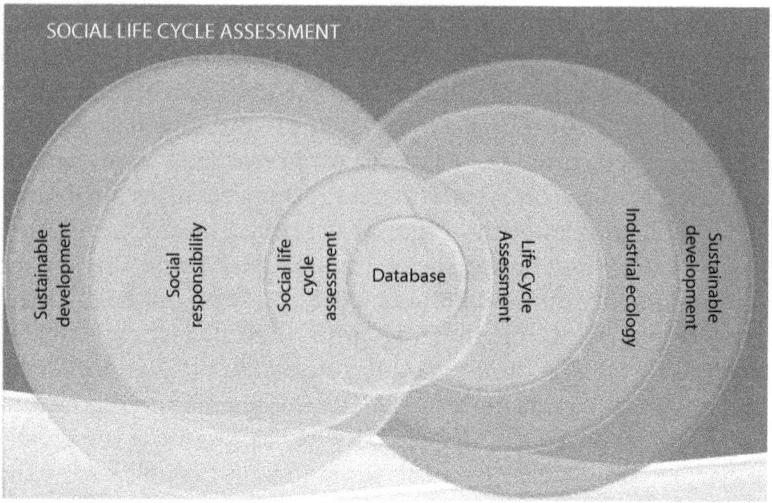

Figure 4.4 A conceptual hierarchy of SLCA. Image courtesy of Catherine Benoît-Norris.

on due diligence in supply chains, without which the life cycle perspective would be lacking an area of application.[90] This narrative, however, tends to confuse two different elements: One is the concepts of sustainability engineering ideologies, which are symbolically portrayed and discursively tested. The other is the actors through which concepts are put into practice. Failure to recognize this distinction implies that concepts assume a life of their own, or that engineering practices are devoid of ideologies. So what is interesting about such distinctions? They enable movement from a historical analysis of ideas and the institutions that have contributed to shaping them towards an appreciation of the ideologies that have sustained those distinctions and conferred them legitimacy.

For some, the key to CSR is life cycle thinking, not sustainability. Former UNEP executive director Klaus Töpfer, for example, is quoted as saying that it is "[l]ife cycle thinking [which] implies that everyone in the whole chain of a product's life cycle...has a responsibility..."[91] In fact, according to SLCA practitioner Evan Andrews and coauthors, the "awareness of managing upstream corporate social responsibility issues has risen due to the growing popularity of LCA."[92] Chemical engineer Michael Z. Hauschild,

---

[90] Interview data.

[91] Life Cycle Initiative (2009): 28.

[92] Andrews, Evan, Pascal Lesage, Catherine Benoît, Julie Parent, Gregory Norris, and Jean-Pierre Revéret. "Life Cycle Attribute Assessment: Case Study of Quebec Greenhouse

who mentored the first two dissertations on SLCA, wrote in 2008 that in particular, the possibility to conduct LCAs "entails responsibility…"[93] Others maintain that CSR drives the inclusion of social impacts in contemporary LCA studies: "the life cycle perspective is key…" a practitioner admitted in personal communication; yet, "the reccurring interest for [SLCA] was alerted from the CSR area."[94] And, "in the academic community," adds an SLCA and engineering practitioner, "sustainable development is like a circle within the bigger circle of CSR and life cycle thinking is like vapors that permeate the whole thing."[95] Similarly, another engineer recalled that "after a Rio+20 meeting…[held in 2012] in the corporate sustainability forum…[it was concluded that]…these big companies…wanted to scale up sustainability along the value chain…[;] so, I think in this sense there was a big space for LCA."[96]

Some SLCA practitioners thus believe that within CSR there is a need for a tool to harmonize views and standardize approaches.[97] To view CSR in this way is to assume a certain ideal of engineering rationality—that of "holism" or "systems thinking"—and then critique it for falling short of that ideal. A few other SLCA authors have maintained that performing life cycle studies is too costly and thus most companies currently conduct environmental "screenings" where climate change is the main focus: "SLCA is for nation states, supra national bodies and NGOs. For companies CSR is enough," said one interviewee.[98] Correspondingly, although the assumption that SLCA would fill a CSR vacuum permeated Andreas Jørgensen's 2010 SLCA dissertation, he was shocked to realize that:

> SLCA was not so attractive as I thought. I made a series of interviews with a list of very CSR-engaged companies asking them about the possibility of using SLCA. Yet, all of [the representatives from the companies I approached] said, 'well, we don't really see the point of that [SLCA],' and basically they said, 'we are not really that interested in the kind of assessment it could provide…' Even though… we are talking about companies in the DOW Jones of sustainability here… And

Tomatoes." *Journal of Industrial Ecology* 13, no. 4 (2009): 565-578.

[93] Hauschild, Michael Z., C. Louise Dreyer, and Andreas Jørgensen. "Assessing Social Impacts in Life Cycle Perspective—Lessons Learned." *CIRP Annals Manufacturing Technology* 57 (2008): 21-24.

[94] Interview data.

[95] Interview data.

[96] Interview data.

[97] Weidema, undated manuscript.

[98] Interview data.

> they also said, 'we don't have the resources you are asking; it's simply not possible for us...'[99]

In other words, when a discipline is in the forming stages, anyone's assumptions can be challenged at any time. This was particularly important in the case of the 2006 SLCA Task Force Committee discussion, where it was noted that although "the aim of the taskforce was to check whether and how social aspects can be integrated into LCA...there was a debate in the taskforce about this preliminary decision."[100] Similarly, while concluding his dissertation, Jørgensen remarked that in the companies that participated in his study, "the use of SLCA tools...[would] be restricted to applications with very limited or no life cycle perspective."[101]

In personal communication, one LCA consultant opined that neither CSR nor SLCA are able to provide the necessary "guidance" to social sustainability at the corporate level. They consider both CSR and SLCA as "approaches," each of which have advantages and disadvantages—SLCA, for example, does not include positive impacts occurring in the use phase of a product.[102] Thus while different actors may, for different reasons, express different conceptual hierarchies, in practice the relationship between CSR (e.g., corporate sustainability reporting based on GRI standards) and SLCA is not hierarchical or natural in and of itself; rather considerable exchange occurs between practitioners of the two fields, with regards both to social sustainability data and social themes deemed worthy of measurement. Apart from such exchanges, as the simultaneous launching in May 2009 of the UNEP Guidelines and the ISO standard for social responsibility (ISO 26000) attests, the institutions associated with CSR and SLCA have also played a role in carving the SLCA landscape.

The history of sustainability engineering may serve as a source to both contextualize and further explore the ideological underpinnings of SLCA, more than a "natural focus."[103] For example, the authors of a 2012 review

---

[99] Personal communication. Jørgensen also admitted that "this is a kind of snapshot of a situation...[meaning that] of course things can change and different companies may think differently in the future."

[100] Grieβhammer, Rainer, Catherine Benoît, Louise Camilla Dreyer, Anna Flysjö, Andreas Manhart, Bernard Mazjin, Andrée-Lise Méthot, and Bo Weidema. "Feasibility Study: Integration of social aspects into LCA." Freiburg, 2006.

[101] Jørgensen (2010): 9.

[102] Interview data.

[103] The idea here is that corporate willingness to protect brands and profits has entailed a "natural focus on...CSR, and [has] broaden[ed] the focus to sustainability rather than just environment." This contention embeds the assumption that researching the social aspects

article found it "surprising" that LCA was *missing* from the discourse at the intersections of engineering and sustainability.[104] Similarly, according to one practitioner, life cycle thinking is the elusive emblem of sustainable engineering: "I would always make the argument that life cycle thinking supports sustainable development, but I don't know...I don't know the history [of how sustainability and life cycle thinking became tied] that well...If I'm doing a paper,... [I] go back to [the 1987] Brundtlandt [report] and you kind of go from there..."[105]

The "natural evolution" assumption expressed above by SLCA practitioners is ideologically rooted in the same technological change premise that sustainable technologies present opportunities to re-embrace the vision of engineering progress in light of challenges posed from outside the profession. This is not to argue that there are no conceptual or institutional links between SLCA, CSR and sustainability. Quite the contrary. Rather, as I illustrate in the next chapter, the "natural evolution" assumption fails to consider how the ideological visions of navigating the unintended consequences of development and of assessing socio-environmental impacts in life cycles have coalesced historically.

---

of products in value chains is a "natural extension" or "natural next development" to linking life cycle thinking with the triple bottom line of sustainability; but it also assumes that corporate social responsibility is a single-dimensional concept. Hauschild *et al.* (2008). Andrews *et al.* 2009; Technical University of Denmark. "Invitation to Seminar: Addressing the social dimension of product sustainability with life cycle assessment," May 31st 2010, 14.00-17.00 Meeting room 1, Building 101A, DTU.

104 Sala *et al.* (2013). The authors note that in "an extensive literature review on sustainability and engineering, life cycle thinking and assessment are not even mentioned" (page 1668). Yet, this assertion was based on reviewing only one article written by an author with an engineering background who himself neglects the engineering-specific literature on sustainability.

105 Interview data.

5

Social Life Cycle Assessment (SLCA) Rationalities

The worldviews of engineers and scientists often voice the epistemological authority of past generations: Walter Klöpffer, the founding editor of the *International Journal of Life Cycle Assessment* referred to LCA as having "scientifically hard and soft components". Hence inventory analysis and impact assessment must meet Karl R. Popper's refutability criterion, Klöpffer believed.[1] He also maintained that the expansion of LCA to include social impacts thus appears to reveal an opening to "pseudo-science."[2] Similar skepticism regarding LCA's ability to "fulfill the high expectations" set by policy makers was expressed in the mid-1990s by Ursula Schleicher of the European Parliament.[3] The *International Journal of Life Cycle Assessment* reflects a rather optimistic blend of technological change and technopolitics despite its early cautionary rhetoric. For example, Andreas Troge, then

---

[1] Klöpffer, Walter. "Editorial." *International Journal of Life Cycle Assessment* 1 no. 1 (1996): 3.
[2] Klöpffer, Walter. "Reductionism versus Expansionism." *International Journal of Life Cycle Assessment* 1 no. 2 (1996): 61.
[3] Schleicher (1996).

president of the German Federal Environment Agency (UBA), argued in 1996 that LCAs "should initially take place without regard to economical or social effects, or even with regard to other political factors."[4] In other words, the technological change vision of a sustainable future framed by LCA boundaries also embodied a variation of technopolitics that challenged the processes by which LCA objectivity is established. In pages of the *International Journal of Life Cycle Assessment*, as well as in pages of other mainstream LCA publishing venues, like the *Journal of Industrial Ecology*, many writers maintained that LCA would not gain legitimacy if LCA boundaries were not breached. The view was that LCA legitimacy was characterized by a transparent discussion of these boundaries. In other words, LCA's rationality was seen as comprised of—and not limited to—the values and normative assumptions implicit in the technique itself.

Less often do engineers and scientists critically engage with contemporary epistemological views and traditions. The article in SLCA literature most explicitly dedicated to the investigation of the ontological, epistemological and methodological bases of sustainability assessment is a 2013 piece by Serenella Sala, of the European Commission Joint Research Centre (EU JRC), and her coauthors, engineers by training, Francesca Farioli and Alessandra Zamagni.[5] The article casts sustainability assessments—and SLCAs particularly—as holistic approaches and treats their ontology, epistemology and methodology as inherently interconnected. The authors describe sustainability as the critical and reflective engineering discourse *par excellence*:

> The discipline [sustainability science] implies a holistic approach, able to capitalize and integrate sectorial knowledge as well as a variety of epistemic and normative stances and methodologies towards solutions' definition....This definition implies that ontological, epistemological and methodological aspects of the discipline are interlinked...[A]t the level of ontology, the main challenges are related to the comprehensiveness of the science in addressing different capitals, values and goals and respective trade-offs, whereas, at the level of epistemology, the key point is knowledge innovation, in other words, how to move to the mode 2 science, through collaboration amongst different disciplines and a broader societal learning...The methodologies have to be designed and developed in a participatory,

---

[4] Troge, Andreas. "Preface by Andreas Troge President of the Federal Environmental Protection Agency." *International Journal of Life Cycle Assessment* 1, no. 1 (1996): 1-2.

[5] Sala *et al.* (2013), Part 1.

> interactive (non-extractive) collaborative way; to facilitate knowledge co-generation through participation processes...from problem framing to strategy implementation, transparently balancing inputs and facilitating knowledge claims, shared control over the process and accountability of results...[6]

The authors fail, however, to discern between the engineering discourse's reflexivity and its potential to lead to unreflective views of technology, thus missing, for example, that the term "participation processes" has varying connotations; therefore a "holistic approach"—as chapter 2 showed—is a broad and precarious term.

The purpose of this section is to examine SLCA practitioners' views about their technique and its social value, and to reflect on how such views might contribute to our understanding of the mutually supportive, ongoing, relationship between the expansion of LCA and the ethical expansion of engineering cultures and practices. Two main engineering ideologies of sustainability have been defined in the introduction. These ideologies enable practitioners to tell stories about their work; each ideology enables stories about sustainability that are construed around a common lexicon and the relatively common perception that technology is autonomous. Where the ideologies of sustainability differ, however, is in regard to questions such as: How does engineering ontology (e.g., the view of systemic interdependence) affect methodology (e.g., the ecosystems approach to sustainability)? How do SLCA practitioners' ideals of—or lack thereof—technology, social justice or participatory democracy affect their views on the nature and purpose of engineering knowledge and vice versa? These ideological elements, their contradictions and their integration, are what enable SLCA studies to claim social relevance.

SLCA Understandings and Verbalizations of Sustainability

In 2008 a leading European advocate of LCA, Walter Klöpffer, implicitly illustrated the interconnectedness between LCA ontology, epistemology and methodology in his definition of Life Cycle Sustainability Assessment (LCSA), which he abstracted from the concept of the triple bottom line:

Life Cycle Sustainability Assessment (LCSA) LCSA = LCA + LCC + SLCA[7]

---

[6] Ibid., page 1662.

[7] LCC stands for "Life Cycle Costing." Klöpffer, Walter. "Life Cycle Sustainability Assessment of Products." *International Journal of Life Cycle Assessment* 13, no. 2 (2008): 89-95.

On one level, Klöpffer's equation is consistent with LCA—and SLCA—visions of holism. Such visions, expressed as early as 1994 by US sustainability authority Jim Fava in an article in *Environmental Toxicology and Chemistry*, cast LCA as "another way of addressing environmental problems from a systems or holistic perspective... [that] gives us new paradigms to...sustain our resources for the future."[8] As the late Ian Boustead, who had collaborated with Fava in the very first LCA conference organized by SETAC in 1990, put it: "...the most fundamental concept that underpins all LCA work is the need to think of systems rather than products."[9] It was this vision that made the articulation of LCA as the technique that operationalized engineering sustainability viable, both at the conceptual and discursive levels. By the mid-2000s, LCA champions were able to extend this vision to argue that three separate assessment tools (environmental, economic, and social) would give the best possible engineering definition of sustainability, drawing from life cycle thinking and a media (input-output) approach.[10]

On another level, for all his holistic rhetoric, Klöpffer's equation also depends on a particular understanding of sustainability. The ontological idea that environmental, economic, and social interdependence "is... a principle of sustainable development" is assumed in the methodological claim that integrating environmental, economic and social impacts within one LCA study would be "rather counterproductive."[11] Methodological tensions capture and redirect ontological conflicts. Rather than assuming, as Klöpffer did, that LCAs and SLCAs will be designed around the same system boundaries, the UNEP Guidelines note that "the reasons why social LCA system boundary may or may not differ from environmental LCA... will become clearer" as researchers accumulate more practical experience.[12] And new ontologies sweep away old methodological commitments: The "functional unit"—the LCA methodological basis that allows for products or services to be compared and examined—may no longer be necessary, some authors argue, in light of SLCA's focus on corporate behaviors instead

[8] Fava, James A. "Editorial. Life-Cycle Assessment: A New Way of Thinking." *Environmental Toxicology and Chemistry* 13, no. 6 (1994): 853-854.
[9] Boustead, Ian. "LCA—How it Came About The Beginning in the U.K" *International Journal of Life Cycle Assessment* 1, no. 3 (1996): 147-150.
[10] Klöpffer (1996).
[11] Hunkeler, David and Gerald Rebitzer. "The Future of Life Cycle Assessment." *International Journal of Life Cycle Assessment* 10, no. 5 (2005): 305-308.
12 Life Cycle Initiative (2009): 56.

of product systems themselves.[13] In calling practitioners to operationalize their sustainability convictions into a life cycle thinking technique, Klöpffer's equation expressly conveys a triple bottom line conception of development. Yet, there is no obvious relationship between sustainability and the life cycle perspective. In a recent essay, the current editor for SLCA in the *International Journal of Life Cycle Assessment* found no evidence for the inclusion of economic impacts (e.g., LCC) in sustainability assessments of technologies.[14]

One of the sustainability engineering discourses where technological change and technopolitics appear as integrative—rather than as opposing—forces is when life cycle thinking is coupled with social impact assessment. This integration becomes workable partly due to the existence of a recurring vocabulary for talking about sustainability in an engineering context. LCA practitioners claimed early on that their systems-based, input-output, approach suggests a "holistic standard" which allows for the understanding of "ecological equilibrium."[15] Similarly, the first attempts to create social auditing schemes viewed the society-economy relationship as part of "rapidly increasing flows" combined with upsurge in entropy.[16] And more recently, SLCAs were described as "integrative processes" meant to inform corporate "integrative reporting."[17] In addition, the same mechanical metaphor—feedback—that has kept the motor of sustainability engineering running for more than two decades (1989 onwards) is currently used by practitioners to understand, explain and rationalize social impact assessment in a life cycle. Yet, as one interviewee pointed out, there is no clear relationship between the meaning of the feedback metaphor advanced by LCA practitioners in varying contexts, and SLCA's intended function as a metric of social sustainability:

> [We] engineers love feedback. What you do is you develop a technology [SLCA] and you show it to the corporate managers and then you get their feedback on it. But… there's nothing social about that in any sense of the word. It's simply allowing the CSR [Corporate Social Responsibility] department to put their best guess forward and then

[13] Zamagni, Alessandra, Oscar Amerighi, and Patrizia Buttol. "Strengths or Bias in Social LCA?" *International Journal of Life Cycle Assessment* 16, no. 7 (2011): 596-598.
[14] Jørgensen, Andreas, T. Ivan Hermann, and Anders Bjørn. "Sketching a Life Cycle Impact Assessment for a Life Cycle Sustainability Assessment," unpublished manuscript.
[15] Troge (1996).
[16] Spangenberg and Bonniot (1998).
[17] Greg Norris, participant observation, "SLCA Short-Course," June 27, 2013.

> finding out if they can get management support for it. What is the sense in which people participate in the development of that SLCA? What is the sense in which they contribute their desires and their needs and their social connections in its development…in the first place? This problem with "feedback" has to be resolved because feedback is utterly an inadequate metaphor of what a social LCA could be.[18]

The interviewee asserted that a common sustainability engineering trope would perhaps become a point of dispute: there was as yet no agreement on whether or not feedback could be an adequate way of conceptualizing to engineering practitioners and their audiences in meaningful terms the intended function of an SLCA study. Correspondingly, SLCAs are depicted as simultaneously "holistic" and "reductionist": Sustainability life cycle assessment, they suggest, "should represent the holistic approach which integrates…the reductionist approach of the single part [e.g., SLCA] of the analysis."[19] And in Greg Norris's account, the concept of "sustainable development" that so many engineers have considered the philosophical basis of utilitarian operationalizations of sustainability, "implies [not a trade off, but] a non-compensatory relationship…"[20] As a result of such discursive and ideological blending, every sustainability engineering metaphor casts every other. Indeed, by virtue of metaphors such as "feedback" not only have technological change and technopolitics become integrated, but SLCA practitioners have also created discursive venues to debate the social and political objective of their work.

As is true in every discourse, however, cultures in power are more likely to establish their metaphors. The ideology of technological change continues to shape the prevailing conceptualizations of sustainable technology in the context of SLCA. The sustainability ideology of technological change is grounded in the belief that technology will provide the means to navigate the unintended consequences of development. On the one hand, its legitimacy is derived from its own rationality, namely that no sacrifice of prosperity is required in the linear path to sustainability. On the other hand, its

---

[18] Interview data.

[19] Sala, Serenella, Francesca Farioli, and Alessandra Zamagni. "Life Cycle Sustainability Assessment in the Context of Sustainability Science Progress (Part 2)." *International Journal of Life Cycle Assessment* 18, no. 9 (2013): 1686-1697.

[20] Norris, Gregory A. "Revisions to LCA Needed to Address Sustainable Consumption." Paper presented at the conference of The Society of Non-Traditional Technology (AIST) Life cycle approach to sustainable consumption: conceptual design of a methodological framework, Tokyo, Japan, 2003.

legitimacy also stems from the engineer's managerial skills and her (quantitative) ability to conduct materials balance analyses. Based on the same utilitarian logic that authorizes trade-offs ("For me, sustainable means... no negative impacts are caused or...[if] caused, negative impacts are compensated"), the prevailing understanding of unintended consequences in an LCA context is that of "problem shifting:"

> [My underlying assumption behind the claim that] life cycle thinking supports sustainable development?... for me it's avoidance of impact shifting... [L]ike, in environmental LCA you can reduce your impact, say, at the manufacturing stage and increase your impacts either in raw materials or traction or in use. So, the life cycle thinking helps you prevent that... That's in my view why life cycle thinking helps support sustainable development, is because you don't want to shift the burden to some other country, some other place, some other time in the life cycle.[21]

The growing emphasis on life cycle thinking approaches—including SLCA—coexists with key elements derived from the dominant engineering ideology of technological change. At the same time though, as technopolitics ideologues of sustainability have lamented the failure of sustainability engineering to engage with social issues and epistemological reflexivity, the dominant engineering cultures they critiqued did not let their colleagues' voices go unheard. The development of SLCA attests to two important, ongoing, transformations in engineering cultures: First, technology's social impacts and corresponding qualitative analyses were viewed, at least by major spokesmen in the LCA community, as precarious, yet legitimate engineering subject matter. Second, parallel to the opening of LCA boundaries to social issues, LCA engineers challenged themselves and their colleagues to be reflective and transparent with regard to what practitioners themselves have viewed as "clearly normative standpoint[s]"[22] in their discipline. European SLCA pioneer Andreas Jørgensen's discussion of the turning point in engineering cultures from making claims to assessing environmental impacts to quantitatively/qualitatively engaging with "human well-being" stresses that it is not an easy, nor a straightforward transition:

---

[21] Interview data.
[22] Finnveden (1997).

> If well-being is understood in its probably most common form as a person's subjective experience of his/her own life, the inclusion of subjective indicators will improve the validity of its assessment... This inclusion of subjective indicators is, however, related to practical, managerial and boundary setting problems...Yet, if well-being is defined differently, as relating merely [to] the world external to the individual...this would eliminate the problem, but would probably also imply that the meaning of well-being would be different from the most common understanding of the term.[23]

Such claims to reflexivity nevertheless do imply the advisability of new kinds of legitimization for the political nature of engineering decision making. The SLCA practitioner, critics from within the community declare, should not set the boundaries of the system or process at hand and then leave the study's results to speak for themselves. Rather, the voice of the modeler should be recognizable, and therefore politically contestable, during the communication of a study's results.

Reflected in questions such as: "Is the payment of the legal minimum wage, fair salary? Is it possible that corruption has positive impacts? Is a high migration rate good or bad?", engineering reflexivity and SLCA are both contingent responses to the problematic nature of engineering identity in the context of sustainable technology.[24]

Entwined Ontological, Epistemological, and Methodological Anxieties

In March 2003, Greg Norris presented a paper to the UNEP-sponsored First International Workshop on Sustainable Consumption held in Tokyo, Japan. His paper imagined LCA and its relationship to sustainable consumption as examples of an ends-based, household focused, engineering intervention. As he quoted the sociologist and father of "conspicuous consumption" Thorstein Veblen, Norris went ahead to challenge the engineering mindset that resonated in LCAs. The (engineer's) view of the product system in LCA, he suggested, does not consider inputs of labor. Within that system, traditional engineering cultures are likely to focus on *how* material and energy flows move through ("attributional" LCAs)—as opposed to examining the *interaction* between such flows and engineering decision

[23] Jørgensen, Andreas, C. H. Lufanna Lai, and Z. Michael Hauschild. "Assessing the Validity of Impact Pathways for Child Labour and Well-Being in Social Life Cycle Assessment." *International Journal of Life Cycle Assessment* 15, no. 1 (2010): 5-16.
[24] Ciroth and Franze (2011).

making. Moreover, by conceptualizing LCAs as addressing questions of "what if" ("consequential" LCAs) Norris found a way to rationalize life cycle thinking methods *independently* of the causal link between assessed flows and impacts. By advocating the extension of "what if" questions to the assessment of social impacts, he implicitly alluded to the idea that, for a complete assessment to take place, LCAs must consider both a realized and a non-realized consumption/production scenario. Moreover, in arguing for an LCA model on the basis of the dynamic relationship between "product-based function and well-being at the individual level," Norris began to envision sustainable consumption not as a purely mechanical-economic construction, but as complex undertakings occurring both within and outside of product life cycles.[25]

For Norris and the rest of the members of the proliferating international SLCA community, the expansion of LCA boundaries poses a set of entwined ontological, epistemological, and methodological anxieties that challenge the conceptual and institutional boundaries of traditional engineering cultures. Having identified social issues as the third pillar of sustainability, how should engineers define "social impacts" in engineering/LCA terms? Even if they agreed upon the notion of "well-being" as the missing "area of protection" in LCA methodology, how would they quantitatively measure that? And even if they decided to merge SLCA into the existing LCA methodological schema, how would they determine the key indicators to describe the social themes they wanted to address? In effect, sustainability engineering discourse can no longer justify itself through invoking technology *per se*: the same social and political issues that were once seen to endanger the engineer's authority now hasten the rise of new engineering disciplines and identities.

In his 2003 article presented at the First International Workshop on Sustainable Consumption, Norris described SLCA engineering cultures as constituting themselves by developing a critique of "the presentday LCA framework"—by revisiting traditional engineering conceptualizations of causality. On the one hand, the common language in LCAs is that of "functional units"—physical quantities (e.g., material flows) per unit of process output—which guide researchers in modeling and comparing product systems. On the other hand, social impacts do not spring forth from technical functions themselves; they do not necessarily depend on product components, but they are shaped by how supply chains are constituted. For example, the production of two otherwise similar PV panels may entail

[25] Norris (2003).

quite different social impacts depending on the manufacturing location and the company's approach to managing these effects. In SLCA studies, then, most practitioners view "impacts" as being dependent on corporate behavior.[26]

In this respect—namely in the absence of a causal relationship between technical process and social impact and the high level of uncertainty associated with modeling social issues in relation to a functional unit—SLCAs reveal their sharpest break with LCA methodology. There are two ways in which SLCA practitioners have dealt with this issue of "allocation," which directly affects the scope and the system boundaries of SLCA studies. One way has been their technological innovation to create software tools to "prioritize" so-called "hotspots," i.e., industrial sectors or parts of a supply chain where data collection will take place. The other way has been their consensual decision on a metric, or "activity variable"—worker hours—that determines the scope of the study by calculating what percentages of a supply chain are more socially risk-prone than others.[27] The two approaches to allocation are interconnected; in combination they define what is considered to be socially relevant in a product's supply chain or a company's list of suppliers.

The assumption that worker hours are the best representation of a company's share in a product's total social impact as well as the development of the social impact databases, have made it easier to structure SLCAs in terms such as: What is the percentage of the population in x economic sectors that is working more than y hours a week? By the same token, this way of demarcating boundary conditions in SLCAs makes it more difficult to address questions regarding the symbolic, aesthetic, or political functions of engineering design. In principle, for example, it is not impossible to use some other activity variable to address more specifically local community impacts, but as of yet no SLCA studies have been applied to the social acceptance of technological projects.

Both allocation and "weighting"—the value-based step in SLCA methodology that places emphasis on some impacts at the expense of others—are examples of bringing values down to the level of the engineer's managerial qualities. They are both consistent with the utilitarian assumption that trade-offs are not only permissible but desirable. Weighting, then, suggests a medium for expressing engineering reflexivity, one that brings, at least potentially, a new level of awareness to the SLCA practitioner. "Potentially,"

---

[26] For example Dreyer *et al.* (2006).

[27] Hauschild *et al.* (2008).

because, according to at least one interviewee, the awareness of how values are embedded in LCA methodology is not always made explicit:

> I think environmental LCA practitioners would argue that weighting is less of a problem for environmental than it is for social. They would make that argument. I think that's BS, but I'm biased because the weighting they use for environmental—I mean it's based on the same kind of subjective assessment that you would have to use for social, too.[28]

There seems to be no predicting what level of society can be relevant to SLCA studies. This was implied by chemical engineer David Hunkeler when in 2006 he distinguished between "societal" LCA—"micro-economic in nature" that concerns itself with labor units (e.g., worker hours)—and "social LCA" which is macroeconomic and is thus designed to cover issues such as government mandates. The point here is not about a correct or wrong definition of an SLCA's system boundaries. The point is that as the number of "societal," or "social" indicators is rather large but also essentially contested, SLCA researchers must set unavoidably arbitrary boundaries around social processes.[29] Hunkeler, for instance, mentions a number close to 300 measures, whereas one SLCA champion claimed in personal communication that they have "analyzed several hundred impact categories and indicators and boiled these down to approx[imately] 30 independent ones that together cover practically all social impacts."[30]

Not only the SLCA system boundaries, but also the "social indicators"—metrics that quantitatively describe the concepts (e.g., "cultural heritage") being assessed, and which are referred to in the UNEP Guidelines as "subcategories"—are set arbitrarily (Table 5.1). In a way that is methodologically unconstrained and dependent on their audience—the UNEP Guidelines themselves do not suppose any method of weighting—[31] SLCA researchers can in theory classify social impacts according to either groups of "stakeholders"[32] or social indicators. In addition, not only the process

[28] Interview data.

[29] Interview data.

[30] Interview data.

[31] Concurrently with the Guidelines, the Life Cycle Initiative also published a set of "Methodological Sheets" which offer examples of social indicators for each of the subcategory defined in the Guidelines' framework.

[32] For that matter, the SHDB can be used to visually estimate the percentage of the supply chain where impacts are more significantly affecting one type of stakeholders instead of another.

Table 5.1 Entwined ontological, epistemological and methodological anxieties in determining SLCA indicators.

1. Should social indicators be expressed in the immediate or close proximity of the industrial system under study (midpoint) or at the broader societal level (endpoint)? "Community participation," for example, if considered in SLCA study, may be an endpoint category. 2. Are there sufficient grounds for quantitatively modeling social impact pathways? *On the one hand, the choice of midpoint indicators requires the establishment of an "impact pathway" between a midpoint indicator (e.g., "incidence of child labor") and endpoint (e.g. "health and longevity"). On the other hand, the concept of an impact pathway can be theoretically bypassed if the defined social indicators do not describe, but immediately reflect the impact or benefit to well-being.[1]
3. Should social indicators be expressed quantitatively, qualitatively, or semi-quantitatively?[2]
4. Should social indicators measure the concept being assessed directly or indirectly or both?[3]
5. Should social indicators be defined at the individual (e.g., manager, worker, etc.) or group or corporate organizational level?[4]
6. Who should be involved in the process of determining the social indicators? 7. Should social indicators be expressed using a top-down (e.g., based on International Conventions), or a bottom-up (e.g., considering local cultural conditions or the context of manufacturer) approach? 8. What forms of data (global generic, process/sector, or site specific) must be considered? [SLCA practitioners have very few generic categories to rely upon]. **These questions relate to issues of accuracy, adequacy, and representativeness in building SLCA databases or reporting social themes of interest (e.g., work-related accidents).[5]

[1] Jørgensen, Andreas, Agathe Le Bocq, Liudmila Nazarkina, and Michael Hauschild. "Methodologies for Social Life Cycle Assessment." *International Journal of Life Cycle Assessment* 13, no. 2 (2008): 96-103; Norris, Gregory A. "Social Impacts in Product Life Cycles." *International Journal of Life Cycle Assessment* 11, no. 1 (2006): 97–104; Weidema (2006); Jørgensen (2010).

[2] Finkbeiner, Matthias, M. Erwin Schau, Annekatrin Lehmann, and Marzia Traverso. "Towards Life Cycle Sustainability Assessment." *Sustainability* 2, no. 10 (2010): 3309-3322; Dreyer *et al.* (2006); Méthot, Andrée-Lise. "FIDD: A Green and Socially Responsible Venture Capital Fund." Paper presented at the conference for Life Cycle Approaches for Green Investment, École Polytechnique Fédérale de Lausanne, Switzerland, 2005; Gauthier, Caroline. "Measuring Corporate Social and Environmental Performance: The Extended Life-Cycle Assessment." *Journal of Business Ethics* 59 (2005): 199–206.

[3] Nazarkina, Liudmila, and Agathe Le Bocq. "Social aspects of sustainability assessments: Feasibility of social life cycle assessment (SLCA)." Paper presented at EDF 2006, Moretsur-Loing, France, 2006.

[4] Ibid.

[5] Jorgensen (2010); Dreyer (2009); Life Cycle Initiative (2009); De Coninck, Pierre, and Carmela Cucuzzella. "A Participatory Approach Seeking Consensus in a Context of Uncertainty. A Sustainable Development Perspective of Life Cycle Analysis. Paper presented at the CALCAS Workshop, Brussels, 27 September, 2007; Swarr, Thomas E. "Societal Life Cycle Assessment—Could You Repeat the Question?" *International Journal of Life Cycle Assessment* 14, no. 4 (2009): 285–289.

of determining social indicators is subject to considerable debate but also the specific definition of a social "impact": Ciroth distinguishes between "impacts," referring to processes, and "attributes," referring to company behaviors; Lehmann and coauthors differentiate between "social impacts" and "social aspects," which "when they undergo a change...may lead to impacts"; Macombe and coauthors talk instead about "social performances" and "social effects." Last but not least, Reitinger and coauthors build on Australian legal scholar John Finis's work to conceptualize impact categories as dimensions of human flourishing, and SLCA subcategories as "capabilities" including "self-expression," "transcendence," and "fairness."[33] Thus SLCA does more than simply providing a quantitative technique to estimate level of risk for social themes associated with labor activity in complex value chains; it also epitomizes a *process of inquiry* into what social attributes matter for whatever is being investigated.

"For people holding a positive view of the market economy," the Swedish chemical engineer and LCA/SLCA researcher Göran Finnveden remarked in 1997, "it may be reasonable to advocate weighting methods which make use of market prices and other types of information derived from the market."[34] Furthermore, the acceptance of metaphors such the "global market" also provides grounds for using analogous language while making certain conjectures about the purpose of SLCAs: "...one of the main added values of S-LCA is to spend the results of [product] evaluation on the market."[35] Indeed, LCAs have become tools that do not challenge, but rather substructure a mixture of pro-growth principles and development policies. SLCA particularly, with its intended emphasis on narrating the conditions of positive social impacts, was envisioned from its inception as an instrument to support the exercise of growth in developing countries. Correspondingly, Catherine Benoît-Norris and economist coauthor Gina Vickery-Niederman advocate that one of the reasons behind SLCA development was the constant growth of ethical consumerism's market share in the Global North.[36] SLCA's very premise, that risks entering the

---

[33] Ciroth and Franze (2011); Lehmann *et al.* (2013); Macombe, Catherine, Pekka Leskinen, Pauline Feschet, and Riina Antikainen. "Social Life Cycle Assessment of Biodiesel Production at Three Levels: A Literature Review and Development Needs." *Journal of Cleaner Production* 52 (2013): 205-216; Reitinger *et al.* (2011).

[34] Finnveden (1997).

[35] Zamagni *et al.* (2011).

[36] Benoît, Catherine, and Gina Vickery-Niederman. "Social Sustainability Assessment Literature Review." Tempe, Arizona, and Fayetteville, Arkansas: The Sustainability Consortium, 2010.

same type of technological process can be very different from location to location or from country to country, means that SLCA databases, like the SHDB, will require a global or regional trade model as the backbone of the technique. In fact, the New Earth team recently signed a Memorandum of Understanding (MU) with the International Trade Center (ITC)—a subsidiary of the World Trade Organization (WTO) that facilitates exports from the developing world—to build a data bridge between the SHDB and ITC's "Standards Map" to indicate certification systems for specific countries and sectors.

Meanwhile, since the beginnings of LCA research in the early 1990s, the systemic approach to environmental impact assessment became linked with a particular notion of the economy—the notion that industrial interdependencies across economic sectors could be captured by linear equations. More recently, however, LCA and SLCA practitioners alike are raising issues—[37] questioning the viability of steady state assumptions that treat economic processes as linear—similar to those that technopolitics ideologues of sustainability advocated in the early 1990s. Deducing the sustainability ideologies' rate of coalescence in the field of SLCA is assisted by such calls, which draw parallels between the "adoption of a societal perspective" in LCA and the integration of cycles of economic activity in its methodology.

Further ambivalence about the relationship between the economy's *modus operandi* and the purpose of SLCAs was particularly evident when the SLCA UNEP Guidelines were officially launched on May 18, 2009: "When you are working with social and economic criteria," the event's organizers surmised, "in fact you have the same intention [The Wealth of Nations] as Adam Smith..." In this fashion it is tempting, they admitted, to assume that an "invisible hand" will regulate the economy for the benefit of "the community as a whole." Yet, "we know from practice," the organizers concluded, that this would be a misleading hypothesis: "this 'invisible hand' has not allowed sustainable production and consumption at all...[;] [thus with] "this publication [the UNEP Guidelines]...we are offering a pair of glasses...*to render this invisible hand visible*."[38]

In both its rationality/lexicon, then, and the relationships between its ontology, epistemology and methodology, SLCA owes a great deal to the converging ideologies of technological change and technopolitics. SLCA

---

[37] For example Norris (2003) and Zamagni (2010).

[38] Mazijn, Bernard. "Methodology of the Social LCA." Presented at the Workshop: Life Cycle Initiative, Brussels, 7 October 2010, emphasis added.

attests to a messy, ongoing tussle between different viewpoints, where the dominant engineering ideology and culture begins to morph into a more open-ended approach.

Sustainability, this section analyzed, has been captured by corporations and is oriented toward market growth. The economistic culture and the dominance of technological systems now exert such weight and power over our imaginations and socialization as human beings that even any attempt to change engineering identity ends up being put within this larger frame. SLCA thus is not really a compromise—it is more that advocates of technopolitics have succeeded to some degree in pushing their agenda and issues into engineering identities by exploiting larger social responsibility concerns.

For most SLCA scholars, therefore, assessing social impacts means not only calculating risk levels for child exploitation or union corruption, but dealing with globalization's unintended consequences as well. In other words, SLCAs have changed and adapted rapidly to the principles of the global economy, much like the value chains that they study.

A logical frame thus arises from SLCAs that equates the effects of globalization with those of technology: both are viewed to be neither good nor bad, and certainly neither of them is considered to be neutral. In other words, the proliferation of social assessments in a life cycle perspective is rooted in an ongoing discursive migration of "unintended consequences" from technological to global economy and globalization contexts.[39] Apart from reinforcing the ontological assumption that sustainability is a problem at the global level—advocated first in like terms by the Club of Rome and then appropriated accordingly by UN officials—[40] the propagation of SLCAs is also closely linked to the question of whether or not the positive social impacts of globalization will outweigh its negative environmental

---

[39] For example, Benoît- Norris and coauthors assert that "[w]ith the complexity of globalized production and consumption, a great deal of transparency is lost, which can result in unintended and overlooked social and environmental impacts." Benoît-Norris, Catherine, Deana Aulisio Cavan, and Gregory Norris. "Identifying Social Impacts in Product Supply Chains: Overview and Application of the Social Hotspot Database." *Sustainability* 4 no. 9 (2012): 1946-1965.

[40] Hauschild and coauthors write that "[t]he globalisation of our economies poses great challenges to our societies but at the same time it presents an opportunity for addressing the sustainability challenge on the scale where it really belongs—the global scale." Hauschild, Michael, Jeswiet, Jack, and Leo Alting. "From Life Cycle Assessment to Sustainable Production: Status and Perspectives." *CIRP Annals Manufacturing Technology* 54, no. 2 (2005): 1–21.

consequences. As chemical engineer and SLCA pioneer Michael Hauschild writes:

> Outsourcing of parts of the production to developing economies will thus normally lead to worse environmental impacts since environmental regulation and infrastructure will be weaker or completely absent in the developing economies. On the other hand, the outsourcing creates jobs and trains local workers, and employment leads to increased welfare in the local economy. These are the two sides of economic growth, and if we only look at the environmental impacts, *we miss the full picture*... The increase in welfare and material security will allow the local citizens to act in a more sensible way rather than erode their local environment in an attempt to meet their immediate needs.[41]

That "full picture" was what Andreas Ciroth had in mind as he turned to the Dalai Lama for philosophical authority while introducing SLCA to a group of engineers and scientists. Later, he claimed that "[social] issues are partly caused by [the] economy, but [the] economy could also reduce these issues."[42]

Once again, the sustainability engineer's argument about SLCAs' potential and effectiveness is rendered plausible by integration of the ideologies of technological change and technopolitics. The field of LCA was expressly conceived as an effort to managing environmental trade-offs. Now it no longer excludes social impacts in its theory, and its advocates also claim greater transparency with regards to the ontological, epistemological and methodological assumptions implied in the management of such trade-offs. The question arises, however, as to how or to what degree SLCA discourse extends its assumed reflexivity to the political operation of SLCA studies themselves.

"Almost Moral": SLCA Limitations

SLCAs are designed to be an ideological compromise between technological change and technopolitics visions of sustainability. Expansion of LCA system boundaries reflects the coalescence of the technological change vision of navigating the unintended consequences of development with that of addressing "problem-shifting" through engineering input-output

[41] Hauschild *et al.* (2008).
[42] Ciroth and Franze (2011).

techniques in industrial product life cycles; but this same expansion of LCA system boundaries also attests to a move towards explicit engineering reflexivity regarding LCA's methodology and the social purpose it aims to address. This chapter details that the development of SLCAs also melds together the engineering cultures for which these studies have enabled convergence between heterogeneous actors. As discussed in the previous section, a further complication is that the discussion topics—for example whether CSR is the "natural" or appropriate framework for the application of SLCAs—are rarely explicitly addressed by researchers in the field. In the rest of this section I introduce other SLCA limitations: the role of International Conventions regarding the choice of social indicators; the political function of SLCA studies; and the exclusion of the use phase from SLCAs.

In 2011, Robert Habib, then scientific director of France's *Centre de Coopération Internationale en Recherche Agronomique pour le Développement* (CIRAD), profoundly challenged the participants of the 2nd International Seminar on SLCA as he argued that, despite their "strong ambition," SLCAs currently include restrictions that hamper their potential:

> [Typically] inventories of...Social responsibility of Industry,... [SLCAs] check whether the inspected company meets international standards. There are two main limitations...of such [a] technique. *The first is almost moral.* The reference to standards, often prepared by the North and imposed on the South seems to be...based on a very Western value system. The second limit is conceptual. Indeed, the investigation doesn't give information on the actual effects of productive activities in...society. Such information should include actual impacts, such as ability to control...[one's] life, freedom of expression, decision, autonomy, etc.[43]

On one level, the idea that SLCA indicators and topics of interest must be in accordance with International Conventions echoes debates in the institutional forerunners of social sustainability. In the late 1990s, Allen White cofounded the GRI with Bob Massie and served as the organization's CEO in its early years. During the World Summit on Sustainable Development in Johannesburg, in 2002, White and fellow GRI colleagues released the second (G2) iteration of the guidelines. At the time, the GRI

---

[43] From author's transcript of Habib's presentation, emphasis added. In 2013 Macombe *et al.* argued that "so called [SLCA] hot spots identify only the cultural and institutional situations that are obviously different from the ones agreed upon by occidental norms."

and the ISO standards were the only two initiatives mentioned explicitly under "changing unsustainable patterns of consumption and production" in the Summit's "Plan of Implementation."[44] On April 4 2002, a few months before the G2's official release in Johannesburg, the inauguration of GRI's Board of Governors was lavishly celebrated at the UN headquarters in New York. During the ceremony, it was stated that the GRI has taken up the cause of transcending state parties to the UN instruments in order to underline its focus on non-state actors and civil society. As Barney Pityana of the GRI Board of Governors put it that day, release of the GRI guidelines indicated that "a philosophical foundation is being laid, founded on international conventions, which would ensure that the principles of equality, human dignity and respect for the environment are honoured in the context of business undertaking..."[45] Twelve years after the inauguration of its Board of Directors, GRI's guidelines suggest the baseline for defining social impact indicators in SLCA literature. Moreover, GRI's Allen White serves on Norris and Benoît-Norris New Earth's advisory board —which, based on International Conventions, has decided upon the twenty "social theme tables" that make up the backbone of the SHDB.

In Louise Camila Dreyer's view, it was these Conventions that constituted the "normative core" of SLCA when she authored the first dissertation in the field in 2009.[46] And, although they recognized that international standards often "define floors rather than ceilings" the SLCA Guidelines made it clear that such Conventions must be the number one repository for social assessment indicators. Yet, despite success in conferring legitimacy

---

[44] United Nations (UN). "Plan of Implementation of the World Summit on Sustainable Development." World Summit on Sustainable Development, 2002.
<http://www.un-documents.net/jburgpln.htm>
Accessed January 26, 2014.

[45] UNISA. "Dr. Pityana's speech at the inauguration of global reporting initiative, UN headquarters, new york." Thursday 4 April 2002.
<http://www.unisa.ac.za/default.asp?Cmd=ViewContent&ContentID=15593>
Accessed January 26, 2014.

[46] Based on the International Labor Organization (ILO's) Conventions and the Tripartite Declaration of Principles Concerning Multinational Enterprises and Social Policy (MNE Declaration) Dreyer *et al.* (2006) propose five "obligatory" impact categories ("Forced Labour," "Child Labour," "Discrimination," "Restrictions of freedom of association, right to organise and collective bargaining," and "Working environment") for the stakeholder category "employees." Similarly, Finkbeiner and coauthors argue that it is the "premise of public data availability" which led them to an approach with social indicators derived from the HDI, UN Global Compact Criteria, and the Gini Coefficient—a World Bank poverty metric system. Finkbeiner *et al.* (2010).

on the SLCA, International Conventions remain ambiguous within the SLCA community:

> One important thing is that we—along with most [SLCA] practitioners—use a "top-down" approach [i.e. premised on broad social and economic issues as defined in accepted documents like International Conventions, UN treaties, etc.] to determining the social indicators... However, oftentimes the affected stakeholders have very different concerns...[;] to give you an idea, one example that comes to mind is the issue of "moral harassment"... This was a main issue that a top-down approach did not take into consideration.[47]

For another SLCA practitioner, International Conventions embody the vision that sustainability "may be reached...[through the application] of [corporate] auto-regulation...by soft laws" which implies the integration of international law into CSR and LCA. This is in line with CSR criteria becoming the standard lens for conceptualizing social sustainability; yet this was simply "not enough to cope with the spirit of LCA" as they defined it.[48]

On another level, the absence of a consistent interface between SLCAs and questions that challenge the political function of such studies reflects the tension inherent in the assumption that technologies are mere "tools" and independent of their judicial-political context. For engineers who embrace the ideology of technological change, SLCA offers the possibility "to quantify sustainability [through a technique] restricted to the assessment of products...[without] treat[ing] any problems which may be connected with...sustainability in large political or macroeconomic systems."[49] It is notable that such a political perspective is lacking for others, however.[50] In the pages of the *International Journal of Life Cycle Assessment*, SLCA engineers read articles such as Klöpffer's influential piece "Life Cycle Sustainability Assessment of Products" in which he lamented the misuse of "sustainability" in the "political discussion concerning global development" and in which he argued that sustainable progress could be made

---

[47] Interview data. These conventions were often negotiated by nation-states in the 1940s to 1970s, so they may not reflect current sentiments among "new" social movements.
[48] Interview data. They are actually "hard laws": if countries have signed up to a treaty, it is binding on them. The UN Declaration of Human Rights is not a hard law as such because it was passed by the UN General Assembly, and is not a treaty. But it is now widely accepted as "customary international law" and is binding on all countries.
[49] Klöpffer (2008).
[50] Zamagni *et al.* (2011).

"without direct political influence…, but not without the economy."[51] Meanwhile, in many ways SLCA marks an extension of the LCA culture reflected in Klöpffer's own career path: after completing his PhD in Physical and Theoretical Chemistry, Klöpffer worked for LCA standardization pioneer Battelle Memorial Institute's subsidiary in Frankfurt and was also a member of SPOLD between 1993 and 1998. Some of his more recent ideas (e.g., "no development can be stable in the long run without social justice"), however, might have been facilitated by broadening tendencies associated with LCA cultures.[52] Similarly, Bo Weidema, another early leader in the field with a traditional LCA background, recently contended that "the unifying concern for all the areas…[of social influence in SLCA] is that of social justice…" although a few lines later he added that "all of these areas can be related to the added value and/or number of working hours per product."[53] The convergence of engineering ideologies via SLCA thus represents the potential of professional engagement with questions about the political nature and application of sustainable technology.

One such potentiality is: Connecting SLCA with problems that consumers and lay people are concerned about, not with the priorities that governments and companies (and technical experts) may have. The politics of trying to make this a reality are highlighted in discussions about the role the "use" (or "non use") phase of a product or service should play in defining the scope and goal of SLCA engineering work.

"In a company application of Social LCA," wrote Dreyer in 2009, "we find that it is not relevant…to consider these impacts [of the use phase for]…these…will be addressed by other…more appropriate tools…[like] decision-making concerning financial investments…"[54] While, however, a preponderance of SLCA studies are business-oriented, a few SLCA practitioners, in personal communications, expressed doubts: a strict focus on business could be an unproductive feature of the field, and SLCAs should be used at various levels within a wide range of applications:

---

[51] Klöpffer (2008).

[52] Ibid.

[53] Weidema, undated manuscript.

[54] Dreyer (2009): 7. Regarding the inclusion of the use phase in SLCAs, the UNEP Guidelines stipulate that "[a]ssessing the use phase represents a major challenge. The accent in method development so far was placed on production, distribution and end-of-life aspects; therefore, use stage aspects require further development. Thus, with regard to the assessment of the usability and satisfaction of the users, it may be better assessed through other tools or become a future field of research for S-LCA." Life Cycle Initiative (2009): 78.

> I don't agree with the perspective [of the Danish school of SLCA]... I think social LCA is not a tool to be used only by companies [;]... [rather,] it should be... a supporting tool for decision makers in... different organizations, administration agencies, or you name it. I also believe it has the potential to be used to assess new technologies....It's not yet clear how we should apply SLCA in the use phase. It's not clear what we should measure, and I don't have an answer for it, not yet. But I think we have to really question whether and how our studies' focus the labor-aspect of the technology can be extended to consumers.[55]

Tina Wegg of the University of East Anglia, who recently completed the first dissertation to apply the SLCA framework as illustrated in the UNEP Guidelines to the field of sustainable energy, and particularly to biofuels' development, argues that a CSR approach means that "only a part of the picture is covered, which is useful to...[the company], but not necessarily of use to the end consumer."[56] In her case study of bioethanol produced in Brazil and consumed in the UK, Wegg found that at the UK end consumers felt that their ethics were compromised. For example, they could not choose at the pump whether or not to consume bioethanol unless they sold their car—which for rural communities in the area of study was not a choice. They also expressed that they would not know whether the impacts at the producer end were any better between different fuels. Thus a closer examination of that particular biofuel life cycle from the perspective of SLCA—and not overlooking the end phase—suggests that the UK government needs to address several issues to prevent opposition to E10 (petrol that is 10% ethanol) policies. Methodologically speaking, Wegg's findings are in line with recent critical voices expressed in the SLCA literature: "[one] reason for skepticism is that the development of SLCA methodology and theory has not, to date, been based on experience from actual case

---

[55] Interview data. Examples of the company perspective's limitation can also be found in the literature. Consider, for instance, "the Fiat Group that announced its intention to shut down the production of the Lancia Ypsilon model in its Italian assembly site of Termini Imerese by the end of 2011 and to move production in the Polish site of Tichy. The social consequences of such a decision are quite evident but can hardly be seized and quantified if we refer to the FU [functional unit] concept or even to a company perspective." Zamagni *et al.* (2011): 597.

[56] Personal communication. Regarding her methodology, Wegg recalled that "I have not seen any yet that have conducted face-to-face interviews with all stakeholder groups to gather qualitative data in the way that I have been doing for my project—but I would very much like to see one if there is and gain advice/thoughts from the practitioner."

studies...[there is therefore] the risk of developing [a] methodology that is neither practical nor relevant to users in real-life situations."[57]

This section discussed SLCA limitations; for example, it showed that in SLCA discourse "social" is equated to universally applicable human rights and labor standards. In addition, the section discussed that SLCA practitioners tussle with each other regarding the scope and goals of their work—some want to focus on products and businesses, and to improve existing technologies, whereas others are more interested in larger systems of application and in evaluating emerging technologies on a more preventive basis.

The next section discusses the expansion of LCA engineering cultures which coincides with, and is facilitated by recent, suggested SLCA methodological requirements to first, assess the scenario of the "non-use" phase of technologies and second, to revisit the role of "stakeholders" in the design of the studies themselves. These requirements—assessing the "non-use phase" of technologies and expanding stakeholder engagement in deciding upon the SLCA boundaries and social indicators—the next section will argue, comprise a political potential as important as the methodological one.

Toward Greater Democracy in SLCA?

The boundaries of LCA engineering cultures do not merely mirror the limitations of LCA approaches to sustainability, but also the ideological space within which lie their current potentialities. This section discusses three related questions regarding SLCA democratic potential. One, whether conducting SLCAs presumes a broader political purpose; namely, to promote

---

[57] Baumann, Henrikke, Rickard Arvidsson, Hui Tong, and Ying Wang. "Does the Production of an Airbag Injure More People than the Airbag Saves in Traffic?" *Journal of Industrial Ecology* 17, no. 4 (2013): 1-11. Other interviewees also critiqued what they called "the Danish school of SLCA": "In Denmark they...want to be more precise but also more consistent with the theory, with the methodology. But...oftentimes this is not that practical for companies" (interview data). Andreas Jørgensen's dissertation from 2010 had recognized the tension between developing SLCA theory prior to or after having conducted case studies: It is...easy to imagine that if the development of SLCA was solely based on the conduct of case studies issues like the potential lack of effect of using SLCA...would...[have been]...overlooked." Jørgensen frames his approach—ascribing a purpose to SLCA first, and then developing case studies—as "top-down," as opposed to a "bottom up" one where "case studies are performed and from which researchers gradually learn about and develop SLCA..." Jørgensen, Andreas. "Social LCA-A Way Ahead?" *International Journal of Life Cycle Assessment* 18, no. 2 (2013): 296-299.

reflexivity not only with respect to methodological assumptions, but also with respect to the technique's political function as well. Two, whether calls by SLCA advocates to extend the boundaries of SLCA to "non-use" life cycle scenarios may permit new modalities of technopolitics in sustainable technology. And, three, to what extent the current development of SLCAs challenges technological change notions of expertise and participation in engineering work. These questions and the requisite SLCA potentialities combined have much to contribute to the—unspoken—inquiry of what is driving the creation of social risks in the first place.

Do SLCAs Have Politics?

SLCA is now presented as a sustainability engineering discourse with increased reflexivity. Yet in both the UNEP Guidelines and in practitioners' arguments, SLCA is seen as simply providing the means for discerning all social impacts in a product's value chain—it "does not provide information on the question of whether a product should be produced or not."[58] Yet the political function of SLCAs essentially depends on the fact that such studies are not just "tools," but that they are part of a larger context within which they quite literally prescribe or legislate engineering outcomes with social and political repercussions. Andreas Jørgensen, the Danish engineer who has subjected SLCA to skeptical scrutiny cautions that assuming "that the developer…by shaping the technology (the SLCA) can decide how it will interact with its surroundings…" may be "too technocratic."[59] Thus there may be a divide between a study's assumed purpose to promote positive social outcomes and the actual effects of its enactment (Table 5.2).

There is, in other words, good reason to question the degree to which sustainable engineering identity assumes and enables an extra-engineering political function for SLCA engineering work. In this connection, SLCA researchers spoke about the drivers behind the proliferation of SLCA studies and their intended application:

> I think the drivers are two: reputation and innovation… When it comes to mitigating risks… there is this crazy questioning coming from different stakeholders, including shareholders and consumers and everybody else asking for transparency. So, they [companies]

[58] Life Cycle Initiative (2009): 13; Traverso, Marzia, Matthias Finkbeiner, Andreas Jørgensen, and Laura Schneider. "Life Cycle Benoît Sustainability Dashboard." *Journal of Industrial Ecology* 16, no. 5 (2012): 680-688; Benoît *et al.* (2010).
[59] Jørgensen (2010).

Table 5.2 What is SLCA *for*?

0. *First-level assumption and overarching aim for conducting SLCAs*: The choice to perform an SLCA is viewed as contributing to the improvement of social sustainability in product life cycles. SLCAs are implicitly construed as tools aimed at not just quantifying social risks in value chains, but also at reducing those risks. *SLCAs are *not* meant to decide whether a product should or should not be produced in the first place*[1]
SLCAs are meant to:[2] 1. *Enhance understanding regarding the totality of supply chain risk by assessing potential and documented social impacts*: As a first step, this includes prioritization of a) areas ("hotspots") where CSR or other sustainability-related management strategies will be implemented, and b) areas where more information needs to be collected. As a second step, SLCAs may be used to inform a "site-specific" assessment in a life cycle perspective where "sensitive topics" can be discussed for the first time: The focus is on the suppliers and business partners influenced by the company that performs the SLCA, and may include practices and behaviors based on so-called Performance Reference Points (PRP).[3]
2. *Provide competitive advantage by integrating sustainability at the company level and by enabling managers and engineers to go "beyond compliance" initiatives and take action in a proactive way.*[4] This integration is viewed by SLCA practitioners as depending on the presence of "the same language" [life cycle thinking] to talk about social and environmental impacts. SLCAs thus help engineers "capture [the] overlap" of impacts which have very specific results on products and would otherwise be lost.
3. *Inform and streamline corporate business management in establishing purchasing procedures, marketing, strategic thinking, and in identifying inconsistencies between written corporate codes of conduct and actual practices.*[5]
4. *Create corporate or social "value added" in a way that makes them readily distinguishable from other social sustainability assessment techniques or approaches.* This implies "going beyond" using reporting to verify positive impacts of business activity reflected in corporate codes of conduct or creating corporate "value added" by applying social labels. The possibility of combining SLCA and social labeling has also been considered.[6]
5. *Combine life cycle approaches with other assessment tools such as Environmental Impact Assessment (EIA) in order to provide a "more comprehensive picture."*[7]

6. *Informing regulatory decisions*
7. *Potentially extend their application in comparative technology assessment or comparative product assessment.*[8]

[1] Finnveden (1997); Traverso *et al.* (2012); Benoît *et al.* (2010). Jørgensen (2013): 298 writes: "On average, basing the decision on an SLCA would thereby lead to more socially beneficial situations that decisions made without the SLCA."

[2] 1, 2, and 3 are similar, but not quite the same as what Jørgensen *et al.* (2012) call "consequential," "educative," and "lead firm" SLCAs respectively, although their categorization is restricted to the mechanisms through which changes in product life cycles relate to social impacts to workers.

[3] Interview data; Benoît-Norris *et al.* (2012); Andrews *et al.* (2009); Webinar on "Combined Environmental and Social Life Cycle Assessment in the Food, Beverage and Agricultural Products Sector." Webinar offered by New Earth on June 26, 2013; Dreyer, Louise C. "A Site Specific Approach to Life Cycle Management of Labour Rights Issues: Using the Social LCA Toolbox." Presented at the International Seminar on Social LCA, Montpellier, France, May 2011.

[4] Interview data; Webinar, June 26, 2013.

[5] Life Cycle Initiative (2009); Dreyer (2011).

[6] Interview data; Gauthier (2005); Zamagni *et al.* (2011); Jorgensen (2013).

[7] Jeswani, Harish K., Adisa Azapagic, Philipp, Schepelmann, and Michael Ritthoff. "Options for Broadening and Deepening the LCA Approaches." *Journal of Cleaner Production* 18, no. 2 (2010): 120-127.

[8] Interview data.

> want to be able to respond—they want to have sufficient arguments when it comes to a question about the social impacts of the product... *they want to be safe.* [The other driver is]... the opportunity for being different. It's an opportunity for performing better than competitors.[60]

We might similarly question just how much SLCA engineers have ceased to regard their work as the specialized activity of a distinct professional group. An important point regarding the intended purpose of SLCAs is that despite popular opinion to the contrary, studies are mostly focused on guiding improvement for a given company or its product—and not on making competing product comparisons. This is because corporate SLCAs, meant to inform managers *internally* about value chain problems such as forced labor, discrimination, or unacceptable treatment of workers, remain by far the most common use of SLCAS.[61] Under the influence of technopolitics, however, social impacts also become entangled in sustainability engineering identity to develop a "common language" to provide consistency between quantitatively-focused internal or external LCA expertise and qualitatively-driven programs on social compliance.[62]

Another factor that challenges the political potential of SLCA work is that, like LCA, SLCA engineers tend to force measurement into monetary terms. This distorts the impacts that are considered worthy of inclusion. Ironically, the political outcome is similar in that the global-economy focus of SLCA also generates the essential limits for the acclaimed critical regulatory implications of SLCA studies and/or the way(s) to translate SLCA data into production and consumption decisions: "In a globalized world, where the inputs come from everywhere and the product goes everywhere, it is hard to have a mandate or rule that pushes companies to be more responsible," recognized one SLCA practitioner.[63]

Despite the high expectations that come with SLCA rhetoric, respondents felt "we are still far from thinking about SLCA as an instrument for decision support."[64] There is, for example, no predicting whether SLCA engineering practices and the information thus assembled could be successfully applied to existing frameworks such as social labeling: "... even if

---

[60] Interview data.

[61] The same can be said for LCAs these days. If we saw significant activity in product comparison for social LCA, interviewees predicted that ISO standards—like the ones applied in LCAs—would need to be decided upon and developed.

[62] Interview data.

[63] Interview data.

[64] Interview data.

we use an SLCA to create a social label, I am not sure whether that would add much to the transparency by which social impacts are presented…," lamented one engineer in personal communication. [65]

What then may be a key problem with constructing a sustainability engineering identity via SLCA viewed as a political mechanism in technology decision making? As one SLCA champion recalls, the actual results of SLCA engineering work may in some sense be considered arbitrary:

> Basically… you can make an assessment and there is no way of validating [that]…assessment…Some [of my colleagues] may opine that I am exaggerating here, but this is what I believe—you can say anything and nobody can check it…You can easily develop…[an SLCA study] and say 'wow, this is really fancy'… but you can't really say anything about its quality…[66]

Although deeply involved in promoting the vision that SLCA can support decision-making processes, interviewees also sensed how premature and ambiguous such suppositions are. On one occasion, an engineer spoke about the "methodological" challenge of establishing "causality" in social impact pathways as the main barrier to applying SLCA to inform regulation, while also pointing out that translating life cycle thinking data into production and consumption decisions is no less precarious in LCA than in SLCA. Another respondent raised the issue of SLCAs being trapped between national regulatory cultures—the US market being more driven by consumer preferences and voluntary industry action and less by public policy, whereas the EU market is configured more by regulation.

In sum, the hypothesis that unambivalently equates sustainability engineering identity with practitioners conceptualizing SLCA as a medium to intervene in technology politics is shaky. As Jørgensen detailed in a 2013 editorial, "…assuming that a [politically] positive effect will come from using…SLCA in decision support may be unsubstantiated."[67] Motivated by the desire to explore the philosophical foundations of the SLCA methodology, Jørgensen shares with early technopolitics advocates of sustainability the idea that political principles (e.g., Is child labor a practice widely condemned as unacceptable? Should child labor be weighted differently in Asia and the US?) and scientifically measured consequences (e.g., the assessed impacts of child labor in a product's life cycle) co-determine the

[65] Interview data.
[66] Interview data.
[67] Jørgensen (2013): 298.

background of sustainable technology. In 2010, he wrote that if companies wanted to completely eradicate child labor, they would commission SLCAs based on "a deontological ethics, instead of a consequential ethics."[68] "Incidence of child labor," as a social indicator of well-being, he claimed, lacked predictive validity: "the mere fact that a child works does not support an accurate prediction of the actual damage (or benefits) to...[their well-being]." On the one hand, he remarks that while trading the totality of human experience for the minimal fraction of an "incidence of child labor," researchers implicitly downplay the importance of other aspects of human experience from which work may be a cultural derivative. On the other hand, when he classifies the SLCA practitioner's reality into two spheres—one scientific and another of "political reality"—he also implicitly lays out the justification that SLCAs partly depend on the acceptance that child labor may be one of the positive unintended consequences of globalization.

The Potential Technopolitics Function of "Non-Implemented" Life Cycle Scenarios

If his subjecting SLCA to reflective, critical, inquiry encouraged such postulations, Jørgensen's analysis also derived from his consciousness of the fact that both SLCA discourse and SLCA studies themselves have, as the previous section discussed, evaded the politics of technology. Consequently, the assertion that SLCAs need to "provide valid assessments of the consequence[s] of the decision that...[they are meant] to support"[69] could be interpreted as an attempt to align questions such as "Should technology x be produced or not?" with a rationalization for justifying SLCA. In his 2003 essay, provocatively entitled "Revisions to LCA needed to Address Sustainable Consumption," Greg Norris had conceded that the "What if?" questions—questions that do not discriminate against causal pathways which cannot be expressed in terms of material or energy flows—are the most difficult to overcome in life cycle-based sustainability engineering, because they are less sensitive to technological fixes.

Seven years later, Jørgensen and coauthors built on that same concept of "consequential" SLCA when they implicitly captured the politics of life cycle engineering as a version of the dilemma of autonomous technology—namely that engineers cannot know in advance whether their

---

[68] Jørgensen *et al.* (2010): 14.
[69] Jørgensen *et al.* (2010): 376.

products (e.g., the SLCAs) will deliver positive or negative consequences to the stakeholders addressed in their assessments.[70] To circumvent this problem, they proposed that SLCA researchers acknowledge the distinction between production/use and non-production/non-use scenarios which they broadly termed "life situations," and which they premised on the acceptance that an SLCA's consequence is "not simply" the enactment of its conclusion. With that in mind, the next step in the process of making SLCAs sensitive to "non-implemented life cycles," they argued, is for practitioners to define the difference that the non-production phase would make to the workers, as opposed to the production phase. This outlook not only effectually recognizes that SLCAs are political—the enactment of an SLCA necessarily changes the power structure in a product's life cycle, making it a political act—but it also construes the ability of practitioners to methodologically address such politics as a constitutive element of the legitimacy of engineering knowledge: While adding a "considerable extra task" to the engineer's practice, the authors contended, "the assessment of the non-implemented life cycle situation increases the validity of the SLCA…"[71]

This counterfactual assessment task will not be easy for engineers to accomplish: On one front, unemployment is an important—but by no means the only—impact on workers of the non-production scenario, yet several issues make its assessment complex and uncertain at a case-specific level. On another front is the impact that the non-implemented cycle has on users, and that Jørgensen and coauthors found "impossible" to rationalize on a generic basis. "Consider for example that we want to assess whether to buy a TV or not," they explained. Indeed, which social impacts may be associated with a technological product's non-use, and how they are translated in the methodological framework of SLCA, reflect variations of technopolitical thought. What if the non-use of TVs is found through SLCA studies to be correlated with happier individuals who have more time to engage in cultural or transcendental activities? Would that be something, as Udo de Haes suggested, that *should be promoted* by life cycle assessments? The clue to understanding SLCA's potentialities is that the field's proclaimed reflexivity makes it conceivable for the politics of technology to become fused with engineering knowledge itself. Echoing the distinction between the "why's" and the "how's" of sustainable engineering, as outlined in the early 1990s by technopolitics ideologues of sustainability,

[70] This is where, for example, "anticipatory governance" can play a role.

[71] Jørgensen *et al.* (2010): 376.

Jørgensen and his northern European SLCA colleagues pronounced the institutionalization of the "'do good' tool": "...only assessing the production/use situation," they urged, "would...be like answering the question 'Will it become better?' with the answer 'It will be good.'"[72]

The promotion of a reflective SLCA tradition suggests that sustainability engineering must renegotiate its standards by expanding its conceptual and professional boundaries. The unreflective pronouncement that through rationally conducted SLCAs, globalization's unintended consequences could be minimized and its positive capacities maximized supports an understanding of SLCAs as mere tools, while the suggestion that studies include an assessment of the "non-implemented" life cycle is a potential critique of that understanding; it is a window to a technopolitical conceptualization of sustainable technology.

The conjunction between efforts to construct the validity of SLCAs in order to legitimize their use in decision support on the one hand, and proposals to include the assessment of the non-use phase in the methodology on the other, opens the first significant democratic potentiality of SLCA: practitioners integrating the idea that SLCAs—and sustainable technologies more broadly—have political qualities into engineering work. Such proposals, the next section will explain, also facilitate the broadening of SLCA boundaries to the potential engagement of engineers with non-engineer users of products.

Widening Sustainability Competencies Beyond the Engineering World

On the whole, the vision sketched in Jørgensen and his coauthors' proposal to include in SLCA the assessment of the non-implemented scenario is fueled, implicitly or explicitly, by persisting calls for sustainability engineering to widen its perimeter of competencies outside of the engineering world. More recently, the UNEP Guidelines acknowledged that "... people trained in different professions tend to view different aspects of the World to be both important and missing from each other's World view and the resulting models,"[73] whereas one interviewee noted that "we are [currently] stuck in the process of moving forward with SLCA because we are all addressing the problems from the same engineering perspective."[74]

In their professional undertakings, SLCA practitioners claim to be faced with the challenge of having to navigate different professional worldviews.

---

[72] Ibid., 381, 382.

[73] Life Cycle Initiative (2009): 56.

[74] Interview data.

Moreover, SLCA authors and practitioners are not a homogeneous lot—although the majority of researchers have LCA backgrounds and, as one engineer put it, "tend to be systems thinkers."[75] For the most part, individuals who practice SLCA have engineering degrees: LCA tools target a user population that is comfortable with engineering-style software. Some of them are environmental scientists, some were trained as economists and are specialized in social impact assessment, others work in the field of agricultural engineering, and a small minority has social science backgrounds. Some social scientists, it seems, are attracted to SLCA because of its potential to expand the breadth of their work into a life cycle perspective.

The second, then, important potentiality here is that of professional experts with non-engineering backgrounds co-defining SLCA plans and strategies of data acquisition and interpretation.

For example, in search of some certainty in interpreting non-engineering cognitive territory, one SLCA project sought refuge in hiring a social scientist expert. Yet participants faced difficulties understanding what was being proposed because "we analyzed the process from the engineer's angle…from input and output, and [the social scientist] was thinking in terms of values, so we weren't able to find a middle ground…"[76] On another occasion, one engineer recalled how in one SLCA conference setting an individual with a very traditional engineering approach to SLCA, which is "'here are the rules, follow the rules, get the answer, and you're done,'" critically addressed a student-presenter who had taken a broader viewpoint.[77] The same conflict and skepticism, however, was amply displayed in one non-engineer's efforts to integrate SLCA into their project proposal: "[My non-engineer supervisors] were really quite negative[;]…they said LCA is a quantitative tool…it's a computer program, you can't put issues like participatory justice in a software tool[;]… and I agree, but what I was trying to do is to supplement that [LCA] information with qualitative research."[78]

Consequently, as SLCA advocates understand the intricacies of global value chains with new comprehension, they have come closer than ever to operationalizing the social pillar of sustainability. But by lending to the "social" the legitimacy of "life cycle assessment," SLCA practitioners soon realized that some of the skills needed to conduct SLCAs—such as designing the data collection instruments, collecting qualitative data and

[75] Interview data.
[76] Interview data.
[77] Interview data.
[78] Interview data.

enabling stakeholder dialogue and engagement—go beyond their prevailing professional expertise:

> People who work on LCA inventories are mainly technicians…working [with] numbers… We need more people trained in the social sciences and the humanities…We face several major challenges in terms of evaluating our qualitative data. For instance, we have managed to phase out child labor in some communities. But the question was, "If they are not working, what would happen to these children?" And we did not know how to evaluate these questions. And this…relates to the fact that we need to be accepted by them [social scientists], otherwise they are saying *this* [SLCA] *is not social at all…* So we are stuck because we are evaluating social aspects…[in a way] that is not accepted by social scientists.[79]

That is, ambiguous steps in SLCAs, such as data evaluation, combined with incidents of clashing professional mindsets, gave rise to the idea that the very legitimacy of the field depends on successfully unraveling its professional boundaries as well. True to their technological change and technopolitical heritage, contemporary SLCA ambivalences echo the challenges of interdisciplinary research that sociologist Martin O'Brien recalls having faced in the mid-1990s, when he collaborated with UK sustainability engineers Alison Doig and Ronald Clift to articulate SLCA:

> [W]e tried to combine basically what was sociology with, what would you call it, thermodynamic engineering… And it failed badly. The sociology was too complicated and the engineering was too simple. And we never really managed to make an effective combination between the two. Partly, *I'm certain this problem still persists, partly because we could not agree on boundaries.* The engineers in the project needed fairly strict…functional units to look through their data sets, but the sociologists on the project couldn't really work with those, and so we had very loose and very vague bounds.[80]

Hence the second potentiality of SLCA mirrors the kind of interdisciplinary collaboration problems faced by contemporary engineering cultures—which by no means are specific to LCA. Academics "don't lend themselves to interdisciplinary work," lamented one engineer and SLCA researcher, whereas "faculty are not rewarded for collaboration" of the kind

---

[79] Interview data.

[80] Personal communication, emphasis added.

that is instrumental in conducting an SLCA.[81] To some extent, the very few and very fragmented SLCA-focused engineering education initiatives are trying to create a novel reward system for engineering collaboration. SLCA is practically non-existent in engineering education, despite the acknowledgement by practitioners that its inclusion in engineering courses would greatly benefit the students and the field itself. At a recent conference presentation, Bo Weidema, who teaches the SLCA course at the International Life Cycle Academy, drew a connection between the "lack of interdisciplinarity," "limited coverage of ethical and social implications," and the "limited scientific rigour and focus" in educational programs focused on sustainability.[82]

The last section highlighted that Jørgensen and his colleagues carried the 1990s technopolitics concern with the "whys" of sustainable technology into the contemporary field of SLCA. In the second decade of the 21st century, the same concerns about participation and commitment to stakeholders that were expressed in the mid-1990s still suggest a "less explored field"[83] in LCA, thus attesting to "the difficulties of free expression of...stakeholders when it comes to evoking the social aspects [of a technology]."[84]

A third, therefore, democratic potentiality of SLCA is that of incorporating user perspectives, rather than expert-defined categories in engineering work.

In principle, SLCA was not designed for any particular stakeholder group—the SHDB can provide data for an entire country's product system if that is what the researcher desires—which means that both the public policy (e.g., its use by governments) and community empowerment (e.g., its use by NGOs or local communities) capacities of the tool are theoretically equally imaginable, along with their corporate application. Nevertheless, although some governments have been looking into using SLCA—the TYGRE project, for instance, has received funding from the

---

[81] Interview data.

[82] Weidema, Bo P., Miguel Brandao, and Jenna M. Watson. "The International Life Cycle Academy: A Quality Network for Teaching in Quantitative Sustainability Assessment: The Why and How." Presented at the SETAC Europe LCA Case Study Symposium, Copenhagen, 27 November, 2012.

[83] Sala *et al.* (2013, Part 2): 1690.

[84] Syndhia Mathé. "Integrating participatory approaches in social LCA: What about functional units and impacts choices? Fish Farming case." Presented at the Conference: Environmental & Integrated Assessment of Complex Systems, Montpellier, France, 2011.

EU to look at the use of SLCA in informing waste policy—the use of SLCA at the community level is practically non-existent.

In the context of SLCA, the notion of stakeholder "participation" is, for the most part, understood either as supply chain actor involvement or as the ISO standard's requirement for a critical review panel—a simple or more elaborate peer review of the LCA study.[85] As a result, like the focus on "well-being," SLCA's conceptual and methodological focus on "stakeholder categories" is also contradictory. On the one hand, practitioners contended that the SLCA framework—as outlined in the UNEP Guidelines—has the notion of stakeholder participation built into it. In studies that actually elaborate on how stakeholder inclusion has been operationalized in the process of conducting an SLCA, the five stakeholder groups proposed in the UNEP Guidelines (e.g., workers, local community, society, consumers and value chain actors) are seen as essential for validating the aspects considered in the assessment.[86] On the other hand, the UNEP Guidelines lack a specific framework for ensuring inclusivity and diversity within the proposed stakeholder categories, and the absence of stakeholder engagement in formulating the purpose of the study itself does not support an assumption that SLCAs address community participation in a structured way. Currently, the UNEP Guidelines speak for the majority of studies when they argue that "due to restricted resources (time and money) the possibility to involve stakeholders in...the goal and the scope phase... [namely the initial planning of the SLCA] may be limited."[87]

With the growth of humanistic professional culture in SLCA, however, professional expertise may become less requisite for conceptualizing sustainability. For example, Arne Wangel, a sociologist by training who teaches at the Technical University of Denmark, recently attempted to remedy the fact that "SLCA does not respond to...the users' needs satisfaction...or the changes in the users' behavior as a result of using the product...". He addressed the viability of participatory processes of developing social indicators and identifying impact categories in an interesting case study that examined three different scenarios of delivering nutritious

---

[85] Sala *et al.* (2013), Part 1.

[86] Lehmann, Annekatrin, Daniela Russi, Alba Bala, Matthias Finkbeiner, and Pere Fullana-i-Palmer. "Integration of Social Aspects in Decision Support, Based on Life Cycle Thinking." *Sustainability* 3, no. 4 (2011): 562-577; Halog, Anthony, and Yosef Manik. "Advancing Integrated Systems Modelling Framework for Life Cycle Sustainability Assessment." *Sustainability* 3, no. 2 (2011): 469-499.

[87] Life Cycle Initiative (2009): 77.

meals to schoolchildren.[88] Wangel's case study, as well as several interviewees' assertions, confirm the idea that expanding SLCAs to the use/non-use phase of a product opens up possibilities for applying SLCA to scenario analyses and for challenging expert notions of "participation" in the design of approaches to assessing social impacts of technology.

On the one hand, this section argued, as "social impacts" become more sound in engineering cultures, sustainability engineering subject matter undergoes a fission between engineering and humanistic expertise. On the other hand, humanists or social scientists are still experts—there are no indications yet that SLCA technical experts are even thinking about relying on lay people as a source of insight and weighting, but also as a source of broadening the scope of engineering work.

Conclusion

With minimal corporate publicity, and reserved enthusiasm among the LCA community, the last decade has witnessed the rise of a common goal on the part of development institutions, CSR managers and sustainability engineers to measure social risks across corporate supply chains.

This chapter examined the historical juncture between life cycle thinking, institutional and corporate forces advocating social impact auditing, and the sustainability engineering vision that the unintended consequences of globalization can be managed and controlled through corporate internalization of development's impacts on workers and the environment. SLCA advocates either claim not to have a particular historical understanding of the connection between "sustainability" and "life cycle thinking," or assume that this "natural" relationship is a recent phenomenon that formed in the 2000s. Yet as this chapter has illustrated, a reading of the emergence of SLCA under the ideological lens of technological change and

[88] Society of Environmental Toxicology and Chemistry (SETAC). "2012 SETAC 18th LCA Case Study Symposium, program and abstracts book." 2012.
<http://lcacopenhagen.setac.eu/embed/Copenhagen/programme_abstracts_book_31102012_v2.pdf>
Accessed February 26, 2014.
The scenarios included home-made lunch, lunch prepared in industrial canteens and lunch prepared and served at the school canteen. The stakeholders of the study included the government, the school board, headmaster, teachers and staff, parents, but also pupils themselves who were expected to have a say in the local "foodscape."

Table 5.3 Change provoked by SLCAs.*

SLCA study of:	Resulted in:
Biobased thermoplastic material manufactured in Brazil, carried by French consultants EVEA in 2012.	The sourcing decision of the firm that commissioned the study, a leading company in the cosmetic industry, began to depend on certification assurance. Also after being advised by EVEA the firm considered the option of a comparative SLCA that assumed a petro-based plastic as an alternative material for the company's intended application.
Canadian Dairy Sector, conducted by a consulting and academic collaborative initiative in 2013.[1]	An SLCA scorecard was developed based on a Performance Reference Points (PRP) method. As a result, the Canadian Dairy Association began to report to its stakeholders about social impacts. Presently dairy producers in Canada are able to transparently communicate the combined results of LCA and SLCA—working towards a comprehensive sustainability strategy. The study also motivated the producers to develop a web calculator to track individual results on an ongoing basis and compare these with practices in the industry as a whole.
European Union Ecolabeled notebook, conducted by German consulting GreenDelta	The EU label considered the certification of social impacts.
Pepsi co products carried by consultants at New Earth	Pepsi co used the results of the study to prioritize engagement efforts with fruit suppliers.
Australian printing company Focus Press[2]	The company began to engage with suppliers of Chain of Custody Forestry to promote social impact assessment in pup and paper processing; the company also began to consider cultural diversity of its workforce in light of national migration trends.

*Interview data. This is meant to be just an indicative list of examples.

1 Consultants Quantis and AGECO with the collaboration of CIRAIG and UQAM.

2 Sack, Fabian, and Louise Pastro. "The story behind the page-social Life Cycle Assessment of Focus Press." Paper presented at the 8th Life Cycle Conference Pathways to Greening Global Markets, Sydney, Australia, 2001.

technopolitics suggests that this relationship is neither "natural" nor did it emerge in the 21st century.

But what happens when SLCA engineering work principles are put into practice? (Table 5.3). For instance, the SHDB of Greg Norris and Catherine Benoît-Norris' New Earth is currently consolidated into the free, open-source, social and environmental index (Higg Index) of the Sustainable Apparel Coalition, a trade organization in the apparel and footwear sectors. In addition, the SHDB was recently tested by the Sustainability Consortium, a unique organization jointly administered by US and European academic institutions, which works with multinationals to promote and conduct sustainability reporting while implicitly maintaining a social life cycle thinking orientation for their research staff. This chapter illustrates that in Europe, as in the US, SLCA and the accompanying enthusiasm have begun to develop in the context of legitimizing sustainable globalization.

SLCA has become the primary motivation for a small group of LCA consultants and academics who have been achieving the assimilation of engineering ideologies of sustainability on a discursive as well as on a practical level. "One dream that the SHDB team at NewEarth has," recalled Greg Norris in personal communication, "is that there will be an SLCA community growing of people applying the [SHDB] data, finding results... presenting case studies at conferences...That we've become a community of users...[in which] each member is strengthened by others."[89] In line with Greg Norris's vision, SLCAs' system boundaries offer a medium to "understand ourselves."[90]

Yet SLCA sustainability engineers, this chapter showed, are also seemingly limited by the connection between SLCA language and engineering identity formation; practitioners mentioned that fellow engineers and people in industry are not "comfortable with SLCA yet. So, they may not label what they are doing at all, other than maybe risk assessment."[91] In addition, LCA engineers have different views on what the use of SLCA

---

[89] Greg Norris, participant observation, "SLCA Short-Course," June 27, 2013.

[90] "We should use this system [SLCA] to understand ourselves...It's a mirror of self-reflection and a live feed on the sustainability impacts from each actor in the global economy." Greg Norris quoted in Tilde Herrera, "Walmart, Seventh Gen Give Free LCA Tool a Trial Run." Greenbiz.com, November 9, 2010.
<http://www.greenbiz.com/news/2010/11/09/walmart-seventh-gen-to-give-freel-lca-tool-test-run>
Accessed March 27, 2015.

[91] Interview data.

language may accomplish, as well as using slightly different jargons for speaking about social sustainability identity: "We have been promoting social as part of what we are as engineers—well, I don't like calling it social LCA…As far as I am concerned, social things are so complex. I used to call it […] I just want to clarify that we don't follow all the…abstractions and recommendations of the ISO…"[92] These claims reveal that many, if not, most engineers are still queasy about using these terms. In turn, several interviewees contended that industrial use of SLCA is still quite marginal and thus many engineers conceptualize SLCA as an academic indulgence. Those interviewees lamented the fact that SLCAs currently suggest tools that academics are "playing around with" and that for the most part either their engineering colleagues or corporations are trying to figure out "where exactly…[the] social fit[s] in."

The vision of SLCA in sustainability engineering identity politics, in other words, faces multiple challenges.[93] SLCA experts claim that they can transparently quantify the level of risk for issues such as forced labor—they do not, however, deal with the level of coercion, i.e., studies do not assess the impacts of labor rights violations; they assume that the labor rights violation is the impact itself. In addition, regardless of whether studies place corporate behaviors or product processes at the core of their analyses, SLCAs predicate the production of a technology as an end in and of itself. Therefore those practitioners using SLCA are smuggling social sustainability work and identity into engineering practice.

In context, this chapter suggested that engaging with the methodological, practical, and discursive structure of SLCAs is a current example of engineering cultures engaging with questions that challenge the traditional understanding of engineering knowledge and identity. On the one hand, within the boundaries of life cycle engineering, managed growth, sustainable production, and "improvement" of the social conditions in value chains have become not only imaginable, but also the subject of ongoing engineering practices. On the other hand, "it seems that some of the problems [engineers] are facing…are not really fixable in a technical way[;]…

[92] Interview data.

[93] The scarcity of corporate social impact data, as well as the time/resource constraints associated with their acquisition also pose significant challenges for SLCA practitioners. The cost and the time scale of SLCA studies vary widely and depend on the goal/scope of the study as well as the priority of the SLCA commissioner. Researchers recently reported costs that ranged from "a thousand dollars to over fifty thousand dollars" [with most studies being close to a $20-$30 thousand spectrum] and time-frames that ranged from six weeks to 4-6 months. Interview data.

[for these] are really deep fundamental problems with the very idea of SLCA".[94]

Committed to "transparently balancing inputs and facilitating knowledge claims,"[95] sustainability engineers expressly blend elements of technological change and technopolitics to forge an SLCA rationality (Table 5.4). In so doing, SLCA advocates use metaphors such as "Version 2.0 social contract" and "rendering the invisible hand visible." Nevertheless, along with the proliferation of new tools to make supply chains transparent comes the casting of shadows over what has been left out of such tools: although SLCA in theory generates democratic potentialities to address the politics of technology, the practice of such politics currently takes place only on the discursive outskirts of the field.

As this chapter has recounted, SLCA is grounded in, and continuously challenges, a historically situated tradition of multiple reflexivities within both engineering and sustainability engineering. Moreover, the boundaries between maintaining professional control over SLCA and practitioners' desire to develop reflexivity about their engineering practices are fraught with ontological epistemological and methodological assumptions.

---

[94] Interview data.

[95] Sala *et al.* (2013), Part 1, page 1663.

Table 5.4 Between Technological Change and Technopolitics: The Rationality of SLCA.

Technological change and technopolitics integrated	Assumptions	Metaphors
Unintended consequences	i) A life cycle approach is the prerequisite for recognizing and managing tradeoffs; engineers use LCAs to avoid impact-shifting between spatial and temporal scales	Footprints; mandalas
	ii) Globalization and the market economy provoke social effects and simultaneously offer opportunities to mitigate them	Human existence as inherent risk
Technological innovation	i) LCAs show that engineering work—through processes of technology and/or globalization—is causing social impacts	Handprints; mindprints
Role of the corporation	A socially responsible actor	CSR
Role of the engineer	i) Values play an important role in the LCA study's outcome	Strong sustainability
	ii) The social element of LCAs is due to the nature of engineering knowledge per se, rather than the context of its application	Engineering with a human face
	iii) Adding qualitative impact assessment to LCAs contributes to the increase, rather than decrease of sophistication level	Fun of the impossible

Role of SLCAs with regards to Society	Co-defining the social contract between business and society	Version 2.0 social contract
Role of SLCAs with regards to the Market	The market creates opportunities that SLCAs can help pinpoint	Rendering the invisible hand visible; mitigating the operation of "unfettered markets"
Presence of SLCA in engineering education	Practically non-existent. Fragmented efforts by researchers in the US and Europe to engage graduate and doctoral engineering students in SLCA. Workshops and training courses on SLCA are offered by software providers (GreenDelta and New Earth) and also through the recently founded International Life Cycle Academy	"Every engineer, as a potential consumer, should be familiar with SLCA"; "SLCA can infuse interdisciplinary collaboration in engineering education settings"
	Name	**Contribution**
Institutional entrepreneurship influence	Simon Zadek	Made measuring social sustainability at the corporate level imaginable
	Allen White	Promoted corporate practices' benchmarking based on life cycle tools
	Bob Massie	Helped integrate social reporting in engineering multinationals
	Raskin	Futurist, pioneered socio-ecological modeling
Institutional forerunners	BCSD, WBCSD, CERES, GRI, Öko-Institut, Wuppertal Institute	

Part III

Case Studies

6

Life Cycle Sustainability, Renewable Energy Project Development, and the Exclusion of Local Knowledge in California's Western Antelope Valley (WAV)[1]

[1] Primary data for this chapter is drawn from forty-five semi-structured interviews with RE project stakeholders in the WAV, including local community members, planners and developers.

Introduction

> "I have never seen anything that I have had to do in my 20 years in the power industry that involved less risk than these projects…[;] it is just filling the desert with panels."[2]
>
> These engineers thought that it was going to be a cakewalk coming up here… getting a tax credit from the federal government by claiming that their panels reduce carbon emissions over the life cycle of their project and then a twenty-year buyout agreement from Southern California Edison... [But] in their fancy engineering methodologies [t]hey forgot that people live in the [Antelope] Valley and… you can't just come in here and have your way.[3]

In a 1997 article published in the *International Journal of Life Cycle Assessment*, Swedish engineer and technopolitics ideologue Göran Finnveden had the foresight to anticipate the question of whether LCA "[w]eighting factors should be] derived from a panel of 'Platonic experts'" and to what extent "environmental justice [would be]…given any consideration" in LCA methodology discussions.[4]

On one level, LCAs have recently been applied within US and international regulatory contexts as the dominant sustainability-integrating engineering methodology, predicated on life cycle thinking and the notion of "holism." Yet the growth of LCAs in technology policy relies on metrics for greenhouse gas (GHG) emission reductions. This reliance, in turn, has become entrenched in the study of climate change mitigation as the convergence point between the two vehicles of engineering and public policy rationale. The US EPA, a key actor in structuring and establishing LCA as a policy instrument, announced in 2011 that it would utilize life cycle methods to assess the environmental impacts of increased amounts of biofuels in gasoline resulting from its

---

[2] NRG Energy's CEO David Crane quoted in the November 11, 2011, edition of the *New York Times*. Crane's NRG Energy received $1,237 billion in federal loan guarantees to build and operate the "California Valley Solar Ranch," a 250-MW PV power plant that was developed by SunPower Corporation in San Luis Obispo, California. Lipton, Eric, and Clifford Krauss. "A Gold Rush of Subsidies in Clean Energy Search." *New York Times*, November 11, 2011.
<http://www.nytimes.com/2011/11/12/business/energy-environment/a-cornucopia-of-help-for-renewable-energy.html?pagewanted=all&_r=0>
Accessed April 12, 2014.

[3] Interview data.

[4] Finnveden (1997).

Renewable Fuel Standard (RFS) program.[5] Similarly, in a publication produced for the California Energy Commission (CEC), the National Renewable Energy Laboratory (NREL) conducted a comparative LCA for several energy technologies in order to inform California's policy on distributed generation.[6] Compatible with engineering ideologies of sustainability—for example, California's Low-Carbon Fuel Standard (LCFC), another LCA-focused CEC initiative, spearheaded in 2007 by the late technopolitics ideologue Alex Farrell—the application of life cycle thinking in energy policy is also consistent with the gauge engineering metaphor of sustainable development as described in chapter 2: closed loop systems that harmonize renewable resources with low environmental impact, material re-use and feedstock recycling.[7]

On another level, the preeminence of LCA as a tool for substantiating sustainability-engineering claims is increasingly illustrated in the field of renewable energy (RE) generating technologies.[8] In a 2012 publication

---

[5] ENVTL. PROTECTION AGENCY, BIOFUELS AND THE ENVIRONMENT: FIRST TRIENNIAL REPORT TO CONGRESS (2011).
<http://ofmpub.epa.gov/eims/eimscomm.getfile?p_download_id=506091>
Accessed April 12, 2014.
The RFS was created in 2005 as part of the Energy Policy Act (EPAct).

[6] Mann, Margaret K., Michael Whitaker, and Todd Driver. "Life Cycle Assessment of Existing and Emerging Distributed Generation Technologies in California." Golden, Colorado, July 2011.
< http://www.energy.ca.gov/2011publications/CEC-500-2011-001/CEC-500-2011-001.pdf>
Accessed April 12, 2014.

[7] Farrell, Alexander E., and Daniel Sperling. "A Low-Carbon Fuel Standard for California, Part 2: Policy Analysis." Berkeley, California: University of California at Berkeley, 2007.
<http://www.energy.ca.gov/low_carbon_fuel_standard/UC_LCFS_study_Part_2-FINAL.pdf>
Accessed April 12, 2014.
Farrell, Alex, and Daniel Sperling. "Getting the carbon out." *San Francisco Chronicle* Friday, May 18, 2007.
<http://www.energy.ca.gov/low_carbon_fuel_standard/FERRELL_SPERLING_SFCHRON_OP-ED.PDF>
Accessed April 12, 2014.

[8] The "life cycle approach to environmental management," contends the newly formed trade organization established by leading solar panel manufacturer First Solar and rival cadmium telluride (CdTe) thin-film module producers, "is regarded as the golden standard in the PV industry."
<http://www.pvthin.org/sustainability-new/>
Accessed April 12, 2014.
Overall, should one conceptualize "sustainability" and "RE engineering" as connected through LCA, that connection has been amplified via the publications of a group of

entitled "Sustainability metrics for extending thin-film photovoltaics to terawatt levels" the world's leading authority on LCAs of solar technologies, Vasilis Fthenakis of Columbia University and Brookhaven National Laboratory, surmised that "[p]hotovoltaic devices…are *inherently sustainable* unless they are too expensive to produce, are manufactured using materials that are depletable, or are environmentally unsafe."[9] Fthenakis' conception of sustainability reflects the nature of metrics (economic, energy/materials related, and environmental) that LCA experts have used to classify photovoltaic (PV) systems.[10] In light of such metrics, Fthenakis' proclamation about the *land use* requirements of utility-scale PV systems is both astute and misleading:

> Concerns have been raised about the land requirements for installing large-scale ground-mount PV systems, but these requirements should be examined within a life-cycle context. In fact, historical data show that ground-mount solar farms often use less land during their life cycle than does coal during its life cycle, and there are plenty of desert lands and rooftops to support many terawatts of PV installations.[11]

The argument is astute because it recognizes the intricate manner in which the numerous impacts of electricity generation manifest themselves throughout an engineered system's life cycle. It is also penetrating because, for instance, it could be taken to mean that in examining RE technologies' environmental impacts from an LCA point of view, land intensity and soil productivity—*how* the land is being affected throughout a project's lifetime and *to what degree* its ecological integrity and fertility can be regained prior or after decommissioning—and not a project's acreage more appropriately encaptures "sustainability." Yet the argument is simultaneously misleading in illustrating the impacts on humans who reside in such "desert lands"—and who, for better or worse, may find themselves in the next five to fifteen

---

engineers and scientists led by Greek chemical engineer Vasilis Fthenakis. See, for example: Fthenakis, Vasilis M., Hyung Chul Kim, and Erik Alsema. "Emissions from Photovoltaic Life Cycles." *Environmental Science and Technology* 42, no. 6 (2008): 2168-2174; Held, Michael. "Life Cycle Assessment of CdTe module recycling." Paper presented at the 24th European Photovoltaic Solar Energy Conference, Hamburg, Germany, 2009.

[9] Fthenakis, Vasilis M. "Sustainability metrics for extending thin-film photovoltaics to terawatt levels." *Materials Research Society Bulletin* 37, (2012): 1-6.

[10] For a sample of PV metrics based on a meta analysis of more than 130 PV LCAs see Mulvaney, Dustin. "Are Green Jobs just jobs? Cadmium narratives in the life cycle of photovoltaics." *Geoforum* 54, (2014): 178-186.

[11] Fthenakis (2012): 2.

years surrounded by "many terawatts of PV installations". "Some think that because [Antelope Valley Solar Ranch One (Solar Ranch One)] is a solar project," expressed one interviewee who resides in close proximity to Exelon's 230 MW farm in Los Angeles (LA) County's Antelope Valley, "it is also inherently sustainable and humanitarian... Yet it's not helping people in the Valley live productive and sustainable lives."[12]

This chapter investigates the tension between conventional sustainability engineering approaches (as evidenced in solar project developers' use of LCA to buttress their sustainability claims) and the local/communal knowledge that is being excluded as a result. This is an intriguing contrast: for example, the solar farms were to be built on what seemed "desert" and "empty" lands, even though the locals were familiar with the local ecology. The RE engineers that developed Solar Ranch One and its adjacent solar facility (Alpine Solar Project) assumed that their projects' sustainability would be self-evident; and that their identities as professional engineers depended on a top-down approach to defining and acting on sustainability. Such culture, this chapter demonstrates was encouraged by the larger federal and state government regulatory and incentive structures that require quantification of sustainability, which LCAs can meet more readily. LCAs, then, played an outsized, "orienting" role in configuring how RE projects in the WAV are defined and designed—because of their quantitative and bureaucratic power.

While they apply LCAs to allegedly intrinsically sustainable technologies, engineering experts also implement concepts of "consequential" LCA as solutions to the "unintended consequences" of technology. Luminaries like Greg Norris and Andreas Jørgensen are working to remove the engineering boundaries between realized and non-realized consumption/production scenarios. Their work is complemented by the contributions of the Nordic industrial ecology expert Otto Andersen. In a 2013 book

[12] Interview data. Likewise, "[j]ust because they're renewables instead of landfills," said one longtime LA environmental activist in a *Los Angeles Times* interview, "doesn't mean they're off the hook."
Hsu, Tiffany. "Wind farms multiply, fueling clashes with nearby residents." *Los Angeles Times*, July 24, 2011.
<http://articles.latimes.com/2011/jul/24/business/la-fi-wind-power-20110724>
Accessed April 14, 2014.
The term "farm" communicates a sense of idyllic, green and pastoral setting—the exact opposite of the industrial scenery connoted by the term "plant" (e.g., solar or wind plant). In this chapter I have chosen to use the terms "project" and "facility" to describe both solar and wind plants.

entitled *Unintended Consequences of Renewable Energy: Problems to be Solved*, Andersen posited that consequential LCA can potentially be used as "a modeling tool for predicting the future environmental consequences of a shift from fossil energy to renewable energy."[13] While engineers focus on studying how flows may change in response to future decisions, the implementation scenarios for consequential LCA in RE, as discussed by Andersen and Fthenakis, are stripped of their human and cultural meaning. The human-life dimensions of solar PV LCAs thus tend to become occluded as attention is focused on energy, environmental, and financial metrics. Such metrics, in turn, are increasingly separated from the lives of those most immediately impacted by the technologies subjected to LCAs. To paraphrase chemical engineer and solar energy/LCA expert Dustin Mulvaney, "there is no life in life cycle assessment."[14]

LCA's self-imposed restriction on its boundaries—and in that sense on its own reflexivity—characterizes the critique Finnveden expressed seventeen years ago. More recently, among those who have considered the role of solar PV technology LCAs from a sociological or political ecology perspective, Mulvaney has gone furthest in detailing how "[i]n the case of cadmium-based PV, LCA was used to construct the environmental impacts of cadmium use, but the results and interpretations obscured EJ [environmental justice] concerns." Almost two decades after Finnveden, Mulvaney called attention to the continuing tensions between LCA and environmental justice: "LCA expertise," he warned, "will increasingly become one of the many narratives that travel with PV modules," thus benchmarking "sustainability" in the design of RE engineering projects.[15]

The way that LCA became embedded in narratives and decision making concerning RE in the US is revealed by the ambiguous case of the Department of Energy's Loan Guarantee Program (LGP)—a public-private financing chain that was intended to bring large-scale PV to market. Established under Title XVII of the 2005 Energy Policy Act, the LGP provided the DOE with the ability to give loan guarantees for "new or significantly improved technologies" to address questions of air pollution or GHG.[16] Four years later, Title XVII was amended by the American

---

[13] Andersen, Otto. *Unintended Consequences of Renewable Energy: Problems to be Solved.* London: Springer, 2013.
[14] Personal communication.
[15] Mulvaney (2014).
[16] Pub. L. (Aug. 8, 2005).
<http://www.gpo.gov/fdsys/pkg/PLAW-109publ58/pdf/PLAW-109publ58.pdf>
Accessed April 14, 2014.

Recovery and Reinvestment Act (ARRA), which added section 1705 to the LGP, stating that the "Secretary may make guarantees under this section… [for] [r]enewable energy systems…and facilities that manufacture related components…[in] projects that commence construction not later than September 30, 2011."[17] The program was designed to require reporting on life cycle metrics.

The LGP rested upon, and at the same time transcended, the vernacular of Silicon Valley venture capital. Jonathan Silver, former venture capitalist and first director of LGP, was quoted in December 2010 as likening his program to a "shadow bank," thus suggesting that it not only carried previously researched and developed technological solutions across what is called "the valley of death"—all the way through to their commercialization—but it also transformed such solutions into game-changing sociotechnical systems.[18] LCA took root in the LGP largely because in order to qualify for the ARRA economic incentives, an RE project developer or "clean tech" company had to prove it was reducing GHG emissions over the life cycle of the technological system. It was the function of LCA to provide this evidence of how the pursuit of "clean technology" would contribute to a sustainable future.[19]

The LGP's main beneficiaries anticipated that the program would be an extraordinary opportunity particularly for solar project developers and a new generation of PV manufacturers. ARRA made solar energy the new and focally visible technological paradigm of the "Green New Deal" and, in the words of NRG Energy's chief executive David W. Crane, this meant

The first three projects supported by the LGP, including Solar Ranch One, were government-funded only.

[17] Section 1705 explicitly mentioned "electric power transmission systems" and "leading edge biofuels projects." Pub. L. No. 111-5 (Feb. 17, 2009).
< http://lpo.energy.gov/wp-content/uploads/2010/09/Sect406ARRA.pdf>
Accessed April 14, 2014.

[18] Kuo, Iris. "DOE loan chief: Where we'll invest in 2011." Venturebeat.com, December 15, 2010.
<http://venturebeat.com/2010/12/15/doe-loan-chief-where-well-invest-in-2011-how-abound-solar-could-compete-with-china/>
Accessed April 17, 2014.

[19] See, for example, Boustead Consulting & Associates, Ltd. "Analysis of Environmental and Energy Benefits of 'AVA Solar manufacturing Facilities' project." 2009. Similarly, the environmental report submitted with Suniva's application for a DOE loan guarantee included a life cycle analysis (LCA) that estimated GHG emissions from the project and decreases in GHG emissions resulting from the use of Suniva's screenprinted ARTisun monocrystalline silicon solar cell technology. Interview data.

that developers "intend[ed] to do as much of this business as [they could] get [their] hands on."[20] In this respect, being guaranteed a loan meant being believed prior to delivery; that is why one vision of RE in contemporary US technology policy focuses on the engineering projects that were financed through LGP (Table 6.1).

Table 6.1 depicts the LGP as a distinctive policy and financial matrix which allowed large-scale solar energy technologies to become imaginable, and bestowed upon them certain attributes. First, the LGP favored "pre-commercial" (ostensibly innovative but still unproven) technologies, namely cadmium-based thin-film PV module manufacturing, which received $1.132 billion as opposed to crystalline silicon PVs (an older form of solar technology), which were not eligible for any of these loans.[21]

Second, the LGP simultaneously implicitly assumed that cadmium-based thin-film PV as well as utility capacity, rather than distributive solar, were well beyond anything like an experimental stage. It aimed to promote utility-scale production. Designed to absorb the risk of scaling up solar PV to 100MW and beyond, the program guaranteed a total of $11.635 billion to solar facilities of unprecedented sizes. About 40% or $4.647 billion went to utility solar PV engineering projects. That the experimental stage had been passed was made plain by the 2011 status of cadmium-based, thin-film PV technology employed in the 230 MW Solar Ranch One, sited in the WAV. In August 2011, the largest operating US projects that employed the same PV technology were just 10 MW and 21 MW in size, with two years' and six months' performing/production/online experience, respectively.[22] In that regard, it is not surprising that a February 2010 letter from then California governor Arnold Schwarzenegger, writing to express his commitment to "expediting" LGP facilities such as Solar Ranch One—projected that distributive generation would account for "no more than 30 percent" of the state's RE needs by 2020.[23] Likewise, the verdict of a 2012

[20] Lipton and Krauss (2011).

[21] See Mulvaney (2014), Table 1.

[22] US Department of Energy (DOE). "Final Environmental Assessment, for Department of Energy Loan Guarantee for the AV Solar Ranch One Project in Los Angeles and Kern Counties, California." Washington, DC: DOE, 2011.
<http://energy.gov/sites/prod/files/EA-1826-FEA-2011_0.pdf>
Accessed April 19, 2014.

[23] Office of the Governor of the State of California, Press Release, *Gov. Schwarzenegger Urges Local Counties to Expedite Large-Scale Renewable Energy Projects*. February 18, 2010.
<http://gov.ca.gov/press-release/14469>
Accessed April 21, 2014.

Table 6.1 LGP financial support to solar energy projects.*

Project name	Project owner/developer	Loan	Technology
Cogentrix of Alamosa LLC	Cogentrix Energy Power Management, LLC (owner and developer)	$90 million	High Concentration PV
Crescent Dunes	SolarReserve Tonopah Solar Energy, LLC (owner and developer)	$737 million	Concentrated Solar Power Tower/Thermal storage
Agua Caliente	NRG Solar, LLC/First Solar	$967 million	PV inverted technology—*First Solar Thin Film CdTe Panels*
Solana	Abengoa Solar, LLC; Liberty Interactive Corporation/ Abengoa Solar, LLC	$1.446 billion	Parabolic trough
Mesquite Solar 1	Sempra U.S. Gas & Power, LLC; Consolidated Edison Development/ Suntech Holdings and Zachry Holdings	$337 million	PV—*Suntech's crystalline Pluto silicon*
California Valley Solar Ranch One	NRG Energy, Inc./SunPower Corporation	$1.237 billion	PV—*Suntech's crystalline silicon*
Ivanpah	Brightsource Energy, Inc.; NRG Energy, Inc.; Google, Inc./	$1.6 billion	Power Tower/Solar Thermal
Mojave Solar	Mojave Solar LLC/Abengoa Solar LLC; Mojave Solar LLC	$1.2 billion	Parabolic trough
Project Amp	Prologis, Inc.	$1.4 billion	Rooftop PV on 750 rooftop buildings owned by Prologis
Desert Sunlight	Next Era Energy Resources, LLC; GE Energy Financial Services; Sumitomo Corporation of America/ First Solar	$1.46 billion	PV—*First Solar Thin Film CdTe Panels*
Antelope Valley Solar Ranch One	Exelon Corporation/First Solar	$646 million	PV—*First Solar Thin Film CdTe Panels*
Genesis Solar	Genesis Solar, LLC/Next Era Energy Resources, LLC; Genesis Solar, LLC	$852 million	Parabolic trough

*Data compiled from various sources.

CEC-led consortium's "interim document," destined to set the regulatory and scientific framework for developing RE projects on as much as 2.3 million acres of California's desert lands, excluded distributive generation from the consortium's agenda.[24]

Third, the emphasis on RE engineering infrastructure at utility scale was grounded in, and legitimated by, the special position of solar PV in providing immediate "jobs for... [Americans'] future with secure, affordable, and reliable energy."[25] For instance, during the peak of the WAV's RE controversy in 2011, the unemployment rates in Lancaster and Palmdale—locations that had recently absorbed proportionately large segments of South Central LA population streams, and were located twenty and thirty miles away from the Solar Ranch One project site, respectively—were reported to have reached 20%.[26] Around the same time, the federal government predicted that a one-year extension of the 1603 Treasury Program, a federal grant finance mechanism designed for RE project developers, would add another 37,000 jobs to the solar sector, in addition to the 100,000 that already existed in the US in late 2011.[27] Recent solar PV job fairs in the WAV were attended by members of the California State Assembly and marshaled close to 1,800 applicants.[28] The PV job fairs propagated the typical discourse of RE metamorphosing California's economic landscape. In other words, as manifested in the LGP, the political function of LCA studies of solar cadmium based thin-film PVs was to provide a framework in the dispute regarding the environmental impacts of cadmium use in PV

---

[24] California Energy Commission, California Department of Fish and Game, Bureau of Land Management, and U.S. Fish and Wildlife Service (CEC, CDFG, BLM, and USFWS). "Description and Comparative Evaluation of Draft DRECP Alternatives." December 17, 2012.
<http://www.drecp.org/documents/docs/alternatives_eval/index.php>
Accessed April 21, 2014.

[25] Pub. L. No. 109-58, Title XVII (Aug. 8, 2005).
<http://www.gpo.gov/fdsys/pkg/PLAW-109publ58/pdf/PLAW-109publ58.pdf>
Accessed April 14, 2014.

[26] Stallworth, Leo. "University of Antelope Valley pays employers for jobs." *Los Angeles News*, Friday September 9, 2011.
<http://archive.today/e357s>
Accessed April 22, 2014.
In neighboring Kern County the unemployment rates in February 2011 were 16%. Meyer, Gary. "Unemployment Dips in Kern County." *The Mountain Enterprise*, March 30, 2012.

[27] EuPD. "Economic Impact of Extending the Section 1603 Treasury Program." Miami, Florida: EuPD, 2011.

[28] Meyer, Gary. "1,000 seek work at solar job fair." *The Mountain Enterprise*, March 8, 2013.

technology, thus empowering the relationship between "big solar" and "green jobs." The LGP provided a symbolic structuring mechanism: the conceptualization of RE engineering projects went hand-in-hand with the vision of GHG reductions constructed through solar PV LCAs. In turn, the pursuit of GHG emission reductions became a powerful force in RE policy, replacing other metrics, like justice, as the standard of sustainability.

Certainly, then, there are parallels between the role played by LCA metrics—particularly global warming potential—in the analysis and modeling of electricity-producing systems on the one hand, and their role in qualifying RE projects on the other. Yet in some cases the parallels are inexact. The conservative think tank Beacon Hill Institute, founded in 1991 by politician Ray Shamie, has lately conducted a number of studies that veer sharply away from an RE vision; the studies have introduced LCAs to actually *undermine* climate change initiatives—for example in the states of Washington, Maine, Arizona, and Nevada. Non-conventional energy sources, argued an April 2013 Beacon Hill Institute investigation into the impacts of Arizona's Renewable Energy Standard (RES), may not be as "environmentally friendly as some proponents claim."[29] Rather, the examination concludes that LCA analyses used to justify RE policy—"a recipe for economic decline" according to the Institute's leaders—are "incomplete" due to a lack of data that address solar and wind resources' intermittency.[30]

Beacon Hill Institute's conclusions are not more legitimate than other LCA-focused research that favors RE. In December 2013, for example, the Institute's host institution, Suffolk University in Boston, renounced one of the think tank's funding proposals, arguing that its "research goals...were inconsistent" with Suffolk's mission.[31] This illustrates the general trend that

---

[29] The Beacon Hill Institute at Suffolk University. "The Economic Impact of Arizona's Renewable Energy Standard and Tariff." Boston, MA: The Beacon Hill Institute at Suffolk University, 2013.
<http://www.beaconhill.org/BHIStudies/AZ-REST/AZ-BHI-REST-2013-0403FINAL.pdf>
Accessed April 23, 2014.

[30] Tuerck, David G., Paul Bachman, and Michael Head. "RPS: A Recipe for Economic Decline, Scheme to cost Nevadans $2.275 billion over 12 years." Nevada Policy Research Institute, April 2013.
<http://www.npri.org/docLib/20130424_RPS-ARecipeforDecline.pdf>
Accessed April 23, 2014.

[31] Goldenberg, Suzanne. "Free-market research group's climate proposal denounced by host university." theguardian.com, December 5, 2013.
<http://www.theguardian.com/world/2013/dec/05/host-university-research-group-climate-suffolk-university>
Accessed April 23, 2014.

whatever researchers' motives may be, they commonly presume that LCAs would offer a holistic framework to address issues of RE economic and environmental viability—more so than would other, less comprehensive policy instruments or methodologies.

Moreover, though clearly dominated by engineers' and economists' rationalities, most discourse on life cycle metrics in RE has not been the exclusive purview of technical experts. Rather, WAV anti-RE project activists have used life cycle thinking language to reject arguments about the "clean" nature of RE systems and justify their opposition to utility-scale solar and wind facilities:

> [The project's preliminary study should] measure the lifetime effects of metal tower and metals in the solar array's surface...and the lifetime effect of temperature changes to habitat, vegetation and microclimate...; How much carbon dioxide will be emitted during the manufacture, shipping, construction, operation, maintenance, and decommissioning of the various components of this facility? ; How much foreign oil will be required to maintain and operate turbines, substations, vehicles? Monthly? Yearly? For the life of the project?[32]

In a similar fashion, the impression that an RE facility must be broken down into "the different phases of the project cycle" so that affected communities and engineers alike are able to operationalize stakeholder engagement has recently been reinforced by constituencies as diverse as the International Finance Corporation—a member of the World Bank group—and the Great Lakes Wind Collaborative, a consortium of wind energy actors based in Ann Arbor, Michigan (Figure 6.1).[33] By early 2013, when the first utility-scale PV projects in California neared completion, the

---

[32] Interview data; Rhyne, Margaret. "PRMDIA Response to Element Power's NOP." December 31, 2011, courtesy of Margaret Rhyne; Zahnter, Susan. "Re: Element Power Wildflower Green Energy Farm, Project Number R2010-00256-(5), RENVT201000063, RCUPT201000121." January 5, 2012, courtesy of Susan Zahnter.

[33] International Finance Corporation. "Stakeholder Engagement: A Good Practice Handbook for Companies Doing Business in Emerging Markets." Washington, DC: International Finance Corporation, 2007.
<http://www.ifc.org/wps/wcm/connect/938f1a0048855805beacfe6a6515bb18/IFC_StakeholderEngagement.pdf?MOD=AJPERES>
Accessed April 25, 2014.
"Community engagement" is, of course, neither a 21st century phenomenon nor a practice that emerged in the context of RE engineering—public health professionals started to involve community stakeholders in developing partnerships already in the 1950s.

"lifetime" of a desert solar facility was viewed simultaneously as a solution to climate change and a contradiction in terms: "[s]olar power projects... have a life expectancy," wrote Chris Clarke, environmental journalist and California desert expert; "it may be 30 years, or 50 years...[;] [p]erhaps [then] we will be able to plant 500-year-old yuccas and 1,200-year-old creosotes from seed."[34]

Sustainability Engineering Downplaying Local Knowledge

On the one hand, then, based primarily on environmental and economic considerations, LCA discourse and requisite metrics have become quantifiable denominators that help organize the meaning of contemporary RE controversies— for example, in the case of a solar and wind project development in Los Angeles County's WAV.[35] On the other hand, in this chapter, I explain that the possible compatibility of solar/wind power with decentralization suggests potential, as yet unexplored, connections between RE engineering, life cycle thinking, and metrics of social justice (defined here as the ongoing struggle of disadvantaged communities to establish equal access to basic services—including energy and a healthy environment—and to overcome many forms of social and/or political oppression).[36]

---

[34] Clarke, Chris. "What Happens When a Developer Abandons a Desert Solar Project." rewire, February 11, 2013.
<http://www.kcet.org/news/rewire/commentary/what-happens-when-a-developer-abandons-a-desert-solar-project.html>
Accessed April 27, 2014.

[35] Another example of a project controversy in the wider Mojave area where LCA metrics have been used to instigate meaning in RE development is NextEra's 250 MW project in Kern County. Fthenakis' studies were used by Kern County planners in supporting the project during an appeal decided upon by the County's Board of Supervisors in October 2012.
Kern County Planning and Community Development Department, Board of Supervisors. "Staff Report." October 30, 2012.
<http://phonyuniontreehuggers.com/wp-content/uploads/2013/07/2012-10-30-Beacon-Photovoltaic-County-Response-to-LIUNA-Appeal.pdf>
Accessed April 27, 2013.

[36] Walker *et al.* (2010) note that "...narratives [of energy policy in the United Kingdom] are clearly predicated on the basis that 'communities' can and do exist in an unproblematic form and within many of the positive qualities with which they are readily associated."

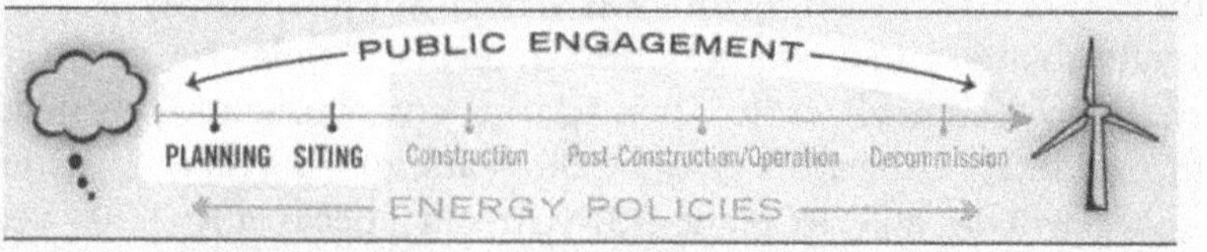

Figure 6.1 Public engagement in wind facility development defined as a continuing process occuring throughout the life cycle of a project. Adapted from "Best Practices for Sustainable Wind Energy Development in the Great Lakes Region."
Great Lakes Wind Collaborative. "Best Practices for Sustainable Wind Energy Wind Development in the Great Lakes Region." Ann Arbor, MI: Great Lakes Wind Collaborative, 2011.
<http://glc.org/files/docs/GLWC-BPToolkit-BP07.pdf>
Accessed April 27, 2014.

Small-scale RE projects in particular *can* solidify community life, redistribute capital and opportunities, and facilitate political self-determination.[37] Over the past forty years, communities in the US have experimented with wind and solar technologies to "break through the political, economic, social, and psychological forces that constrain and oppress [them]."[38] More recently, the example of the Maple Ridge Wind Farm, one of the largest wind farms in the northeast, located in Lewis County, New York, provides an illustrative sketch of effective communication between RE project development, economic sustainability

Such considerations, however, are in fact problematic: "[w]hilst appearing inclusive, community can also be deeply exclusionary, marginalizing those who are seen as not fitting." Also, what in each case counts as "affected community" may be contingent on the very type of RE technology at stake. I am thinking particularly about off-shore wind projects. Does "community," in that case, simply mean "coastal community," or must the term be broadened to include the mainland, too? Walker, Gordon P., Patrick Devine-Wright, Sue Hunter, Helen High, and Bob Evans. "Trust and community: exploring the meanings, contexts and dynamics of community renewable energy." *Energy Policy* 38 no. 6 (2010): 2655-2633.

[37] Schumacher, Ernst F. *Small in Beautiful: A study of economics as if people mattered.* London: Abacus, 1974. Lovins, Amory B. *Soft energy paths: toward a durable peace.* San Francisco, CA: Friends of the Earth International, 1977. Winner, Langdon. *The whale and the reactor: a search for limits in an age of high technology.* Chicago: University of Chicago Press, 1986. Harding, Jim. *Social policy and social justice: the NDP government in Saskatchewan during the Blakeney years.* Waterloo, Ont: Wilfrid Laurier University Press, 1995. Gordon, Uri. *Anarchy Alive!: Anti-Authoritarian Politics from Practice to Theory.* London: Pluto Press, 2008.

[38] Boyle, Godfrey, and Peter Harper, eds. *Radical Technology.* London and New York: Pantheon/Random House, 1976.

and preservation of agricultural land along with traditional knowledge and rural lifestyles.[39] Nevertheless, however promising the emancipatory potential of RE, there is no energy panacea. All energy technologies involve environmental costs and risks to public health. Wind technologies, for instance, have environmental and health impacts such as land and materials use, biodiversity, and wind blade recycling. They also may face community opposition.[40]

"[T]here are economies of scale you get by going big," reflected one energy analyst in a 2012 interview examining the fate of massive solar projects in the California desert, "but communities can just make your life miserable."[41] Indeed, unmet local concerns may entail costly project delays or cancellation. Thus strong political and financial incentives encourage state authorities and RE developers to address issues of social acceptance. For example, in response to a sharp decline in wind project approvals in the UK—from 70% in 2008 to 35% in 2012—the British government took drastic measures to ensure that local concerns were better incorporated into the RE planning process. "It is important that onshore wind is developed in a way that is truly sustainable—economically, environmentally and socially," argued Energy Secretary Edward Davey in the summer of 2013, after publicizing a decision that supported a five-fold rise in the economic assets provided by developers to communities hosting wind projects.[42] Correspondingly, intense public concerns are being expressed with regard to siting large solar and wind facilities in the southwestern US.[43]

---

[39] This is a story where RE helped save family farms as well as maple syrup production traditions and practices. Unknown. "Tapping Maple Ridge. A Film About Maple Syrup, Wind and Community." Tappingmapleridge.com, undated.
<http://www.tappingmapleridge.com/02About.htm>
Accessed April, 27, 2014.

[40] The notion that RE might empower individuals but can also generate durable contention becomes apparent in the size and design of projects themselves. For instance, after its original owner's bankruptcy and an ongoing dispute over Native American cultural resources, the size of the Blythe Solar Power project—once celebrated as the world's biggest solar facility—was reduced from 1000MW to 485MW.

[41] Koren, James Rufus. "Cooling Off? Solar ranch nears launch as large projects lose sizzle." *Los Angeles Business Journal* 34, no. 35 (2012): 1 and 50.

[42] Harrabin, Roger. "Local communities offered more say over wind farms." bbc.com, June 6, 2013.
<http://www.bbc.com/news/business-22791815>
Accessed April 27, 2014.

[43] A 2009 press release by the British Wind Energy Association (BWEA) was lamenting over the finding that "local council approvals of wind farm applications have fallen to a shocking new low of just 25%." In addition, according to a 2011 news report (Derbyshire

With more than thirty proposed RE projects for the WAV, rural Los Angeles County residents have likened their emotions to "what John Muir must have felt watching the Hetch Hetchy Valley fill with water."[44] They believe they are "completely powerless[;]... not empowered to do anything other than to ask 'please' nicely."[45] One local, who fell victim to a car accident during the intense April 2013 dust storms that Antelope Valley natives associate with Arizona-based First Solar company's on-site grazing practices at Solar Ranch One, put it as follows: "[n]ow we live in Hell Valley...[and] I am in too much pain to [even] be polite."[46]

2011), data obtained by the British law firm McGrigors showed a growing percentage of wind projects being turned down by planners (29% in 2005, 33% in 2009 and 48% in 2010). The fact that the increase in the percentage of opposition is partly due to the increase in the number of project applications does not negate that project opposition suggests a significant concern in RE development internationally. I could not find a good source of aggregate data showing combined (solar and wind) opposition against RE projects in the US. The best source for community resistance to wind energy in North America is Phadke's work, which uses GIS technology to plot the most recent wind oppositional movement in the US. Of course, many proposed projects do not get built, and may not go forward for financial or other reasons. According to NREL researcher Eric Lanzt, the country's leading RE agency has very few data on the number of RE project failures and "that's been one of the things we've been talking to with developers lately." In the early and mid-2000s, the same source details, the sentiment in the NREL community was that for every successful project there were ten failed projects or ten projects that were stalled or disabled "for some reason or another" (personal communication). The lack of a "social acceptance" database is also due to the industry's competitive pressures, namely that developers have to be strategic about when they announce their project, otherwise their competitor is likely to come in and sign up all the land that they are trying to sign up. BWEA. "Wind farm planning approvals by local councils slumped to a record new low of 25%." 2009.
<http://www.bwea.com/pdf/press/PR20091020_25pc_approval.pdf>
Accessed November 23, 2011.
Derbyshire, David. "The wind turbine backlash: Growing public opposition thwarts green energy drive." *Mail Online*, July 11, 2011.
<http://www.dailymail.co.uk/news/article-2013233/The-wind-turbine-backlash-Growing-public-opposition-thwarts-green-energy-drive.html#ixzz1jUo8QMZa>
Accessed December 8, 2011.
Phadke, Roopali. "Resisting and Reconciling Big Wind: Middle Landscape Politics in the New American West." *Antipode* 43 no. 3 (2011): 754-776.

[44] Rhyne, Margaret. "Blue Sky Wind Energy Project NO. R2011-00408 EIR Scoping Meeting." Letter to the Los Angeles Department of Regional Planning, October 6, 2011, courtesy of Margaret Rhyne.

[45] Interview data.

[46] Trabish, Herman K. "Construction Halted at First Solar's 230 MW Antelope Valley Site." *Greentech media*, April 22, 2013.

The unprecedented, and often contradictory, challenges and opportunities brought about by RE in the California desert do not predict uniformly negative or positive consequences for WAV residents. A few private landowners in the Valley are seizing the profit-making opportunity by either selling or leasing thousands of acres to RE companies. State and federal mandates have increased the pressure to convert farmland into RE project and "mitigation" sites. WAV residents have admitted in personal communications that a few farmers in the area have stopped working their property and have high hopes of selling it to developers. Others opine that RE represents the "lesser of two evils," endorsing solar and wind projects to prevent the area from being used for residential development. In electronic media and on local community boards, some long-time residents—many retired, some not—have raised their voices in opposition to RE developers, citing a multitude of concerns, including the cumulative loss of land across the Valley for conversion to non-agricultural uses.[47] Opponents of RE projects in the WAV strongly identify with their activism and describe their intervention as a "second job"—one went so far as to contend that they "have not been personally invested in most [of their professional career] as [they are] in this situation."[48] The majority is still trying to grasp the implications of tens of thousands of acres of mountainsides and valley floor being turned into an RE laboratory.

North of Los Angeles County, surrounded by the Tehachapi Mountains and the San Gabriels, lies the Antelope Valley portion of Kern County. In December 2011, San Diego-based wind project developer Helo Energy LLC abandoned a proposal for a facility after strong opposition from Kern County officials and local residents. The company reported that it was defrauded by a local real estate agent and two other firms about the efficiency of wind resources in the area before investing $9 million into

---

<http://www.greentechmedia.com/articles/read/Construction-Halted-At-First-Solars-230-MW-Antelope-Valley-Site>
Accessed April 28, 2014.

[47] Trabish, Herman K. "Element Power, First Solar, and Intense Local Politics." *Greentech media*, November 29, 2011.
<http://www.greentechmedia.com/articles/read/element-power-first-solar-and-intense-local-politics>
Accessed April 28, 2014.

[48] Interview data
<http://www.avhidesert.com/forum/showthread.php?tid=2030>
Accessed April 28, 2014.

buying land and commissioning an environmental impact study.[49] Only a few months later, the Kern County Sheriff's Department was confirming that "at least eight incidents of vandalism at wind farms in Southern California"—presumably by anti-wind activists—had been reported.[50] Likewise, at least three incidents of vandalism—copper wire, but also other expensive equipment theft and destruction—have marked the tumultuous history of Solar Ranch One.[51] "When you have dissatisfied people in the community, this sort of thing happens," one local board officer was reported to have remarked in the aftermath of a $100,000 worth of vandalism at Exelon's Solar Ranch One in July 2012.[52]

Similarly, wind developer Terra-Gen's decision to drop its plans for a 7,000-acre project in the Tehachapi Mountains "was a result of several important development concerns, including local opposition."[53] A recent proposal (Clearwater-Yakima) to build two separate, yet adjacent, 20-MW PV facilities right on the edge of West Mojave has drawn "some of the fiercest opposition I have seen in 60 years living in the area," according to Mojave Chamber of Commerce board member Bill Deavers. The proposal was initially approved by a 4-1 vote during a Planning Commission meeting that started in the early morning hours of January 24, 2014. The perception was that the proposal was a threat to health and safety, as well as to local commercial development and residents' "sacred right to express their perspective." Mojave dwellers, community leaders and representatives from the Mojave Air and Space Port have requested that the Mojave Specific

---

[49] Nelson, Jill Barnes. "Wind Company to Sue." *Mojave Desert News*, Friday April 6, 2012. <http://www.desertnews.com/news/article_84f204f0-7f7c-11e1-a45b-001a4bcf887a.html> Accessed April 28, 2014.

[50] Trabish, Herman K. "Is Wind Becoming a Victim of Ecosabotage?" *Greentech media*, August 18, 2011.
<http://www.greentechmedia.com/articles/read/is-wind-becoming-a-victim-of-eco-sabotage>
Accessed April 28, 2014.

[51] Hedlund, Patric. "First Solar Seeks 'New Beginning' for 2012." *The Mountain Enterprise*, January 27, 2012.
Trabish, Herman K. "$100,000 Vandalism at the First Solar/Exelon Antelope Valley Solar Ranch One." *Greentech media*, July 18, 2012.
<http://www.greentechmedia.com/articles/read/100000-vandalism-at-the-first-solar-exelon-antelope-valley-solar-ranch-one->
Accessed April 28, 2014.
For a timeline of Solar Ranch One's Siting, Permitting, Financing and Pre-construction phases see Appendix 1.

[52] Ibid.

[53] Hsu (2011).

Plan—a 2003 Kern County planning document for the development of land uses—be modified to incorporate community concerns with regard to expanding solar and wind facilities. Mojave natives have also advocated that boundaries be set around their town, confining future RE project development.[54] To this purpose, in the summer of 2011, Tehachapi officials requested that Kern County enforce a temporary moratorium on wind projects.[55] And two years later, in June 2013, county supervisors from San Bernardino—bordered by LA and Kern Counties—unanimously passed a 45-day industrial-scale solar project moratorium, thus expressing the "need for a thoughtful and balanced approach to locating" projects by consciously engaging local constituencies.[56]

With its focus on realizing utility-scale engineering projects, the problem of including underrepresented viewpoints in RE project siting in the southwestern US has become apparent in various ways: the protests, the proposals for RE moratoriums as well as other forms of community activism, and the contents of politicians' speeches.[57]

Solar Developers and Local Communities in the WAV: An Unsustainable Relationship

The California pride in being a nation-leading state shone during the first panel of the sixth Intersolar North America trade show at San Francisco's InterContinental Hotel on July 8, 2013. Governor Jerry Brown and San Francisco mayor Edwin Lee were joined by Bernadette Del Chiaro, the new director of the California Solar Energy Industries, and National Renewable Energy Laboratory (NREL) director Dr. Dan Avizu, in an address before

---

[54] Personal communication.

[55] Hsu (2011).

[56] Weiner, Terry, and Miriam Raftery. "San Bernardino Board of Supervisors OK's 45-day solar moratorium." *East County Magazine*, March 15, 2012.
<http://www.vvchamber.com/index.php?module=corp_news&post_id=66>
Accessed April 27, 2014.

[57] The adversarial nature of utility-scale RE project development across the desert southwest was considered in March 2012 when rooftop solar proponents and desert activists joined forces to protest that distributed generation was missing from the annual Imperial Valley Economic Development Corporation's RE summit. Weiner, Terry, and Miriam Raftery. "Advocates of rooftop solar raise voices in protest over industrial-scale desert solar and wind projects." *East County Magazine*, March 2012.
<http://www.eastcountymagazine.org/advocates-rooftop-solar-raise-voices-protest-over-industrial-scale-desert-solar-and-wind-projects>
Accessed April 27, 2014.

a full auditorium of solar enthusiasts from North America and elsewhere. The panelists referred to Intersolar 2013 as "the meeting of those who want to save the world and those who want to make money generating electricity." The utopian view of California's RE destiny undoubtedly moved many attendants. Governor Brown explained that between his first and current political appointments he "had to expiate [his] many political sins and... spen[d] some time in the wilderness" to reflect.[58] On that note, this chapter emphasizes that the governor's—and many others'—views on community opposition to utility RE projects need to be reconsidered: In 2011, Governor Brown was reported to have said "some kind of opposition you have to crush" and "any old fool can object to anything".[59] On July 8, 2013, he added that "[he] is going to move aside all of the obstacles, whoever—whatever they are—get [them] out of the way."[60]

The Intersolar 2013 inaugurating panel may have missed an important point when it did not consider the conflict and irresoluteness associated with many RE projects. Challenging the "inherent" sustainability of RE projects does indeed have a local community referent. These challenges also amplify the dissonance between the rise within contemporary US engineering and public policy of a movement to create, communicate and distribute RE expertise, and the decline in public trust in RE experts, planning authorities and project developers.[61]

---

[58] Participant observation.

[59] Kahn, Debra. "Calif. Governor Vows to 'Crush' Foes of Renewable Energy." *New York Times*, July 26, 2011.
<http://www.nytimes.com/gwire/2011/07/26/26greenwire-calif-governor-vows-to-crush-foes-of-renewable-22698.html>
Accessed November 21, 2011.
Trabish, Herman K. "Will NextEra Energy's Wind be Part of the New California Gold Rush?" *Greentech media*, July 26, 2011.
<https://www.greentechmedia.com/articles/read/will-nextera-energys-wind-be-part-of-the-new-california-gold-rush>
Accessed April 28, 2014.

[60] Participant observation.

[61] NREL'S energy analyst Joyce McLaren, who has done original work on the social acceptance of wind energy in the UK, confirmed in personal communication that it is indeed questionable whether "a lot of the [US] federal and state policy makers...have thought about what the [social acceptance] challenges [of RE] are." In some states, for example, like Oregon, Washington and Wisconsin if an RE project is of a certain size it has to go through a state siting process, whereas if the project is below a certain threshold, it is required to go through a local process. This, basically, means that the developer is able to choose whatever they think will be "easiest." Yet that still misses the problem of social acceptance of RE.

This chapter shows that those developers who used carefully balanced aesthetic and engineering considerations in their verbal *assertions* to capture specific RE project design intent in the WAV found themselves blocked by local opposition—a reflection of the deep mistrust on the part of community members toward the corporate engineers and Los Angeles County planners. Ironically, such assertions were used by residents to expose the discrepancy between promises made prior or during project permitting, and construction realities.

This chapter's analysis reveals that in Los Angeles County's WAV, RE project development is perceived as a potential threat to agricultural land, local wildlife, community water-rights, property values and landscape aesthetics—to name just a few casualties. Contrary to federal and corporate visions which assumed the inherent sustainability of large-scale facility planning in rural unincorporated areas, this chapter argues that the fractured relationships between stakeholders and the disparity between rural and urban mechanisms of governance has facilitated the diminishing fairness in RE project dispute resolution. In particular, the limits imposed on the design, planning, and permitting processes of Los Angeles County's inaugural utility-scale RE projects are nonsensical if clashes between communities and developers and between communities and planning authorities are not considered. Nor one can apprehend the history and future fate of RE in the WAV separate from the severe misalignments between urban and rural politics.

Beginning in late 2010, some rural communities in Southern California changed their assumptions about RE, not because of the discovered intermittency and lower reliability of wind and solar sources of energy compared with the base-load power currently furnished by coal, natural gas or nuclear. Rather, by the time Los Angeles County's premier utility-size solar PV facility had secured its permits, RE was widely viewed by WAV natives as a problem: it seemed that RE development was synonymous with corporate/planning authority distrust and socio-environmental injustice.

During the construction phase of Solar Ranch One, and particularly between early 2012 and mid-2013, the relationship between affected community members and solar developers became oxymoronic and unsustainable: While First Solar was unable to meet local concerns, Solar Ranch One meanwhile proceeded as anticipated by institutions financially and politically invested in the facility's success. Solar Ranch One engineers wanted to demonstrate that a new mechanism for transparent engagement with locals was emerging, with transformation of Solar Ranch One from a project plagued by slow progress to one that would justify its loan guarantees and become an emblem for the economic and social potentialities of RE.

In early February 2012, First Solar company representatives attempted to make it seem that their most diaphanous claims might be honored. Solar Ranch One engineers were "roundly applauded for crisp, factual answers to issues of [local] concern" as they "vowed to be more forthcoming with information of concern to the community."[62] First Solar executives admitted that "with the benefit of hindsight... [laying off workers without prior notice is] probably something we could...have done better on."[63] A more apparent shift in corporate attitude was the developers' determination to include engineering experts in Town Council meetings, such as First Solar's Sustainable Development director Parikhit Sinha, who briefed WAV residents on health and safety issues pertaining to CdTe encapsulated in Solar Ranch One's solar panels. Such initiatives required community engagement in and of themselves. To say that Solar Ranch One had triumphed over "hearts and minds" would not be unfair. But it was a transitory juncture, and to say otherwise would have been unjust and unrealistic.[64]

By the time First Solar engineers and executives attempted to put community engagement visions into practice, Solar Ranch One's planning and permitting reality prevailed. For instance, in January and February 2012, Los Angeles County withheld the facility's construction permit until planners could approve a change in on-site grading practices—"not [an] unusual" procedure, according to one DRP employee.[65] Consequently, Wall Street speculators worried about the fate of the controversial project's $646 million federal loan guarantee. Solar Ranch One's planning uncertainties and continuous conflict between developers and WAV communities had amplified the risks for economic sustainability. Indeed, after having secured Los Angeles County's permission to remove some extra 280,000 cubic yards from the facility's 2,100-acre site, and oblivious to regional conditions, First Solar personnel were unable to control the blowing sand, due to the very high winds that hit the Valley in late April 2012. This scenario

---

[62] Hedlund, Patric. "First Solar and Rural Councils Inch Toward 'Good Neighbors' Goal." *The Mountain Enterprise*, February 10, 2012; Hedlund (January 27, 2012).

[63] Currier, Craig. "Solar firm ups community gift to $280,000." *Antelope Valley Press*, January 16, 2011.

[64] Hedlund, Patric. "Change on Display in the Antelope Valley. Solar Begins to Wind Hearts and Minds." *The Mountain Enterprise*, March 9, 2012.

[65] Sweet, Cassandra. "Los Angeles County Approvals Boost $1.36 Billion First Solar-Exelon Deal." www.4-traders.com, February 17, 2012.
<http://www.4-traders.com/FIRST-SOLAR-INC-37008/news/LA-County-Approvals-Boost-136-Billion-First-Solar-Exelon-Deal-14029228/>
Accessed May 24, 2014.

was eminently predictable by local residents who were familiar with both the weather conditions and the friability of the sand. Their experiences and thoughts had been ignored in the planning of the project.

In response, local residents expressed their fears about being exposed to the dust. Yet, "First Solar had no actual intentions of meeting with… concerned citizens," argued a May 2012 protest letter.[66] This community claim was buttressed by the company's decision to discontinue the meetings of the "Landscaping Committee," for which locals had been recruited. Moreover the Los Angeles Department of Water and Power (LADWP) had already granted a waiver for the Solar Ranch One CUP by acquiescing to the developer's plans to plant 109 instead of the promised 4,700 trees along the project's perimeter, which could have stabilized the sand or reduced the windblown dust. As this chapter subsequently explains, RE engineers failed to consider the extent to which engineering work depends on local knowledge. As a result, both the outline and execution of their utility-scale solar project did not engender trust on the part of WAV residents. Engineers from First Solar "have expressed many times their eagerness to be open and cooperative with the public," remarked one Valley native; yet, "that appears to be nothing more than rhetoric, a way to put the public off over anything they do not want to discuss or do."[67]

By early 2012, mistrust and incredulity had become consolidated into the web of relations between Los Angeles County, First Solar, and rural communities in the WAV: at the same time, Supervisor Antonovich's field deputy Norm Hickling told rural Town Council leaders and First Solar representatives that he was "no poker player" and that his desire was to "start fresh."[68] One of the most telling examples of such embedded distrust is the controversy over a seemingly straightforward issue: the electrical certification of Solar Ranch One's solar panels, which ultimately delayed construction on site for more than three months.[69] What began as a routine visit to

---

[66] Friends of Antelope Valley Open Space and Concerned Citizens of Westside Antelope Valley (FAVOS and CCWAV). "Letter." May 22, 2012, courtesy of FAVOS and CCWAV.
[67] Ibid.
[68] Hedlund, Patric. "Meeting Held to Explain First Solar's Future Goals." *The Mountain Enterprise*, January 13, 2012; Hedlund (January 27, 2012).
[69] Another, equally telling, example involves a US EPA complaint referral received on November 6, 2012, by the Los Angeles County Fire Department, Health Hazardous Materials Division (HHMD) Investigations Unit alleging that, on "more than one occasion a First Solar foreman had instructed an employee to dispose of buckets containing diesel contaminated soil in the desert [approximately two miles away from the Solar Ranch One site]. The alleged toxic dumping was reported to the EPA by members of the Oso Town Council. According to a February 2013 report by HHMD, there was no "sufficient

First Solar's site at 170th Street turned into sharp disagreement, reflecting participants' uncertainty about the technical adequacy of First Solar's panels, and signifying an intensification of the quandary about the symbolism of the project itself. Engineers, planners, workers and local community disagreed sharply over the meaning of the technical, economic, procedural and symbolic dimensions of a prolonged RE construction freeze in the WAV.

Operations at Solar Ranch One hit a stumbling block when Los Angeles County inspectors noticed a discrepancy between First Solar panel electrical connector compliance with International Electrotechnical Commission (IEC) certification standards—commonly applied in European countries—and the requirements of the US-based certification entity, Underwriters Laboratory (UL). "We don't believe there is any safety issue here," public relations director of First Solar, Alan Bernheimer, told an RE news reporter in June 2012. He further observed that conformity to UL standards had not previously raised any polemic against the company in any of its US installations, including a 21-MW facility located in nearby Riverside County:[70] "The issue is which electrical code is appropriate for a utility PV power plant," Bernheimer suggested.[71] Yet according to Los Angeles County Department of Public Works spokesman Kerjon Lee, his agency was not "focusing solely on the UL approval issue…[;] [the county's] focus…[was] on compliance with… health and safety standards" at Solar Ranch One as a whole.[72] First Solar's becoming mired in the solar module standard dispute was situated in the panel manufacturer's desire to tame public opposition in the WAV, and simultaneously to recover from a twelve-month economic hiatus during which the company's shares had dropped from \$140 to \$14 (Figure 6.2). Solar Ranch One became a local political and labor policy violation battlefield, reportedly laying off close to 240 blue-collar employees

evidence at this time to verify the complaint"; the former First Solar employee who initially communicated the alleged dumping to the Oso Town Council members waited seven months to file an official complaint "because he feared retaliation" from the solar developer. Los Angeles County Fire Department, Health Hazardous Materials Division (HHMD), Special Operations Section, Summary of Events, February 1, 2013, Log Number 12-38.

[70] This is the Blythe Solar Project owned by NRG-Solar.

[71] Trabish, Herman K. "More Furloughs at First Solar while Dispute with Los Angeles County Drags On." *Greentech media*, June 18, 2012. <https://www.greentechmedia.com/articles/read/more-furloughs-at-first-solar-while-dispute-with-la-county-drags-on> Accessed May 26, 2014.

[72] Hedlund, Patric. "First Solar Lays off about 240." *The Mountain Enterprise*, June 1, 2012.

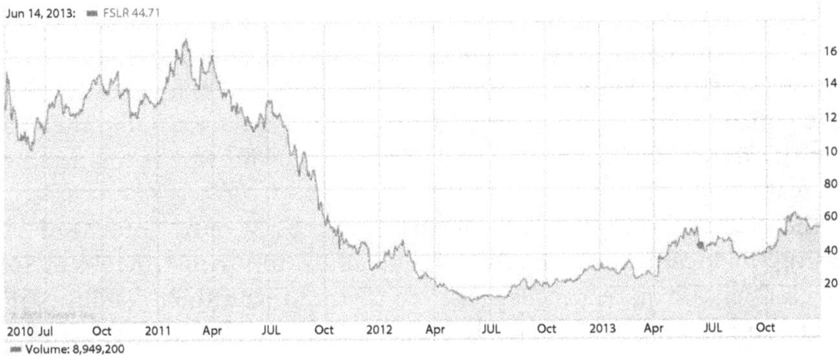

Figure 6.2 First Solar stock chart from the acquisition of Solar Ranch One until the termination of the project's construction phase in late 2013.
Finance.yahoo. "First Solar, Inc. (FSLR)." Undated.
<http://finance.yahoo.com/echarts?s=FSLR+Interactive#symbol=FSLR;range=1d>
Accessed May 28, 2014.

between April and June 2012.[73] In May 2012 First Solar announced an extensive round of layoffs, two weeks after company executives had told Congress the solar developer was "financially strong."[74]

Interview material:

A lot of us surmised that these [furloughs of local workers at the Solar Ranch One site occurring on multiple occasions] were First Solar's leverage in its negotiations with the County… [Workers] had been laid off because of some kind of political finagling that was going on between the Supervisor, using the levers of county government, and First Solar Executives. So, they were playing hardball behind closed doors. They said it had to do with not complying with electrical standards for these kinds of [utility scale PV] installations in the United States. There was something else going on…

Engineering personnel out there [in the Solar Ranch One site] has changed.[75] Someone from First Solar struck a person from planning after a dispute, so Solar Ranch One managers moved him out of there as soon as possible… He probably just blew it. You know… the stress from having to deal with community affairs while trying to meet construction deadlines

---

[73] Ibid.

[74] Stiles, Andrew. "First Solar Furlough." Freebeacon.com, May 30, 2012.
<http://freebeacon.com/national-security/first-solar-furlough/>
Accessed May 28, 2014.

[75] For example, in January 2012, First Solar replaced Solar Ranch One's project manager.

got to him and he just struck the inspector. The dispute between First Solar and County was over the panels but its repercussions for everyone involved were more far reaching than the technicalities of a certification standard. [The panels] weren't UL approved for a 1,000 volt system... So, they have those kinds of problems because permitting in downtown LA was a straight out scam; or a hoax—probably both...

The electrical installations controversy actually came to a head over taming the local town councils... They had to stop [construction at Solar Ranch One] for quite a while. Then, to catch up with their deadlines they had to hire all kinds of people; they were asking for all kinds of extensions. They wanted to work around the clock...It was... what local community? Which town councils? We have to get this thing built no matter what.

But the electrical certification "dispute" goes back to the working condition violations [at the Solar Ranch One project site]. [The violations] are very blatant. They are very basic. You don't give breaks in the afternoon? No, we don't give breaks in the afternoon. Oh, okay. And let's see, we're required to be there and service our equipment and everything, and be ready to work at 7:00, but we do this at all time. Okay, great. Do you pay overtime for that? Oh, by the way, no we don't. Let's see. You need us to service your vehicle in the evening time, okay—Do we get paid for it? No, we don't get paid for that. You have people out there working without any dust masks?... We are talking about several state and federal law violations, and also paycheck work violations, which are very basic...

If you talk to other members of the workforce they will too describe the job as neck-down. Even if you are highly familiar with that area, even if you have been working in the electronics field for years and years, you are not allowed to make a suggestion. Take the panel certification issue, for example. You are not allowed to make a comment on there being a better way of doing things. You are not supposed to think. You are just supposed to do what you are told to do by their engineers, even if your experience tells you that that is the dumbest thing possible... They [Solar Ranch One managers] don't want to hear it. We've had crews here that have worked ten days in a row...

Setting:

First Solar's director of Sustainable Development, Parikhit Sinha, at a background briefing for WAV residents about CdTe based thin-film solar PV panels used in Solar Ranch One. March 6, 2012, William J. Fox Airport in Lancaster, California, fifteen miles away from the Solar Ranch One project site:

CdTe is still a potentially toxic material. But the way it is encapsulated in glass in First Solar panels, Sinha said, sharply limits the possibility of inhalation or ingestion exposure... It would take a very rare and daunting set of circumstances to both free the CdTe and release cadmium.... [Moreover,] [i]n keeping with SVTC [Silicon Valley Toxics Coalition]-set aims, First Solar has, Sinha indicated, a pre-funded end-of-life take-back and recycling program.... But Antelope Valley residents, who know their desert to be a rare and daunting place, were not convinced even when Sinha talked about the accelerated life cycle testing the panels are subjected to and the international certifications they have earned. Speaking for those who will have to live with the panels, a local noted that First Solar has been in business less than half the 25-year term of the panels' warranty. The accelerated life cycle testing, another pointed out, covered only ten years. And First Solar's real-world experience with the panels, he added, is even less than that...Others seemed convinced. "Dr. Sinha went a long way toward allaying my fears," a councilman said. He added that he now feels "much more confident" that the community "can live with this technology."[76]

For most WAV residents, however, questions regarding the compatibility of human lives with utility-scale, thin-film solar PV systems "should have all been asked before [First Solar] ever turned a shovel of dirt."[77] Much like the corporate language of "engagement," then, the rhetoric of life cycle analysis only temporarily camouflaged the physical and psychological realities for rural Los Angeles County residents who experienced firsthand between 2011 and 2013 the construction of Solar Ranch One in the WAV:

> And then, one day we had their sustainability guy [Sinha] paying us a visit here in the Valley. Oh that was fun! We are not dumb you know...I know what a life cycle assessment is. It's the reason why these PV subsidy suckers exist in the first place... He [Sinha] threw us a bunch of numbers and spoke about efficiency... [In that regard,] [y]ou should have seen that engineer's look when I asked him what his numbers said about things beyond emissions to the environment...But let's talk about trust: What I feel is what I see... And what I see everyday is millions of CdTe panels—me and my community are dealing with uncertain

[76] Trabish, Herman K. "The Lowdown on the Safety of First Solar's CdTe Thin Film." *Greentech media*, March 19, 2012. <http://www.greentechmedia.com/articles/read/how-safe-is-first-solars-cdte-thin-film> Accessed May 22, 2014.

[77] Kings Canyon resident quoted in Hedlund, Patric. "Company Says Solar Panels of Cadmium-Telluride Harmless." *The Mountain Enterprise*, March 9, 2012.

> technology risks...He also spoke about panel recycling and how socially responsible their company is—yeah right, we learned the other day that First Solar discontinued their recycling program in the US...[78]

Even as they praised their panels' efficiency, RE engineers integrated life cycle metrics discourse with their health-risk claims about thin-film PV modules. However, such claims have not gone unchallenged within the solar PV industry. Quite the contrary: "efficiency is not the key," a former VP of Sharp Electronics Corporation, a First Solar competitor, said in a 2010 interview, adding that his company has "a premium for not being carcinogenic."[79] To that effect, engineer and PV LCA analyst Dustin Mulvaney recently concluded that "it is well within reason" to consider CdTe thin-film PV technologies "as posing cadmium pollution risks and exposures at various points in the product life cycle."[80]

Moreover, the argument commonly advanced by thin-film PV manufacturers, according to which converting Cd—a by-product of refining copper and zinc—into less toxic CdTe panels to make "clean energy" represents sustainability engineering *par excellence*, can end up reinforcing grievances.[81] In other words, the production of life cycle metrics in some cases can obscure questions of justice, partly because LCAs quantify environmental performance in a way that masks ecological and social effects other than pollution and climate emissions. LCAs may ignore questions of distributive justice such as the displacement of other electricity generating technologies.

Similar to the requirements of the loan guarantee program, the role of LCAs in mitigating the toxicity controversy associated with thin-film PV panels was about plugging in "a bunch of numbers" on environmental emissions. LCAs thus facilitated disembodied and context-less decision making. First Solar—once the sole PV panel manufacturer that compelled its clients

---

[78] Interview data.

[79] For example, Advanced Energy and Solar Frontier are two thin-film PV manufacturers that have phased out cadmium in their products' buffer layer. Wesoff, Eric. "Sharp Solar and Solar Frontier Ready to Reclaim Japan's Leadership in Solar." *Greentech media*, October 19, 2010.
<http://www.greentechmedia.com/articles/read/Sharp-Solar-and-Solar-Frontier-Ready-to-Reclaim-Japans-Leadership-in-Solar>
Accessed May 27, 2014.

[80] Mulvaney (2014).

[81] Pvthin.org. "Sustainability. Sustainable energy from beginning to end." Undated.
<http://www.pvthin.org/sustainability-new/>
Accessed May 27, 2014.
Participant observation.

to recycle their end-of-life modules—recently discontinued its recycling arrangements everywhere but in European countries where solar PV recycling is mandated by law. "We were told [by First Solar engineers] that the so-called accelerated life cycle testing of their panels covered about 10 to 12 years," one interviewee said. "Yet ironically, those engineers' allegedly real-life experience with those panels is less than 10 years".[82] And like Los Angeles County and First Solar's political manipulations during the construction phase of Solar Ranch One, the life cycle testing performed on First Solar panels was perceived as being subject to diffusion of social responsibility.[83]

As plans for RE project development in the WAV proliferated, the disconnections between expert-produced life cycle knowledge and the knowledge contained within local community networks increased.

Mistrust and hostility: Perceptions of Engineering and Community Engagement in RE project development in the WAV

> So, I don't have a target on my shirt, but you are welcome to throw whatever you want at me[,] [said one project manager as he was introducing himself to the attendants of a Fairmont Town Council meeting]. I hope it's Guinness because it is Saint Patty's Day.[84]
>
> We ate that [community liaison person] and spit [them] out for lunchmeat. We had [them] scared running.[85]
>
> What is your framework? What can you offer [to us] for making such a huge change in our living and natural environment?
>
> We bring green energy and jobs—I cannot offer or promise anything else.[86]

[82] Interview data.

[83] According to the Silicon Valley Toxics Coalition, "First Solar, for years the only major company with a fully funded EPR program, has eliminated its EPR program for most of the volume of its sales." See Silicon Valley Toxics Coalition (SVTC). "2013 SOLAR SCORECARD." 2013.
< http://www.solarscorecard.com/2013/2013-SVTC-Solar-Scorecard.pdf>
Accessed May 25, 2014.

[84] Nat Parker speaking at the Fairmont Town Council on March 17, 2011, author's transcript.

[85] Interview data.

[86] March 20, 2013, Antelope Acres TC meeting: A Town Council member persistently asks SunPower's "community liaison" about the company's "framework to engage" local groups, participant observation.

Things got off on the wrong foot between long-term WAV residents and RE engineers when First Solar executives placed the inevitability of Solar Ranch One at the center of their community engagement policies. Over the course of a turbulent Fairmont Town Council meeting in the summer of 2011, First Solar development director Jack Pigott "first looked aloof and then slightly ill," as a local reporter captured his facial expressions, when the developer realized that neither was his facility going to be an easy sale, nor could he slide Solar Ranch One through with his characteristic, "we've got our permits," style.[87] What the directors of the foremost US solar company had in mind when they assigned the management of one of LGP's showcase projects, $680 million, to Jack Pigott remains a mystery, but what the developer's approach did to the perception of RE engineering in the WAV does not. Pigott's work attitude—"arrogant and dictatorial"—captures the tone of community "engagement" in an RE tale of illegitimacy, mistrust and hostility.[88]

Setting an "Engagement" Precedent, Breaking with Transparency

Like their proclamations about openness and law abidingness in planning Solar Ranch One, RE engineers' language and attitude made a universal rhetorical and attitudinal standard emblematic which, nevertheless, lacked meaningfulness and substance for their audience. To RE developers, presenting their projects in rural Town Council meetings in the WAV served as both a solution to checking out "local engagement" and a

---

[87] Hedlund, Patric. "First Solar Offers Jobs to Neenach Area—From Fire to 'Hire.'" *The Mountain Enterprise*, July 15, 2011. A very similar incident took place in Durham, Ontario, Canada in 2010 when after they had been "humiliated, embarrassed and… dysfunctional" a group of NextEra representatives were prompted by a 400 audience to "leave town." According to one anti-wind activist, "[w]hen one lady started to ask some tough questions (you know, the same ones that are never answered or are skirted around over and over regardless of company or area), the NextEra spokesperson looked ever so sad and said "It's very intimidating to be in front of this crowd." The woman asking the question didn't miss a beat! She replied clearly and sternly, (I'm paraphrasing): YOU are coming to my community to destroy my life, my health and my home. I'm not concerned about YOU being intimidated."
Unknown. "Nextera told by 400 at meeting to Leave Town!" Ontario-wind-resistance.org, February 5, 2010.
<http://ontario-wind-resistance.org/2010/02/05/nextera-told-by-400-at-meeting-to-leave-town/>
Accessed May 26, 2014.

[88] Interview data.

mechanism to broker the encounter between an allegedly inherently sustainable technology and its social context:

> The engineer told us that [the palliatives used at the Solar Ranch One site] are environmentally sustainable—but he did not answer my question of what are they. "Don't worry they are qualified along the life cycle'"—that's not what I asked you…So, it just seems like [developers] are doing their dog and pony show…[;] but there is honestly absolutely nothing meaningful going on, because they do simply what they're formally mandated to do… Their traveling circus slide show's predominant feature is that permitting and development processes are transparent. And that we are privileged because green energy is going to create a lot of jobs and reduce emissions. But this is a joke…[89]

The break with transparency associated with Solar Ranch One was no less sharp with regard to First Solar's stance of avoiding recorded sessions with locals. During the first months of Solar Ranch One developers' presence in the Valley, the company's VP James Woodruff demonstrated his unwillingness to open up to public debate the roots of contention between Solar Ranch One and local communities: "We do not…believe that it is in anyone's interest," he wrote in a letter to the Fairmont Town Council members in October 2011, "for the media to participate in these meetings or for the meetings to be recorded in any manner."[90] One local resident's remark echoed the Solar Ranch One managers' approach: "[the] problem with dealing with Woodruff is his huge talent for…[turning]…the profound, heartfelt and dedicated commitment [into] absolutely nothing."[91]

The principles of transparency were invoked to help frame both developers' rhetorical positioning in the WAV, and long-term residents' polemics against RE corporations and planning authorities. In the wake of publicized incidents involving private negotiations between Town Council members and corporate managers, a leading member of the Three Points-Liebre Mountain Town Council, Suzan Zahnter, observed in her January 2012 letter addressing Los Angeles County's DRP: "Please require transparency and public availability of all information with regard to offers

---

[89] Interview data.

[90] "First Solar control of the Press, questions for 11/2/11 meeting agenda." avhidesert, November 1, 2011.
<http://www.avhidesert.com/forum/showthread.php?tid=2059>
Accessed May 22, 2014.

[91] Currier, Craig. "Locals battling windmills." Antelope Valley Press, November 8, 2011.

of money or 'community benefits packages' in return for support of" RE facilities."[92]

Indeed, most concerns regarding RE projects and developers expressed by WAV residents suggest petitions for greater transparency in another guise: "It would be easy to conclude," reads another letter to the County's planning department regarding the West Antelope Solar Project developed by TUUSO Energy in collaboration with Canadian Solar, "that the thirty-day review period without easy public access to documents equates to obfuscation[;] I was personally supplied with a few documents from public entities, but only when I asked, so, no other member of the public has had access to this information."[93] The same approach that brought developers close to "transparency" marked the beginning of contention, according to WAV residents' testimony and publicly available sources: the origin was private meetings between Town Council officials, and as a result of such meetings, a series of accusations pertaining to conflicts of interest.[94]

As the last section analyzed, the attenuation of legitimacy in solar PV project permitting determined in part the fate of wind facilities in Los Angeles County. By pursuing a distorted notion of "community engagement," this section argues, developers operating in the WAV helped construe the notion of engineering as being the authoritative organization of unfair siting—in other words, the engineering aspect of project development displayed contempt or animosity towards local expertise and, as a consequence, toward procedural justice.

---

[92] Zahnter (2012).

[93] Zahnter, Susan. "RE: West Antelope Solar Project, R2012-01589-(5) / CUP 201200086/ ENV 201200158." November 20, 2013, courtesy of Susan Zahnter. For an indicative list of community concerns see Appendix. 2.

[94] First Solar's vice president James Woodruff was reportedly asking that a private meeting be held between his company's representatives and members of the Fairmont Town Council, with Fairmont residents suggesting that "a public meeting is a good idea." Hedlund, Patric. "Fairmont Town Council's Solar Merry-Go-Round Ride." *The Mountain Enterprise*, June 24, 2011. In another instance, one town council president accused a member of a neighboring Town Council for planning a private meeting with First Solar: "Regarding your phone call to me earlier today during which you insisted that today's meeting with First Solar be closed, Oso Town Council's residents and I couldn't disagree with you more...So if Fairmont insists on having closed meetings—that's your business... but that's not the way we do things here." (Manuscript provided to the author dated July 14, 2011.)

Conclusion

Savvy Desert Rats versus Arrogant Corporate Suits: Planning Paradoxes and Perceptions of RE Engineering in the WAV

> [A]nd we conducted a poll,...it was across the whole Antelope Valley...[I]t was over 1,000 registered voters, and the results in the poll were overwhelmingly in favor of [our] project.[95]

Some of the most significant tensions due to RE project development in the WAV arose when developers perceived their own work as being inherently beneficial. "Investors and developers think of themselves as the good guys; they believe every project is going to be wonderful, because *it is in the nature* of their green business," bewailed one Valley local. "You could really tell, when their eyes were scanning our faces, that they were actually thinking: 'What's wrong with you ignorant NIMBYs? What is it *exactly* that you can't understand?'"[96]

In this section, I maintain that the assumption of support for renewables among "the general public" has legitimized the choice of permitting

[95] The author was refused access to the company's poll and, thus, could not verify how distance from the proposed facility affected the alleged support, or whether the pollster included residents of the Valley's rural unincorporated areas. Likewise, in February 2012, San Francisco based non-profit organization "The Vote Solar Initiative" released a public opinion poll on what communities in the desert counties of Southern California think about solar–project development. Vote Solar says respondents "overwhelmingly" support solar project development in the desert (75% voted in support of solar projects). However, the study does not account for the proximity of respondents to proposed projects. Finally, according to a July 2012 survey executed by the Public Policy Institute of California, 78% of Californians favored "increasing federal funding to develop wind, solar, and hydrogen technology," whereas 77% supported the state's Renewable Portfolio Standard (RPS). Jackson, Rosalind. "California desert communities say yes to solar, no to climate change." votesolar.org, February 9, 2012.
<http://votesolar.org/2012/02/polls-california-desert-communities-support-solar-development-care-about-climate-change/>
Accessed June 7, 2014.
Baldassare, Mark, Dean Bonner, Sonja Petek, and Jui Shrestha. "Californians & the environment." Public Policy Institute of California, July 2012.
<http://www.ppic.org/content/pubs/survey/S_712MBS.pdf>
Accessed June 6, 2014.

[96] Interview data, emphasis added. Wolsink (1994) notes that in some cases RE proponents have likened NIMBY with a social disease (e.g., "NIMBY syndrome"). Wolsink, Maarten. "Entanglement of interests and motives: Assumptions behind the NIMBY-theory on facility siting." *Urban Studies* 31, no. 6 (1994): 857-866.

services and RE engineers to refrain from consciously engaging with local opposition to solar and wind technologies.[97] Some developers believed that the difference between the perception of the "public" being supportive of RE on the one hand, and individuals or local groups resisting projects on the other, was that people endorse RE only as long as facilities are not located "in their backyard." Investing the idea of NIMBY with connotations of "emotional" or "irrational" rejection, critics assert that NIMBYism is "the first insult that big developers throw at their opponents...[arguing further that NIMBYs are] selfish, short-sighted enemies of progress, prepared to put their narrow interests above those of a wider society."[98] As one Vermont wind project co-owner had it, "It's time for...[projects' opponents] *to grow up and get real on the future*, and the future is renewables."[99]

It may be that the notion of NIMBYism leads to distortion, sacrificing nuance and complexity with the mentality of "getting real on the future." For example, social scientists consider communities' resistance in terms of what they call a "social gap"—a disparity between high levels of public support for RE and high levels of local opposition—arguing that the idea of NIMBYism oversimplifies opposition to RE.[100] This section uses the WAV case study to further verify that the language of NIMBYism is unlikely to lead to a resolution of the complex socio-technical dilemmas at the source of RE opposition.[101]

---

[97] Has "[a]lternative energy lulled...[stakeholders] into a sense of complacency?" wonders one Antelope Valley resident. Hsu (2011).

[98] Australia's Community Engagement Manager, for example, reported in 2011 that local opposition to wind projects is "based on fear and not fact. I guess it's a challenge to face issues that are based on emotion and not reason." Kingsnorth, Paul. "Nimbys are the true democratic heroes." *The New Statesman* 17, (2004): 22-24.

[99] Gram, Dave. "Vermont loves renewable energy, except when it arrives." reformer.com, March 2, 2014.
<http://www.reformer.com/morelocalnews/ci_25257813/
vermont-loves-renewable-energy-except-when-it-arrives>
Accessed June 10, 2014.

[100] Bell, Derek, Tim Gray, and Claire Haggett. "The 'Social Gap' in Wind Farm Siting Decisions: Explanations and Policy Responses." *Environmental Politics* 14, no. 4 (2005): 460-477.

[101] Burningham, Kate. "Using the language of NIMBY: a topic for research, not an activity for researchers." *Local environment* 5, no. 1 (2000): 55–67; Wolsink, Maarten. "Wind Power and the NIMBY-myth: Institutional Capacity and the Limited Significance of Public Support." *Renewable Energy* 21, no. 1 (2000): 49–64. Wolsink, Maarten. "Invalid theory impedes our understanding: A critique on the persistence of the language of NIMBY." *Transactions Institute of British Geographers* 31, no. 1 (2006): 85-91.

Like the opaqueness of Solar Ranch One and Alpine's permitting processes, the language of NIMBY shaped the perceptions of RE engineering in the WAV and was reflected in developers' notions of "community engagement." Even though RE practitioners believed that their activities in the Valley were inherently just in light of the "clean" aspect of RE, this section argues that because local knowledge was assumed to be irrelevant, "community engagement" did not constitute an inherent part of RE engineering.

The theory and WAV practices of community engagement in RE project development derived from different notions of knowledge: planners and developers held a vision of short-term, qualified expertise founded on LCA metrics of GHG emissions, whereas WAV communities premised a locally acquired set of skills and know-how based on a long-term view of the Valley's cultural values and rural desert habitat. Some developers perceived WAV natives as unsophisticated rural citizens whose absentminded NIMBYism provoked unfounded apprehension. Those developers lacked a genuine notion of engagement which regards laypeople as bearers of local learning, and of a form of awareness that non-experts attain in the daily course of their lives:

> We got treated either like kids or desert rats—the suits came in and told us what was going to work *for* us. We felt that they thought we didn't understand and our suggestions were ridiculous...[In other words,] they believed because we are rural... [that] we fell off the turnip truck...[For instance,] we warned them that if they stripped the land we would have major dust issues; now we have sandstorms that are almost as bad as they were fifty years ago... The local weather was another example: When the clouds are in the notch of the Tehachapi Mountains, you better batten down the hatches because wind is coming. Did they ever ask that?... You couldn't talk to somebody about the sunrise or watching the wildlife out there and how these things affect the projects. They wouldn't get the idea...I asked [one of the RE engineers] "what makes you an expert?" and he got up and started going on about all these degrees. They didn't listen to us, so you'll now see all these little pine trees that are dead in the periphery of Solar Ranch One... And we sit there and laugh because we said, "hadn't worked for a hundred years, but go ahead"...Well, guess what, they still got the dunes and the bare land just like everybody else.[102]

This chapter showed that LCAs are interweaved not only with the solar energy systems that RE engineers perceived as the embodiment of

---

[102] Interview data, emphasis in the original.

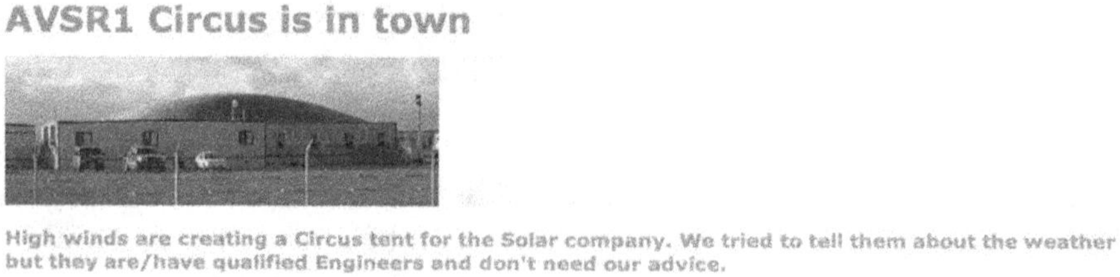

Figure 6.3 "Solar Ranch One Circus is in town." Oblivious to local knowledge, Solar Ranch One developers promulgated a misconstrued model of community engagement which, in turn, influenced how WAV locals perceived RE engineering.
"Solar Ranch One Circus is in town." January 21, 2012.
<http://www.avhidesert.com/forum/showthread.php?tid=2178>
Accessed June 7, 2014.

"sustainability," but engineering experts' assumptions about the local environments and communities in the WAV. For example the tensions between technical expert and lay expert knowledges that swirl around LCAs, the previous section revealed can engender local community opposition. LCAs have the potential, this chapter argues, to warp the vision of sustainable engineers and prevent them from appreciating the complex ecological and social landscapes that they are intruding into. RE engineers used LCAs to legitimize their vision of large-scale solar project deployment, yet in doing so they subverted two of sustainability's most important dimensions: respecting and preserving local natural and social ecologies.

First Solar personnel at Solar Ranch One ignored local knowledge regarding local weather conditions safety standards and the facility's environmental impacts at their peril (Figure 6.3). In the summer of 2011, a few months after it became the model of contested RE permitting, Solar Ranch One emerged as proof of heavy-handed arrogance. On July 2, a non-local surveyor wearing shorts and tennis shoes walked in Solar Ranch One's project site—an area where one commonly finds venomous Mojave green rattlesnakes—without a snakebite kit.[103] Driving across dry brush without a spark suppressor shielding the exhaust portions of the truck from two-and-a-half-foot high dried-out grass, the surveyor set a

---

103 Solar Ranch One's final EIR does not specify whether "soil sampling" had to be performed by a local entity/expert or not; Jack Pigott, however, had promised that the site's jobs would be given to locals. Los Angeles Department of Regional Planning (LADRP). "Findings of fact regarding the final environmental impact report for the AV SOLAR RANCH ONE PROJECT COUNTY PROJECT NO. R2009-02239 VESTING TENTATIVE TRACT MAP NO. TR071035 CONDITIONAL USE PERMIT NO. RCUPT200900026 ENVIRONMENTAL REVIEW NO. RENVT200900027 STATE CLEARINGHOUSE NO. 2009041145." December 2010, 119.

full-scale fire. Within a matter of weeks, RE engineers were facing a sea of problems instead of the envisioned sea of solar modules, as they resented community claims that Solar Ranch One's construction phase be opened up to local input. For instance, one former Solar Ranch One employee had to repeatedly fix First Solar's motor pumps because, being too large, the machines lowered Fairmont's water levels while causing septic issues for the project's neighbors. He recalled, in personal communication, how First Solar "did not figure out and did not care that not everybody out there was hooked up to the city water system...as each of those homes has a well going into a common aquifer." As a result, the developer built four substantial wells with large pumps that could not withstand the speed at which First Solar was extracting the aquifer's water for the site's dust abatement, and consequently burnt out their own well motors. Local knowledge was, in short, dismissed with a "disdainful point of view."[104]

Despite their internecine struggles, one concern that WAV residents shared was the solar PV facilities landscaping that was intended to block the blowing dust and at the same time address the projects' viewscape. At its core, Solar Ranch One landscaping reflected disdainful perspective and poor planning; this time, however, there was a persistent ambivalence caused by a dissonance between Los Angeles County's supposedly therapeutic role of assuaging conflict in the WAV, on the one hand, and its ultimate refusal to actually settle RE disputes, on the other.

Around the time that representatives of the Greater Antelope Valley Economic Alliance (GAVEA) were celebrating Solar Ranch One's owner company for having set "the standard for community participation," the project's August 2010 final EIR articulated a solution for installing an area of Joshua trees and other native vegetation along the facility's fence lines.[105] The EIR also premised that the landscaping plan would be completed before solar workers were expected to break ground in Fairmont.[106] Ironically, in linking local flora to Solar Ranch One, County planners failed to reveal the bigger picture of Solar Ranch One's development life cycle; they thought that indigenous plants covering the project's perimeter would act as windbreaks and habitat connectors. As rural residents belatedly became familiar with the EIR for the 230-MW solar PV project, locals contested many of the document's presumptions: a fairly technical debate resulted between RE engineers, planners and community, set

---

[104] Interview data.

[105] LADRP (2010). According to the EIR's proposal, the developer was going to intersperse pine trees, transplant Joshuas, use some conifers and other local vegetation.

[106] LADRP (2010).

up on desert landscaping grounds. "You could not imagine or create as a fantasy something more silly than putting up some Joshua Trees," one interviewee bitterly asserted, illustrating the paradoxical elements of the proposed viewscape conception: a large percentage of Joshua trees die when transplanted, with most taking approximately sixty years to mature, implying that the landscaping plan's lifetime was twice as long as that of the solar project's. Blatantly contradicting the local economic alliance, the planning department further confirmed its inability to work with locally available information: it relied upon the assumption that seeds for trees and grasses indigenous to the WAV would be provided by the Antelope Valley Resource Conservation District (AVRCD). But the AVRCD, having never been engaged by the developers, could not fulfill that expectation.

The "community engagement" formulations of RE proponents obscure the fact that the foundation of developer policy was the depreciation of local knowledge and procedural justice (Table 6.2). In the summer of 2012, the Solar Ranch One landscaping conflict finally became critical as County, developers and local community members were summoned to set up a viable plan, and First Solar's Jim Woodruff repeated Pigott's "I've got my permits" legacy. With the silent approval of County representatives that were supposedly facilitating settlement between Town Councils and RE companies, Woodruff rejected Oso and Fairmont Town Council members' suggestions.[107] While the perception is that the facilitator for "community engagement" in RE project development in the WAV is the Los Angeles County department of regional planning, that facilitation was undermined, paradoxically, by the department's own efforts to sustain the false prerogatives of an inadequate and non-participatory permitting procedure.[108] Thus participatory design of RE projects became WAV residents' paradoxical problem as some locals' perception of community engagement is riddled with mistrust and suspiciousness: "You want the engagement…but if you tell [engineers] how to control the impacts to the community…its like giving them the key to the kingdom."[109]

All the while, WAV residents' sense of attenuated participatory experience in mitigating the dust emanating from Fairmont's two graded solar PV

---

[107] To use Middle East pomegranates and Afgani pine trees, respectively.

[108] "I spoke with [one Los Angeles County zoning department staff person], one interview reported, and asked him: 'if the corporation agrees to it and the community agrees to it, would it be possible to get [an operational] landscaping plan?' And [they] said, no, that's already done." According to the interviewee the planner "brought the issue of water and the fact that they had already settled that it's going to be Joshua trees."

[109] Interview data.

Table 6.2 RE developers' notions and the resulting paradox of "community engagement".

Community engagement *means*: offering construction jobs, having a good communication strategy, organizing obligatory "open houses," hiring "community liaisons" to attend Town Council meetings;
Community engagement *is viewed as*: potentially creating a barrier to moving along with construction or securing permits for a facility;
The goal of community engagement is: securing a Conditional Use Permit (CUP), convincing stakeholders to write a letter of support, making as few changes in the project's design/RE company's construction plans as possible;
The paradox of community engagement in the WAV: even if local knowledge is required to mitigate RE projects, sharing that knowledge with engineers arouses contention.

project sites went hand-in-hand with the need to designate a culprit—the technicians—who were arrogant about their expertise, founded as it was in specialized engineering education (Figure 6.4). Solar Ranch One itself, in the words of a rural resident, reflected an "uncaring, selfish and single-minded technician; green as the almighty dollar." First Solar executives did face up, at least rhetorically, to the question of how local knowledge might be used as an authoritative source for solar PV project mitigation measures. "There is no knowledge like local knowledge," exclaimed the company's director of Environment and Safety during a Town Council meeting after the July 2011 fire.[110] On the other hand, the developers continued to underestimate how negation of the local expertise alienated rural Antelope Valley communities. For instance, engineers' assertions that dust problems in Fairmont predate the advent of RE companies in the Valley are partly true. Yet, they are also partly wrong: their view downplays photographic evidence (Figure 6.5) that clearly depicts the source of the airborne dust to be the Solar Ranch One and Alpine project sites. This chapter provides evidence that local community members could have helped design the solar facilities in the WAV to have a lesser environmental impact (although this does not address the question of whether they should be supportive of projects that are actually damaging).

[110] Trabish, Herman K. "Was the First Solar Fire at Solar Ranch One a Blessing in Disguise?" *Greentech media*, July 11, 2011. <http://www.greentechmedia.com/articles/read/was-the-first-solar-fire-at-solar-ranch-one-a-blessing-in-disguise> Accessed May 7, 2014.

Figure 6.4 A representation of RE engineering steeped in irony; First Solar accused of having "EIN-GIN-EERS that have edumacation."
"Saturday Feb 23 Solar Ranch One Dust storm RADICAL." avhidesert, February 23, 2013. http://www.avhidesert.com/forum/showthread.php?tid=2650
Accessed June 16, 2014.

Figure 6.5 A view of dust storms emanating from the Solar Ranch One entrance, March 2013. Photo, courtesy of Robert Kerekes.

In the summer of 2013, shortly following the injury of six WAV residents in dust-related traffic accidents and twenty-eight reported cases of solar workers coming down with valley fever, Los Angeles County belatedly sprang into action, two years after the soil grading controversy between First Solar and Valley locals had begun at Solar Ranch One.[111] On May 14, 2013, Supervisor Antonovich made a motion to instruct the directors of Planning, Public Works, and Public Health departments to explore potential code amendments to require RE companies to employ alternatives to mass scraping and the earth stabilizers used by developers in the Valley. Yet Los

[111] Valley fever or cocci is the disease also known as coccidioidomycosis caused by inhaling the spores of coccidioides immitis—a fungus found in desert soil. According to locals' testimonies, the cases of valley fever in the WAV have increased by over 500%.

Angeles County's measures against fugitive dust were generalized, abstract, and long delayed, in much the same way as were the DRECP drafts, which were supposed to set the regulatory standard for RE project development in the southwestern US. Speaking at a Town Council meeting in Antelope Acres in May 2013, Vickie Rausch, the lead specialist with the Antelope Valley Air Quality Management District (AQMD), admitted that RE project permitting had implicitly entailed an accommodation to fugitive dust originating from Solar Ranch One and Alpine, as planners' main concentration had been on calculating the particulate matter in diesel fuel attributed to the facilities' construction phases.[112]

As in Solar Ranch One's landscaping controversy, during the dust mitigation debate in the Valley, developers, planners and locals argued ineffectively about hydroseeding and hydromulching until the County rejected the locally proposed solutions to control fugitive dust particles.[113] Similar to Solar Ranch One's EIR that prescribed unsubstantiated action regarding local vegetation, First Solar's insistence on using rye—a winter crop—to perform hydroseeding in the summer exemplifies an assault on the legitimacy of RE permitting procedures. And like the LGP's limited notion of LCA metrics, the planning focus on authorizing Solar Ranch One by calculating vehicle emissions—while neglecting to account for fugitive dust policies—led to a diffused and unfocused perspective. The use of "green metrics", then, can—seemingly paradoxically—lead to greater mistrust and rejection by local peoples. But, as this chapter explained, it is not surprising or paradoxical—it is embedded into the nature of LCAs themselves and the practices of engineers in using them to cast expert visions of sustainable energy futures. Overall, by limiting the scope of local expertise during the planning and permitting phases of utility-scale solar PV projects, Los Angeles County and RE developers set the stage for more emerging paradoxes during the construction and operation phases of the facility.

---

[112] Antelope Acres Town Council meeting May 15, 2013.

[113] Hydromulch/Hydroseedconsists of the mixing fiber mulch, grass seed, fertilizer and/or other agriculture approved additives (one is a tactifier or bonding agent) to water. This mix is placed in a machine to form a homogeneous slurry. The slurry is sprayed under pressure to form a uniform application over the soil. Hydromulching is, a term used by many contractrators in the Southern parts of the US, when this slurry is 35 lbs. of mulch material or more per 1000 SF (1500 lbs. per acre). Anything less is called Hydroseeding in Southern US. Hydroseeding is used in the North for all thickness of Hydroplanting. Hydroplanting is a new term to represent all hydrolic planting. The solutions included using decomposed granite or woodchips.

7

Through an Ethnographic Lens: Local-Regional Politics of Utility Solar and Wind Siting in California

The Cinderella of Utility-scale Renewables

> The story of Cinderella is a well known fairy tale. An East of West rendition is presented for your entertainment and enlightenment. The role of the evil step mother and sisters are played by the Los Angeles politicians. The prince is played by the solar and wind corporations. One day…[the prince] stumbled upon the far west Antelope Valley which became a top ranking CREZ (Crank out solar and wind projects here Real EZ) location. The proverbial slipper fit our region perfectly.[1]

[1] Tuszynski, Jack. "Cinderella of the Far West Antelope Valley." July 7, 2011. Courtesy of Jack Tuszynski.

An unparalleled, utility-scale RE experiment took place in California's Western Antelope Valley. Mushrooming wind farms featuring enormous numbers of the newest wind turbines along the ridges of Kern County's Tehachapi Mountains have blazed the way, while more than 1,000 megawatts of solar energy have been built or are currently under planning or permitting processes in the area. Under construction since February, 2013, MidAmerican's Antelope Valley Solar Projects (AVSP)—the world's largest PV facility with 579 MW of combined solar capacity, developed by SunPower Corporation—encompasses 3,230 acres of privately owned land, and spans two counties (Kern and LA County) in the Mojave Desert. California's first utility-scale PV projects, Exelon's Solar Ranch One and NRG Energy's Alpine Solar Project (Alpine)—a 66 MW facility built by First Solar—came online in 2013. Element Power, NextEra Energy, TUUSSO Energy as well as Silverado Power and Beautiful Earth have engineered solar and wind projects in the region (Table 7.1).

In this chapter I focus my ethnographic lens on the local politics of such an unprecedented, utility-scale, solar and wind siting experiment in the California desert. I discuss how informal rural "Town Councils" in the WAV ended up disguising a pervasive lack of accountability and communal transparency; how local residents and groups fought with each other over their welcome of RE projects; the strategies of solar and wind companies in dividing and ruling as well as in manipulating local institutions through money; and the egregious acts of Los Angeles County leadership. This chapter argues that technical LCA of RE is often inadequate in capturing the finely granular socio-political and ecological conditions of project development. Two connections between LCAs and RE project development in the WAV, then, will become evident from this chapter: one that LCAs do not recognize such complex politics and therefore do not prepare firms for potential conflicts; and, two, that LCAs in no way guarantee that RE projects proceed democratically and fairly. In this chapter I use ethnographic data to draw contrast between the fates of utility-scale solar and wind facilities in California—tracing the evolving history of reception and oversight of successive waves of RE projects, particularly in the same area and within a few years. As a whole, this chapter provides evidence for the claim that the story of RE project development in the US has a profound internal rift. Around 2009, solar and wind farms were seen very positively—the development of so-called RE "zones" in the Southwest US was a concrete historical grounding of that positive outlook. Yet the ensuing experience

---

<http://www.avhidesert.com/forum/showthread.php?tid=1823>
Accessed March 1, 2014.

Table 7.1 Indicative list of RE projects in the WAV, multiple sources and interview data.

Project name/status	Owner/developer	Area (acres)/MW
Alpine Solar Project/ Completed	NRG Solar/First Solar	835/ 66 MW of solar PV
Antelope Valley Solar Projects/ Under development	MidAmerican Solar/SunPower Corp.	3,230/ 579 MW of solar PV
AV Solar Ranch1/Completed	Exelon/First Solar	2093/ 230 MW of solar PV
Blue Sky Wind Project/ Cancelled	NextEra Energy	187/ 38 MW of solar PV
TA-High Desert/Completed	TUUSSO Energy LLC. Co- development with Akula Energy Ventures	180/ 20 MW of solar PV
West Antelope Solar Project/Completed	TUUSSO Energy, LLC. Co-development with Canadian Solar (USA) Inc.	216/ 20 MW of solar PV
Western Antelope Blue Sky Ranch A and B/ Completed	Silverado Power	157/ 40 MW of solar PV (20 MW each)
Wildflower Farm/Cancelled	Element Power	4,197/ 300 MW wind/solar hybrid

described in this chapter leads us to be more cautious in building a new socio-technical system founded on RE.

The Fabrication of a "Competitive Renewable Energy Zone"

There is precedent for the WAV's accommodation of RE-induced technological change. This change was made possible by the notion that the Valley was uniquely positioned—due to the area's locational proximity—to deliver its plentiful RE resource to LA and the Central Valley. During the early 1980s, federal and state tax incentives were approved which paved the way for the US's original utility-scale RE projects—the wind facilities in Altamont Pass (Alameda County), Tehachapi Pass (Kern County) and San Gorgonio Pass (Riverside County), respectively. By 1986, according to statistics of the American Wind Energy Association (AWEA), the Altamont Pass alone numbered 6,200 turbines, totaling 583 MW.[2]

Then in the mid-1990s, California ran out of RE transmission infrastructure. The wind industry and Southern California Edison (SCE) spent about twelve years planning the Tehachapi Renewable Energy Transmission Project (TRTP), designed specifically to provide this missing infrastructure. The question arose regarding how an enterprise initially conceived by SCE to accommodate energy produced by the already mature wind industry might be recommissioned for a utility-scale solar PV industrial infant:[3]

> The solar industry in 2007–2008 was like wild catters, they didn't have anybody to speak for them... It was dominated by a few big players like BrightSource [Energy], who had their own [solar thermal] agenda. It took me [the Director of Kern County's Planning Department] literally 128 meetings on solar proposals—on which thirty of them were real. The rest of them...had no idea what they were doing...They would come down here and say, 'where should I put the project'... They didn't understand the permitting process. They didn't understand CEQA [the California Environmental Quality Act of which public participation is an essential part].[4]

Construction of the TRTP began in 2008, with the utility solar PV industry in its infancy at exactly the same time that planning for California's inaugural, large-scale RE projects began to take shape. Concurrently,

---

[2] Interview data.

[3] Kern County, for instance, has people in the wind industry who are third generation.

[4] Director of Community & Planning Development, Kern County Lorelei Oviatt, personal communication.

there were indications of compelling engineering and policy ambition, an imperative for an RE future, exemplified by the invention of the "Tehachapi Wind Resource Area"—a construct that was used in TRTP's Environmental Impact Report (EIR) and was useful for wind investors as a device for legitimizing the new infrastructural regime.

"The desert is a big part of California's renewable energy future," explicated Roger Johnson, the current deputy director of the CEC's Siting Transmission and Environmental Protection Division, in a 2011 interview.[5] Originally instituted in 2002 under Senate Bill 1078 and expanded in April 2011 by Governor Jerry Brown, California's Renewable Portfolio Standard (RPS) is the cornerstone of such a desert-based RE future; it requires that, by 2010, the state receive 33% of its electricity from RE sources like solar and wind.[6] In the WAV, public officials have called the RPS' repercussions for local communities a "mini land rush"—and it has proven to be a relatively short-lived one, particularly as regards large-scale RE projects.[7] In 2012, energy analysts were quoted in the *Los Angeles Business Journal* as saying that most California desert locations upon which utility-size projects could be sited had already been claimed.[8]

Equally important as the TRTP, in December, 2009, through its Renewable Energy Transmission Initiative (RETI), the CEC baptized the Valley's Fairmont area a "competitive renewable energy zone" (CREZ), namely a zone that features both the high quality renewable resources and a great enough resource density to make building transmissions to that area economically worthwhile.[9] Like the fabrication of the "Tehachapi Wind Resource Area," the construction of Fairmont's CREZ sheds light on the

---

[5] Trabish, Herman K. "Is this the new California Gold Rush?" *Greentech media* July 1, 2011. <http://www.greentechmedia.com/articles/read/is-this-the-new-california-gold-rush/> Accessed March 4, 2014.

[6] The first US state renewable energy mandate—Iowa's 1983 Alternate Energy Production law—was signed by Republican governor Terry Branstad and required the state's utilities to buy an average of 100 MW of wind electricity. The law, nevertheless, was not enacted until 1997.
Institute for Local Self-Reliance. "Renewable Energy Mandate—Iowa." January 20, 2009. <http://www.ilsr.org/rule/renewable-portfolio-standards/2564-2/>
Accessed March 5, 2014.

[7] Interview data.

[8] Koren (2012).

[9] Pletka, Ryan. "RETI Phase 2 Update Workgroup." December 10, 2009. See also, the Renewable Energy Transmission Initiative (RETI) Phase 2B Maps found in the California Energy Commission website:
<http://www.energy.ca.gov/reti/documents/>
Accessed April 15, 2014.

California Energy Commission's role in authorizing favorable criteria for RE project development in the WAV, such as physical location and access to the state's populous urban centers' electricity markets via an existing and developing high-voltage transmission infrastructure.[10] Similarly, CREZ implements the federal government and the LGP's vision to encourage

---

In Element Power former project manager Nat Parker's words, RE developers began flocking to the WAV because it is "a place where the trees grow sideways and the sun burns bright." Quoted in Hedlund, Patric. "'Where the Trees Grow Sideways and the Sun Burns Bright.'" *The Mountain Enterprise*, April 29, 2011. Indeed, according to NREL, the Antelope Valley, comprising the Western tip of the Mojave Desert is home to one of the highest quality insolation (solar radiation) in the United States.
Roberts, Billy J. "Photovoltaic Solar Resource of the United States." September 19, 2012. <http://www.nrel.gov/gis/images/eere_pv/national_photovoltaic_2012-01.jpg> Accessed April 20, 2014.
In addition, the WAV is located in the nexus of the San Gabriels and the Tehachapis where the winds happen to peak in the late afternoon, which is when the power use in Los Angeles is the greatest. This explains why utilities, in particular, were interested in the Valley.
The CREZ's must not be confounded with the Department of Interior's Solar Energy Development Program (SEDP's) policy that delineated "solar energy zones" throughout the southwestern United States. CREZ's were originally defined as a "group of projects that when combined has improved economics." Black and Veatch. "Renewable Energy Transmission Initiative Phase 1A Work Group Meeting 6." March 6, 2008.
[10] In other words, instigated by RPS and RETI, the assimilation of the concept—and, of course, the reality—of utility-scale RE projects in the WAV depended on transmission infrastructure transformations that increasingly began to incorporate a greater load of renewables. One example of such transformations is the California Independent System Operator (CAISO)—the state's electricity management board—analysis of the potential integration of intermittent energy sources like solar and wind into California's electric generation mix. Through what became known as the "Duck Chart," CAISO maintained that the trend to produce electricity at projected large-scale solar projects combined with energy brought about by "conventional" sources such as coal or nuclear would result in over-generation sometime around 2015 and beyond. Since current electricity transmission infrastructure was designed to accommodate the concept of base load generation, CAISO detailed that more solar energy is likely to plunder California's grid supply and demand balance—"one of those 'good problems to have,'" according to Rocky Mountain Institute's analysts. Utilities like Southern California Edison (SCE) reacted to the reliability challenges associated with projected electricity load growth in the WAV by developing the Tehachapi Renewable Transmission Project (TRTP), projected to be completed in 2015. TRTP's northwest direction leads to a Southern California Edison substation called Whirlwind (adjacent to 170th Street West and Rosamond Boulevard) that Solar Ranch One and Alpine Solar were designed to connect to. Smith Owen, and Mathias Bell. "Renewables' Bird Problem. It's not what you think. Meet the Duck Chart." Blog.rmi.org, October 29, 2013.
<http://blog.rmi.org/blog_2013_10_29_renewables_bird_problem>
Accessed April 20, 2014.

positive investor sentiment for utility solar and increased competition amongst the country's biggest developers.[11]

Part of the reason why the CEC Fairmont Parcelization Map (Figure 7.1) shows such a sprawling purple splotch for Fairmont is because of its particular socio-geographical characteristics: it featured a less dense population and was an unincorporated rural area. A sparsely populated locality, according to LA County's Department of Regional Planning (DRP) the Antelope Valley consists of "a wide variety of very low density, rural villages… [whose] residents express a sense of community pride and local identity."[12] (Figure 7.2 and Table 7.2). One longtime RE project developer seemed to confirm the notion that WAV communities are notably identifiable; but simultaneously, WAV resistance was associated with a rural variant of the "Not in My Backyard" (NIMBY) syndrome, and, developers argued, both uniqueness and NIMBY attitudes had to be sacrificed in the drive for RE expansion:

> There's a general psychological and sort of societal identification in this valley as being a very different place than what they [locals] refer

---

[11] Political leaders add to and further legitimize the mix of favorable conditions for RE project development in the area. Of all public administrators intimately engaged with promoting renewables in Antelope Valley, LA County Supervisor Michael (Mike) D. Antonovich conveyed in 2011 the sense of an imaginable cultural advancement which all involved constituencies can and must participate and which, as a prototype, might serve the needs of RE-oriented communities more broadly. The Antelope Valley, he argued, "has the potential to become the nation's leader in green, alternative energy innovation and production." This standard of green leadership was also applied to Kern County: Our economy, "is a star in California, competing with Silicon Valley for high marks in recovery" boldly announced Mike Maggard, chair for the Kern County Board of Supervisors in January 2013. See Simmons, Ann M. "Antelope Valley residents not fired up over green energy projects." *Los Angeles Times*, July 4, 2011.
<http://articles.latimes.com/2011/jul/04/local/la-me-wind-turbines-20110704>
Accessed March 18, 2014.
Hedlund, Patric. "Black gold and renewable energy make Kern County a recovery star." *The Mountain Enterprise*, February 8, 2013.
Referring to Beautiful Earth's Del Sur Project, Lancaster Mayor Rex Parris—leader of the first US city, since January 2014, to have made residential solar mandatory—has said it could bring his jurisdiction "one step closer toward the goal of becoming the alternative energy capital of the world." Unknown. "Beautiful Earth Group breaks ground on large California solar plant." CleanEnergyAuthority.com, January 3, 2011.
<http://www.cleanenergyauthority.com/solar-energy-news/
beautiful-earth-group-breaks-ground-on-large-california-solar-plant-010310/>
Accessed March 23, 2014.

[12] Antelope Valley Area Plan (AVAP), Community Recognition, Section III-1.

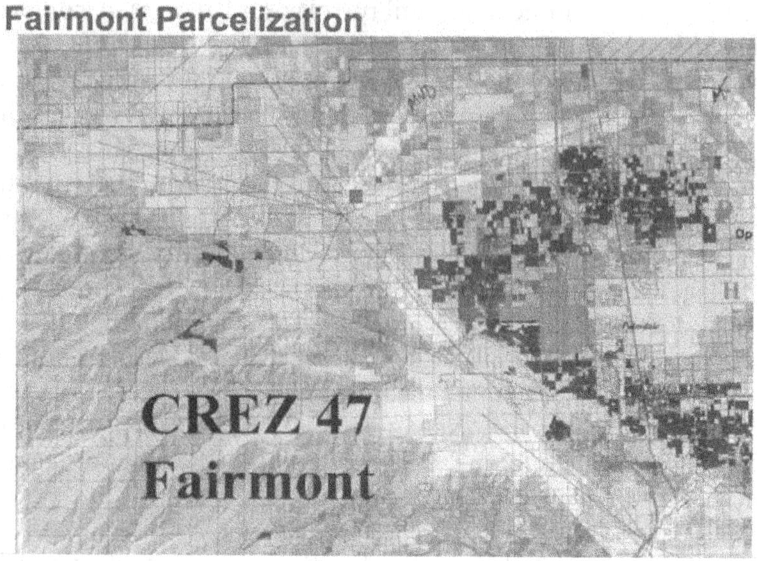

Figure 7.1 "Fairmont Parcelization," adapted from Pletka, Ryan. "RETI Phase 2 Update Workgroup," December 10, 2009.

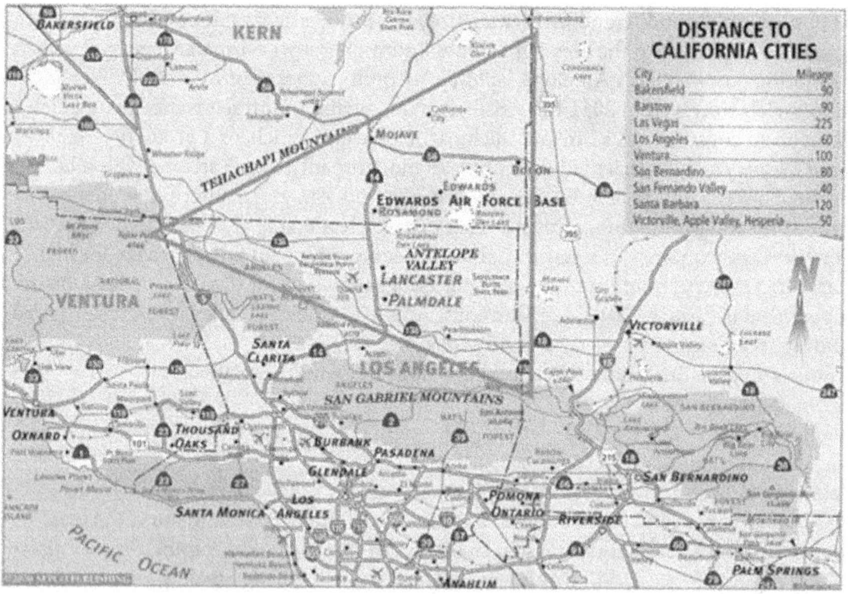

Figure 7.2 Antelope Valley Map.
Senn, Evan. "Antelope Valley Art Outpost: Building Community in the High Desert." KCET.org, March 13, 2015.
<http://www.kcet.org/arts/artbound/counties/los-angeles/antelope-valley-art-outpost-high-desert.html> Accessed May 5, 2015.

> to as down in the basin of the LA metropolitan area. [Locals] also have something of a rugged, individualistic, western identity—it's a predominant view of the NIMBY movement. But RE development is happening here anyway, so some of them will stay and enjoy, the others will have to go…[13]

It was thus tempting to associate the high RE potential, the low parcelization, and the abandoned agriculture lands in the Valley with the notion that its residents were also expendable. The RE fate of the WAV, as discussed in the next section, interestingly contrasts with the story of Riverside, California, which, despite its similar RE "zone" status, is an agricultural area and is more densely populated than Fairmont.

The links between RE project development in the WAV and environmental/social justice became apparent when Exelon's Solar Ranch One was discussed in the 2011 Department of Energy (DOE) Environmental Assessment (EA). After quoting the US Census Bureau, which described the Valley as accommodating "urban expansion needs for Southern California," the study asserts that the Solar Ranch One area is itself "planned for substantial population growth."[14] Integrating environmental justice into the project's evaluation, the report continued, requires assessing whether "the potentially affected community includes minority and/or low income populations." If poverty and minority rates calculated near the project area are lower than those of LA County and if they do not exceed 50%, they constitute—according to the DOE—grounds for assuming that Solar Ranch One "does not have the potential" for disproportionally affecting communities in the WAV.

RETI's failure to bring legitimacy to the planning of RE infrastructure in the California desert prompted one of the earliest critiques, which expressed a connection between the solar project permitting process in the WAV and a loss of procedural justice: "It appears your self-imposed comment deadline only serves to stifle public participation," lamented a March 2009 letter from the Center for Community Action and Environmental Justice (CCAEJ), addressing the unfolding of RETI's Phase 1.[15] The unfolding trajectory dismayed longtime WAV residents such that

---

[13] Interview data.

[14] DOE (2011): 76.

[15] Center for Community Action and Environmental Justice (CCAE). "RE: Black and Veatch Renewable Energy Transmission Initiative Phase 1A." March 29, 2008. <http://www.energy.ca.gov/reti/steering/workgroups/phase1A/comments/Center_for_Community_Action_and_Environmental_Justice.pdf> Accessed April 14, 2014.

Table 7.2 History, Geography, and demographics of the WAV.

History	The area "was a trade route for Native Americans traveling from Arizona and New Mexico to California's coast. A later exploratory period starting in the 1840s led to the valley's first permanent settlement during the following decade, fueled by California's Gold Rush and new status as American territory." The Antelope Valley suffered from a "drought that began in 1894-the worst in southern California's recorded history… The 1913 completion of the aqueduct spanning 233 miles between the Owens Valley and Los Angeles…revived the valley's economy. Today the Antelope Valley retains elements of its agricultural past but its economic base is now supported by aerospace and defense industries."[1]
Geography	Located in northern Los Angeles County and the southeastern portion of Kern County—between the Tehachapi and the San Gabriel Mountains—the Antelope Valley suggests the westernmost part of the Mojave Desert having an average elevation of approximately 900m. It consists of a series of "closed basins," that do not provide an outlet for its surface streams. Along the southern slope of the Valley runs the San Andreas Fault.[2]
Demographics	The area in close proximity to Solar Ranch One and Apline projects has approximately 20,000 households with a median household income of $68,230. A significant percent of the population is retired: "Most of us that live in the Fairmont area have large parcels of ground and we live in a rather isolated fashion," recounted one resident in personal communication. "We are all pretty much independent. You know, we are not a lot of welfare folks and things like—I mean, there are some, but for the most part, people that live out here have a little bit of income, a little bit of money. We are not struggling, necessarily. My husband and I are both retired."[3]

County of Los Angeles Public Library. "Antelope Valley Community History." Undated.
<http://www.colapublib.org/history/antelopevalley/>
Accessed May 5, 2015.
Michaelsen, Joel. "Mojave Desert Region Physical Geography." Undated.
<http://www.geog.ucsb.edu/~joel/g148_f09/readings/mojave/mojave_desert.html>
Accessed May 5, 2015.
18 California Hometown Locator. "Lancaster, CA 93536 ZIP Code Profile." Undated.
<http://california.hometownlocator.com/zip-codes/data,zipcode,93536.cfm>
Accessed May 5, 2015.
Interview data.

they did not even use the name CREZ 47 to distinguish their locality.[16] The "Fairmont" zone encompassed—in effect green-lighted—all possible locations for land development, inescapably transforming RE corporations into monopolized industries in the Valley.

In March 2011, a project manager working for an RE corporation bluntly conveyed the sense of inevitability that marked the planning of RE facilities in the Valley:

> [Y]our valley is being inundated with companies like my own…But, I don't think it's going to stop, and I'm going to give you an example. If you guys light your torches and you run [my company] out of town. What do I do? What does my company do with the millions of dollars that we sunk into the community, into the project so far? We have to sell it. Well, guess who we sell it to?... We are going to have to flip it to another developer…[17]

By late 2011, longtime WAV residents had adopted such attitudes, and the sense of inevitability marked their response to tactical problems and potential consequences and benefits:

> We [rural residents] have a choice of fighting them as far as we can and then getting as much as we can or fighting them farther than we should and getting absolutely nothing and them coming anyway.[18]

By June 2013, more than thirty RE projects were planned and/or underway in CREZ 47, according to data from CEC's Renewable Energy Action Team (REAT)—a consortium of federal and state agencies that includes the US Bureau of Land Management, California's Department of Fish and Game and the US Fish and Wildlife Service (Figure 7.3).[19] The grants and loans from the government, the tax concessions from the counties, the inevitability conveyed by CREZs. What is currently happening

---

[16] Interview data.

[17] Nat Parker speaking at the Fairmont Town Council on March 17, 2011, author's transcript.

[18] Currier, Craig. "Money, power tangle in west Valley." *Antelope Valley Press*, November 19, 2011.

[19] California Energy Commission. "2013 Renewable Energy Action Team (REAT) Generation Tracking Projects." 2013. <http://www.energy.ca.gov/33by2020/documents/renewable_projects/REAT_Generation_Tracking_Projects_Map.pdf> Accessed March 27, 2014.

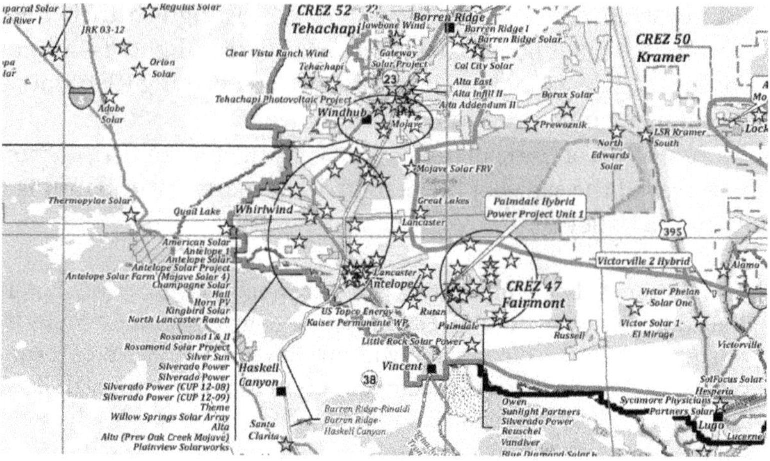

Figure 7.3 More than thirty RE projects are planned and/or underway in "Fairmont's" CREZ 47. Adapted from "2013 Renewable Energy Action Team (REAT) Generation Tracking Projects."[1]

[1]Ibid.

in the WAV resembles a big business redevelopment industry—except it is the Wild West. In fabricating and licensing CREZ 47, federal and state authorities inevitability resorted to the RETI's "top-down" planning process which reverberated in the mindsets of RE developer; meanwhile, as the next sections describe, the contentiousness of unstructured public engagement procedures became tied with the political disempowerment experienced by communities residing in rural, unincorporated areas.

Rushed RE Project Development: Planning After the Fact in the Californian Wild West

In addition to the subtle shift towards RE as an inevitability, another gradual shift occurred in the creation of utility-scale solar facility assemblages in the southwestern US. The key elements were the instigation of RE planning and siting-related regulatory innovation, including providing a land-use framework for the conservation of natural communities in the state of California. The drive for RE regulatory change has sought to incorporate a process of legitimate transfer between federal, state and county entities; the major institutional players have been the REAT, the Significant Ecological Areas Technical Advisory Committee (SEATAC), and LA County's DRP. Each player represents a particular initiative to mediate between RE project development, regulatory principles of California's land use, and mandates of public participation in technology planning: crafting California's

Desert Renewable Energy Conservation Plan (DRECP); extending and reconfiguring Significant Ecological Areas (SEAs) in LA County; and fusing a Renewable Energy Ordinance (REO) into Antelope Valley's Town and Country Plan, respectively.

Authorized by Governor Brown's RPS, DRECP development can be traced to the 2008 establishment of REAT under the aegis of California's Natural Communities Conservation Act and the federal Habitat Conservation Plan section of the Endangered Species Act. With its draft version long overdue and intended to "advance state and federal conservation goals in...[California's] desert regions while also facilitating timely permitting of renewable energy projects [and transmission corridors]," the DRECP has built two important dimensions into the planning of RE engineering projects in the WAV.[20]

First, DRECP may be based on a problematic foundation: Natural Community Conservation Planning (NCCP), whose recorded history attests to continuous efforts towards "fix[ing]...gaping problems" while walking "a fine line between encouraging creative cooperation and being too permissive."[21] An example of such problems is DRECP's non-consideration of the joint EPA and National Renewable Energy Lab (NREL) "RePowering America's Land" program—a database of more than fifteen million acres of private and public land that suggest either Superfund or other "brownfield" sites and which could imaginably host RE projects. Second, while the DRECP leadership encouraged cooperation and contributions for public comment, it also provided limited seats at the

---

[20] According to REAT's initial planning the DRECP draft plan should have been available by 2012.
California Energy Commission, California Department of Fish and Game, Bureau of Land Management, and U.S. Fish and Wildlife Service (CEC, CDFG, BLM, and USFWS). "Planning Agreement by and among California Department of Fish and Game, California Energy Commission, United States Bureau of Land Management, and United States Fish and Wildlife Service for the Desert Renewable Energy Conservation Plan." May 2010, 5. <http://www.energy.ca.gov/2009publications/REAT-1000-2009-034/REAT-1000-2009-034-F.PDF>
Accessed April 17, 2014.
The DRECP's overarching goal is to provide a multidisciplinary framework and a public engagement platform for identifying locations for utility-scale RE project development—in October 2011 the initiative identified 3.5 million acres in five locations (West Mojave, Barstow, Blythe, Imperial Valley, and Owens Valley) that were designated as preliminary Renewable Energy Study Areas (RESAs).

[21] Pollak, Daniel. "Natural Community Conservation Planning (NCCP), The Origins of an Ambitious Experiment to Protect Ecosystems, Part 1 of a Series." Sacramento, CA: California Research Bureau, 2001, 32.

decision-making table. Furthermore, the DRECP eliminated the WAV's western section from its defined planning area; this removal meant that solar or wind projects could be developed in the future without oversight of their mitigation by the DRECP overseeing their mitigation. In these respects, the DRECP Independent Science Panel (ISP) 2012 concluded that its 2010 recommendations[22] were ignored and that the initiative "is unlikely to produce a scientifically defensible plan without making immediate and significant course corrections.[23] The panel's work led to some of my interviewees critiquing the hierarchical organization and repudiation of expertise within the overall planning process:

> The independent science panel to the DRECP was actually made up of brilliant people. It's the only reason that those of us who were excluded from the stakeholding process did not blow our tops…[;] because people that we knew and we trusted were on that independent science panel and we felt that if their message gets through, it will be okay. [But] their message didn't get through and they didn't have us to back them up and all of the letters of comment in the world don't compare to the political contributions that they are getting from PG&E, and Southern California Edison….That's why what is happening in the Western Valley is just such a disaster because they didn't let science guide them. They thought that they were going to be better than Nature, and…you can't be better than Nature.[24]

By taking formal and normative shape *following* the approval and siting of the first—and most controversial—utility-scale solar PV projects in California, the DRECP was deprived of its overseeing force, and of the monitoring power that is essential for the entire process.

The 'planning after the fact' model also characterized DRP's outreach work to address concerns related to utility-scale RE projects in the WAV.

---

[22] DRECP Independent Science Advisors (ISA). "Recommendations of Independent Science Advisors for The California Desert Renewable Energy Conservation Plan (DRECP)." October 2010. According to the report, "[t]he advisors recommend obtaining additional scientific input as soon as possible…" in order to address DRECP's "significant errors of omission and commission…" ISA (2010): 86 and 21 respectively.

[23] Independent Science Panel (ISP). "Initial Recommendations of the Desert Renewable Energy Conservation Plan (DRECP) Independent Science Panel Based on Review of Draft DRECP Materials." August 2012.
<http://www.drecp.org/documents/docs/Independent_Science_Panel_2012_Initial_Recommendations.pdf>
Accessed May 1, 2014.

[24] Interview data.

As part of its efforts to update LA County's Town and Country Plan, the DRP organized a "Renewable Energy Meeting" in the Lancaster Regional Library on June 18, 2011, ten months *after* Solar Ranch One's Final EIR was made public and three months after NRG Energy's application for Alpine's Mitigated Negative Declaration (MND) was approved. Belatedly, the DRP realized that its "map approach" was inadequate for doing justice to the concerns of the Valley's longtime residents. As a result, the Department's perspective "changed to focus on developing an ordinance [REO] to regulate" project development.[25] The DRECP and the REO thus share a capricious attitude toward the role of public participation. The REO process exemplifies how the turbulent nature of RE projects becomes aggravated rather than assuaged when scientific and local community input is nullified by bureaucratic agencies downplaying or delaying it:

> The DRECP and the County's [RE] ordinance are supposed to be this combination of... initiatives for putting together what's best to site all these renewable energy "farms". Yet, they are still in the planning stage. Meanwhile, projects are going forward everywhere. So, what the hell—...they are way behind what is actually happening. They're pushing everything as fast as possible. And since the DRECP hasn't come out with what should or shouldn't be done, they can still do whatever they want until that time...[26]

A similar exacerbation of controversy around RE development is true of LA County's December 2012 draft SEA ordinance. This rule is a proposed amendment to the Antelope Valley's area-wide plan (i.e., Town and Country Plan), which notably increases the total acreage on the west side of the Valley, presently regarded as SEA. LA County's SEA program has relied on biological science for almost half a century to designate special management areas—these are lands documented in a series of research reports as hosting rare and endangered species, and these attributes grant them the strongest protection possible. Simultaneously, the DRP is amending the county's zoning ordinance to make RE projects particularly suitable for areas located in zoning code A-2 (heavy agriculture). As a result, these new codes are redefining how WAV residents subdivide their land and how they use it.

---

[25] LADRP. "Renewable Energy Meeting." Presentation, October 26, 2013. <http://planning.lacounty.gov/assets/upl/project/energy_presentation_oct2013.pdf> Accessed May 3, 2014.

[26] Interview data.

Drawing on the DRP's vision to provide stakeholders with guidance regarding permit processing in rural unincorporated areas, the Town and Country Plan thus amplifies both local constituencies' and developers' struggles to ascribe environmental and socio-economic authority to notions of "development" in the WAV. For example, the Blue Ribbon Committee (BRC), which is a consortium of farmers, real estate and business people, private citizens and WAV community leaders, has proposed the creation of residential and industrial zones in the Valley, and is reinterpreting the DRP's outline for zoning and land use in rural unincorporated areas of LA County.

A closer look at the planning history of RE projects in the WAV reveals that, in addition to the BRC, RE developers are also inserting themselves into the Town and Country Plan: "[L]ocated within areas designated in the TCP [Town and Country Plan] as a Renewable Energy Priority Area," reads one developer's Notice of Preparation of a Draft Environmental Impact Report (NOP), "[the project]... is on the leading edge of implementing... critically important TCP planning programs..."[27] The developer seems to be using environmental sustainability as a reference point in the suggested Town and Country Conservation and Open Space policies. But this obscures the complexity of interactions between County regulation and local conservation concerns. As Suzan Zahnter of the Three-Points-Liebre Mountain rural community puts it: "If they wish to promote their project with proposed, unapproved planning policies and goals, then they should correspondingly acknowledge in the NOP and EIR the presence of their project in [the proposed] SEA 21."[28]

Like the DRECP, the perceived validity and effectiveness of the SEA ordinance depends to a large degree on whether or not scientific proficiency is relied upon and becomes the center of the planning enterprise. In her July 2012 letter to the DRP, Margaret Rhyne, president of the Poppy Reserve/Mojave Desert Interpretive Association, urged that the planning authorities be guided in their RE projects in the WAV by the "scientific expertise of the Significant Ecological Areas Technical Advisory Committee."[29] It

---

[27] LADRP. "NOTICE OF PREPARATION AND NOTICE OF SCOPING MEETING. WILDFLOWER GREEN ENERGY FARM. Project No. R2010-00256-(05)/Conditional Use Permit. 201000121/Environmental Impact Assessment 201000063. November 4, 2011.

[28] See Zahnter (2012).

[29] The Poppy Reserve is a 1,800-acre reserve and one of California's premier Parks, established in the 1970's, visited by hundreds of thousands of people every year. Rhyne, Margaret. "June 2010 SEA Ordinance Summary Draft SEA Ordinance Change Comparison Chart 2012." July 30, 2012, courtesy of Margaret Rhyne.

is precisely the SEA program's authority that has become the subject of controversy.[30] Commenting on one preliminary draft version of the SEA ordinance, locals articulated their opposition to empowering the DRP to waive public review regarding a SEA conditional use permit.[31] One possible outcome of updating SEAs based on new science is that development—especially that of RE projects—will be subject to higher scrutiny. On the one hand, potential extension of one particular SEA, the San Andreas SEA 21, would cover a very large part of Element Power's property, initially intended for the "Wildflower" RE project: the "Wildflower" met with fierce resistance from locals until the DRP blocked its development in early 2012. On the other hand, remodeling of the regulations may, if they result in fast-tracking development in those areas, make the expansion of SEAs meaningless.

As regulatory embodiments of conservationist principles, planning documents like the DRECP, the REO and the SEAs contribute to defining what counts as compatibility between RE and environmental sustainability. Yet decisions to approve projects at the expense of scientific and local community input has cast a cloud of mistrust around the legitimacy and transparency of any efforts to determine such compatibility.

How might the siting of utility-scale RE projects preserve the WAV's rural character so that, as the TCP surmises, "[n]ew development…[becomes] consistent with a rural environment"? How can WAV's rural communities contribute to the definition of appropriate concerns regarding an RE project's biological constraints, when final review of the potential project site by SEATAC has not been completed within the NOP comment period set by the DRP? To what extent can local citizens meaningfully engage in steering the DRECP process if the first time representatives from the Bureau of Land Management (BLM) and CEC visit the California desert to address public comments three years[32] after the inception of the DRECP process?

RE regulatory innovation becomes a double-edged sword, potentially safeguarding conservation innovations, but also promoting the proliferation

---

[30] Interview data.

[31] Antelope Valley Conservancy (AVC). "Re: Preliminary Draft Significant Ecological Area and Hillside Management Area Conditional Use Permit Ordinance." February 25, 2012, courtesy of the Antelope Valley Conservancy.

[32] Schmidt, Kati. "National Parks Conservation Association, San Bernardino County Convene Desert Renewable EnergyPlan Workshop with CA Energy Commission, BLM." www.prweb.com, September 5, 2013.
<http://www.prweb.com/releases/2013/9/prweb11095185.htm>
Accessed May 6, 2014.

of what interviewees have called "green scars" as an imaginable future of the California desert. This is especially true in cases where RE planning documents simply add to the contention around the siting of projects. Almost three years after Solar Ranch One broke ground in California's WAV, conservation values found a home in draft RE planning documents like the DRECP and the TCP. "In a way, processes like...the state's Desert Renewable Energy Conservation Plan," stated retired Superintendent of Mojave National Preserve Dennis Schramm in a September 2012 interview, "are what we were crying out for five years ago...[;] [nonetheless, a] lot of the most problematic projects were proposed before...[such federal and state regulations] were drafted."[33] Schramm's reflections have been mirrored exactly by longtime WAV residents; ironically, however, the political structure invented by LA County to supposedly enable the voicing of such concerns fostered factionalism, thus making procedural justice in rural RE project development a prominent problem in and of itself.

WAV Rural Town Councils: A Lack of Due Diligence While Creating Animosity and Division

Davis Mars, a professor of public administration at the University of Southern California, recalled that in the early 1990s, the rural residents of the WAV participated in electing advisory boards under the auspices of LA County supervisor Mike Antonovich. They were referred to as Town Councils, and were a "reflection of the disenchantment with politics as usual" and a "way to gain control over their own lives."[34] Supervisor Antonovich envisioned the town councils as a leverage structure to "alleviate a lot of conflict and build trust"; they were constructed as an indication that LA County rural empowerment was on the rise.[35] Nonetheless, a closer look at the language employed by rural citizens reveals that in their opinion, public officials' empowering rhetoric can also serve as a tool to control rural communities.

---

[33] Clarke, Chris. "Group Calls for Strict Limits on Solar Power Near National Parks." *Rewire*, September 18, 2012.
http://www.kcet.org/news/rewire/government/solar-peis/parks-group-calls-for-solar-limits.html
Accessed May 6, 2014.

[34] Kaplan, Tracey. "Town Councils Give Communities a Chance to Be Heard: Government: The advisory boards reflect a 'grass-roots uprising' created to empower and safeguard rural northern L.A. County." *Los Angeles Times*, June 14, 1992.
<http://articles.latimes.com/1992-06-14/local/me-949_1_first-town-council>
Accessed May 7, 2014.

[35] Ibid.

Consider what a local Town Council officer said in a 2011 address to the Association of Rural Town Councils (ARTC):

> Perhaps you don't really know what a "Town Council" actually is, despite your lofty position as the Director of the ARTC. Town councils *have no power or borders and only exist because of a consensus of the local residents. There are no laws or regulations to govern them—there are only suggestions and guidelines.* Maybe some of the "officers" of our neighboring town councils have become a bit power-mad and think they're something that they're not, but we know exactly what we are—an informal group with no officers, no elections, no scheduled meetings, and no secret handshakes, uniforms, or ceremonial hats...[36]

The officer expressly mentions the ambiguous stature and non-official standing of rural Town Councils: Legitimate oversight of Town Council bylaws does not occur and the Town Councils, lacking official legislative power, are not authorized to commend or condemn a particular project. He also shows concern about the Town Councils' potential roles as actors in the construction of RE facilities in the WAV. There are at least two potential Town Council roles, and they appear to be interrelated: One level is the non-structure of the Town Councils, which deprives WAV residents of a set of formal 'local research' codes, practices and symbolic means to legitimately seek and operationalize community engagement during the planning, siting, and erection of RE projects. A second level is the antagonism resulting from the tension between expressed citizen desires for empowerment and government tendencies toward control and manipulation. The creation of secrecy then gives the appearance of conflicting interests.

Town Councils "are just discussion forums," laments one WAV resident. "They are a sick joke...[part of] a cruel hoax... that Mike Antonovich has played on the far West Antelope Valley."[37] Created by Supervisor Antonovich and not by action of the County's Board of Supervisors—namely not by charter, ordinance or other formal action of a legislative body—the Town Councils are, in the words of assistant head deputy district Jennifer Lentz Snyder, "not subject to the Brown Act as a matter or law."[38] The Brown Act mandates that all local government meetings and their agendas be announced publicly and be open for residents to attend

---

[36] Manuscript provided to the author.

[37] Tuszynski, Jack. "OpEd: Richard the Fourth?" *The Mountain Enterprise*, September 23, 2011.

[38] Manuscript provided to the author.

and participate; Fairmont Town Council officials were pointedly reported in April 2011, just a few months after the Council's initiation, to have issued a proposal to remove the Brown Act from the Council's bylaws.[39] Equally troubled by the fate of biological mitigations involved in potential exclusive contracts between Town Council representatives and RE corporations, is the Director of the Antelope Valley Conservancy (AVC), a local public benefit entity preserving habitat and watershed resources in northern LA and southeastern Kern counties. In her July 2011 communication to the Supervisor's Deputy, she outlined the dialectical and structural contradiction at work in virtually every aspect of siting and planning RE projects in WAV:

> ...if Los Angeles County supports the District Attorney's position that town councils are not subject to compliance with public Sunshine Laws such as the Brown Act because they are not representative or legislative councils then the Supervisors must not rely on them as representative or legislative councils.[40]

Initially proclaimed to be golden carriages of democracy, when their input in structuring public engagement in the development of RE projects in the WAV became critical, rural Town Councils turned into pumpkins. This is in keeping with the "Cinderella of the Far West Antelope Valley" analogy.

Opaque and Contentious by Design: LA County's Inaugural Utility-scale Solar PV Projects

> My running joke is that they don't need a communication consultant, they need an exorcist... [Adam Eventov, First Solar Solar Ranch One community liaison, quoted in Hedlund, "First Solar Lays Off 165"].[41]

The Letter of the Law and a Double Standard

On June 30, 2011, the federal Department of Energy publicized offers of conditional commitments for a total of $4.5 billion in loan guarantees for three cadmium-based, thin-film PV projects that First Solar was involved with either as owner or developer. One was a potential $680 million

---

[39] Hedlund (April 29, 2011). In her critical commentary, the local newsperson lamented over the fact that Fairmont's draft bylaws document "does not appear to put in place any protection against private meetings with [RE] developers to make private deals."
[40] Manuscript, courtesy of Wendy Reed.
[41] Hedlund, Patric. "First Solar Lays Off 165." *The Mountain Enterprise*, May 25, 2012.

loan guarantee for Solar Ranch One, sited in the heart of the WAV.[42] One day earlier, the first trade news article to detail the early months of Solar Ranch One's life expressly characterized the then world's largest approved PV project as "*unexpectedly* knocking heads with some irate neighbors."[43] In this section I am building a case showing that while the California Environmental Quality Act might have been met on paper, local communities in the WAV were disenfranchised.

Setting:

June 27, 2011, Fairmont Market, Lancaster, California, six miles away from the Solar Ranch One project site.

"When you're this giant corporation," David Kerr, President of the Fairmont Town Council, told [Project Development Director at First Solar Jack] Pigott, "and your project is deep in our area and you don't respond to our requests for a meeting, it can only mean one of two things: laziness or that's your policy." It was neither, according to Pigott. Since First Solar took on the project, he explained, the company has had over 100 meetings with 40 different groups, listening to and addressing concerns. At that time, he said, there was no Fairmont Town Council. The company was told Antelope Acres was the governing community...Kerr began requesting meetings with First Solar that December [in 2010]. His requests went unanswered. Pigott said there was no response because the company was on Christmas season hiatus. "By December 2010," Pigott said, "the plan for our project was done and mitigation measures were in place." ...Antelope Acres, some twelve miles from the Solar Ranch One location, was granted [mitigation funds].[44] First Solar also donated mitigation lands to Antelope

---

[42] The projects were Solar Ranch One (230 MW)—which finally received $646 million in loan guarantee—Desert Sunlight (550 MW) and Topaz (MW) (see also Table 7.1). In September 2011, however, First Solar announced that it "would not meet the statutory deadline to receive a federal loan guarantee" for Topaz. See: Osborne, Mark. "First Solar fails to gain US DOE loan guarantee for 550 Topaz solar farm." www.pv-tech.org, September 22, 2011.
<http://www.pv-tech.org/news/
first_solar_fails_to_gain_us_doe_loan_guarantee_for_550mw_topaz_solar_farm>
Accessed April 23, 2014.

[43] Trabish, Herman K. "Is the World's Biggest PV Project About to Get Stopped Cold?" *Greentech media*, June 29, 2011, emphasis added.
<http://www.greentechmedia.com/articles/read/
is-the-worlds-biggest-pv-project-about-to-get-stopped-cold>
Accessed May 7, 2014.

[44] See below regarding the fate of Solar Ranch One's mitigation funds.

Acres' chosen desert conservancy [Desert and Mountain Conservation Authority].[45] Fairmont, whose residents live within 500 feet of the site, has been granted nothing. When First Solar's Pigott told Fairmont's residents he considered them to be part of the Antelope Acres settlement, the room erupted, according to an attendee.[46]

Setting: June 28, 2011, Riverside County Board of Supervisors, Riverside, California, 106 miles away from the Solar Ranch One project site.

The [Riverside] County supports solar energy and acknowledges its benefits. The County also recognizes *solar energy production can have adverse, unavoidable impacts on communities where it occurs*, including impacts on visual, cultural, historic, agricultural, recreational, and biological resources...[In recognition of the "sudden influx" of RE companies in our area] the Board unanimously amended the County's 2011 state legislative platform...to ensure *the County does not disproportionately bear the burden of renewable energy production*...[D]evelopers claim their solar power plants will bring...jobs...However, the majority of these are short-term construction jobs, and there is no requirement to employ local area residents. [In addition, the state and federal governments that have provided the financial incentives for these loan guarantee supported projects] will not bear the brunt of the impact...[that local residents will]. [Riverside County] *residents must be compensated*...[47]

Less than twenty-four hours after the first formal encounter between Solar Ranch One developers and Fairmont's longtime residents, the Riverside Board of Supervisors approved—despite vocal opposition from local trade unions—the "Board Policy B-29 pertaining to solar power plants"—according to which a franchise fee of 2% would apply to the gross annual revenues on utility-scale solar projects that require county permits.[48] The "Sun Tax" would apply directly to the 550 MW thin-film PV "Desert Sunlight" project—Desert Sunlight and Solar Ranch One would be developed concurrently by First Solar. Contrary to the vision expressed in Riverside's Sun Tax, LA County's inaugural utility-scale RE projects were

---

[45] The project owner at the time of the settlement of the mitigation agreement was NextLight, not First Solar. Also, the author's claim that Antelope Acres Town Council *chose* the Desert and Mountain Conservation Authority is misleading—see below.
[46] Trabish, Herman K. (June 29, 2011).
[47] Executive Office, County of Riverside. "Submittal to the Board of Supervisors County of Riverside, State of California." June 23, 2011.
<http://rivcocob.org/agenda/2011/06_28_11/03.112.pdf>
Accessed February 2, 2014.
[48] Ibid.

conceived, permitted and developed under thoroughly different economic, planning, and socio-political conditions.

"[Riverside County's] much-vaunted solar intensity, considered among the best in the world, will keep companies coming here... They're not going to run," Michelle DeArmond, chief of staff for Supervisor John J. Benoit, remarked confidently in the summer of 2011, in response to objections raised against her County's Sun Tax. "And if they do," she added, "there are 10 [other RE companies] behind them."[49] According to state and federal agencies, Fairmont's CREZ 47 is no less an RE "Mecca" than Riverside County's Desert Center, adjacent to Joshua Tree National Park. Riverside County officials convincingly countered LA County's misinformation about WAV community representation: the economic prosperity promised by RE developers would be ephemeral—lasting only through the one to two-year construction phases for each project. First Solar VP James Woodruff made a plea to Riverside County that the "2-percent fee is dramatically out of line for the kind of right-of-ways and easements" his company had hoped for.[50]

In contrast, Los Angeles County, as the mediator in siting Solar Ranch One in the WAV, neglected the possibility of productive and meaningful engagement with, and expansion upon, local needs and human resources. Los Angeles County also dismissed the idea that rural residents might "disproportionately bear the burden" of numerous solar facilities. Solar Ranch One's project director brought WAV communities into a public consultation process that vacillated between arrogance and mistrust. "We will never pay that," he was reported to have laughingly replied to Fairmont residents when—during the confrontation of June 27, 2011—locals argued that neighboring Riverside County was benefiting more from its solar projects because of the Sun Tax.[51] Predictably, Solar Ranch One's development director embodied the double standard in the relationship between RE project mitigation and a perverted notion of community engagement. The sole feedback Pigott felt obliged to offer Fairmont's residents was included "in the project's environmental impact study... which was done publicly and transparently according to state and county guidelines."[52]

---

[49] Kaufmann, K. "Riverside County may levy fee on solar firms." Mydesert.com, June 17, 2011.
http://www.ccpda.org/documents/solar-issues/ca-model-photovoltaic-ordinance/94-riverside-county-may-levy-fee-on-solar-firms-june-17-2011/file
Accessed June 24, 2014.

[50] Ibid.

[51] Hedlund, Patric. "Sabers Rattle as Fairmont Chides First Solar for 'Bullying.'" *The Mountain Enterprise*, July 1, 2011.

[52] Trabish (June 29, 2011) emphasis added.

Table 7.3 California's inaugural utility-scale solar PV projects.

Project name **Area (acres)** **MW** **Electricity buyer**[19]	**Technology**
Solar Ranch One 2,100 Private 230 PG&E	The project employs FS Series 3 PV thin film CdTe modules and also features a utility-scale deployment of inverters with voltage regulation and monitoring technologies that were new to the US solar market in 2011. The inverters enable the project to provide more stable and continuous power, increasing the reliability of large-scale solar power plants. Each module is 1,200mm in length, 600mm in width and 6.8mm in thickness.
Alpine Solar 835 Private 66 PG&E	Alpine Solar used the same technology.

Electricity produced in both projects is sent to Southern California's Whirlwind substation, located about three miles away from Solar Ranch One.

Indeed, the RE engineers that designed California's inaugural utility-scale PV project (Table 7.3) were firmly grounded in the state's foremost environmental law (California Environmental Quality Act or CEQA): They issued a notice of preparation and conducted at least one scoping meeting to incorporate public input regarding the confines and content of Solar Ranch One's draft Environmental Impact Report (DEIR). Subsequently, they watched their project proposal twice go through the hearing procedure of LA County's Regional Planning Commission (RPC) where federal, private and public actors commented on the design of the DEIR before the latter was finalized and a Conditional Use Permit (CUP) was granted to then-NextLight's solar electricity generating facility.[53] During the public hearings in the summer of 2010, for example, RPC staff decided, that based on DEIR project alternatives, Solar Ranch One's transmission lines should be laid underground. The RPC certification of Solar Ranch One's EIR was likewise based on Governor Schwarzenegger's authoritative recommendation to the LA County Board of Supervisors that

[53] For a documentation of the project's change of owners see "Appendix 1".

they subject the project to a "thorough yet expeditious review."[54] Thus in September 2010, the Commission concluded that no recirculation of the project's EIR was necessary, since no "significant new information" had been added to the EIR after stakeholders were notified of the availability of the DEIR for public review.[55] Correspondingly, to comply with the National Environmental Policy Act (NEPA) at the federal level, the DOE published an August 2011 Environmental Assessment (EA) that did not identify any significant impacts for Solar Ranch One.[56] The result was that no Environmental Impact Statement (EIS) was ever conducted for the project.

The LA County Board of Supervisors decided during its November 23, 2011, meeting to reject, on two grounds, army contractor Northrop Grumman's appeal against Solar Ranch One's planning permit decision. First, LA County's planning department maintained that the appellant's comments—"received after close of the duly noticed draft E.I.R. [DEIR] public comment period"—did not warrant recirculation of the project's DEIR. Second, central to the Board's ruling was a public response from the Department of Defense (DOD) that the Solar Ranch One-Northrop Grumman dispute was "a land use issue" for which the DOD had no position. Similar to how they dealt with Solar Ranch One's social and environmental effects upon Fairmont's residents, LA County planners assumed that the project's EIR did not need to analyze impacts on Northrop Grumman's radar testing facilities. And like most WAV natives, the military contractor, in the words of its attorney, "did not learn about the existence of" Solar Ranch One until after the LA County planning commission had approved it.[57]

---

[54] Schwarzenegger, Arnold. "Letter to LA County." February 2010.

[55] See LADRP. "Project No. R2009-02239-(5) AV SOLAR RANCH ONE, LLC. Vesting Tentative Tract Map No. 071035 Conditional Use Permit No. 200900026 Environmental Assessment No. 200900027 Agenda Item No. 7." Regional Planning Commission Public Hearing September 15, 2010, page 4. The Commission's decision that a recirculation of the EIR was not required was—unsuccessfully—challenged by the California Unions for Reliable Energy (CURE).

[56] DOE (2011).

[57] Minutes of the November 23, 2010, Meeting of the Los Angeles County Board of Supervisors. Northrop's decision to appeal Solar Ranch One was based on the supposition that the "[solar] project will adversely impact the support of the military mission." To the Board of Supervisor's claim that its decision was influenced by the lack of alarm from the Pentagon, Northrop Grumman engineering director Leonard Figueroa replied that "our military customers are in classified program offices and they cannot take public positions." The Board of Supervisors. "The Meeting Transcript of the Los Angeles County Board of Supervisors. Regional Planning Committee Hearing." November 23, 2010.

Not surprisingly, following the letter of the law—in this case, the formal requirements of California Environmental Quality Act guidelines—was not quite the same as following the "spirit of the law": to integrate as much public input as possible with environmental governance decisions and processes.[58] Engineering an RE project may be consistent with developers' rhetoric that processes unfold "publicly and transparently," in the words of First Solar director Jack Pigott, but it must go beyond such rhetoric. As an example, transparency requires changing engineering and planning practices which are blind to problematic assumptions that may entail injustices. For instance, communication of design intent by LA County through rural mailing was "ridiculed" by WAV residents who also claimed that scoping meeting notifications to property owners within a "1000-foot radius from Solar Ranch One's project site" consisted of unjust, selective, disclosure in rural, unincorporated areas.[59] Furthermore, of the alleged 100 public outreach meetings that Pigott referenced to the Fairmont Town Council on June 27, 2011, only three actually engaged local communities, and none was specifically targeted at households residing within close proximity of the Solar Ranch One site.[60]

---

[58] Hedlund (July 15, 2011).

[59] Hedlund, Patric. "Fairmont Town Council Asks for Emergency Planning Meeting on First Solar Intentions." *The Mountain Enterprise*, June 28, 2011. In the article Pigott is reported to have said that First Solar was "instructed by Los Angeles County Supervisor [Michael] Antonovich's office to have the scoping meeting and [EIR] hearings at the Antelope Acres Town Council."
During the June 30, 2010 Regional Planning Committee hearing on Solar Ranch One it was mentioned that "the community was appropriately notified for the public hearing by mail, newspaper, property posting, library posting and Department of Regional Planning (DRP) website posting. Newspaper notices were published on May 27, 2010, in the Antelope Valley Press and La Opinion. Mailing of notices to land owners located within a 1,000-foot radius of the property boundaries and to five local libraries, were sent on May 24, 2010…[A DEIR Notice of Completion (NOC) was] published in the Antelope Valley Press and La Opinion newspapers [on June 16, 2010]." Another example, then, of planners/developers following the letter, and not the spirit, of the law is the fact that Solar Ranch One notices were published in the "Antelope Valley Press" which lacks general distribution in rural WAV. The Board of Supervisors (2010).
Interview data.
In January 2012, three years after First Solar arrived in the WAV to hold its first public sessions with locals in Antelope Acres, Jim Lamon, the solar company's senior Vice President of engineering, procurement, and construction (EPC) admitted that the developer "had not understood that zip codes in this area can be misleading." Hedlund (27, 2012).

[60] Besides, it was by Fairmont's neighboring Oso rural Town Council's president himself, that First Solar managers were informed in March 2011 regarding the status and location

Thus although certainly predicated on law, RE engineers and LA County planners failed the operational test of AVSR's legitimacy—its consistency with procedural justice.[61] Ironically, to meet DOE's construction deadlines between 2011 and 2013 and to receive federal loan guarrantees, First Solar's engineers had to develop an *ad hoc* framework and methodology to address a multitude of local knowledge and procedural justice-related contingencies, such as on-site dust mitigation and landscaping—the very elements that Solar Ranch One's planning, siting and permitting processes downplayed. Following the spirit of regulatory procedural provisions requires more than formal compliance; it also involves a radical shift of inward perception in engineers. Fulfilling their ethical obligations challenges RE engineers to construct their professional identity *in response* to the marginalization of their role in progressing community engagement. They feel that they must overlook local concerns if they are to complete their design and construction, and thereby secure greater power within a RE industry setting.

Besides mounting 3.7 million Cd-Te thin-film PV modules—a large-scale toxicity experiment in itself—the chronicle of Solar Ranch One includes erupting Town Councils, opaque negotiations, contending military contractors, bottoming corporate stock, furloughed workers, construction managers literally fighting county inspectors—even a plea to an exorcist.[62] However complicated the interweaving of animate and inanimate actors in shaping the fate of Solar Ranch One, the project's contentious character can hardly be considered "unexpected." In fact, this section details how Solar Ranch One's design, permitting, and construction "consensus" was dubious and based on shaky foundations; the project's proponents saw legitimacy in it, whereas its critics constantly challenged the grounds upon which it was maintained.

---

of the totality of the Town Councils in the area adjacent to their project. Richard Skaggs, personal communication. According to its members, the Fairmont Town Council was formed partly to help longterm WAV residents "avoid being steam-rolled by all these [RE] companies" and partly in opposition to a proposed racetrack at Fairmont Buttes (see below for the racetrack project). Interview and participant observation data.

[61] The lack of procedural justice in Solar Ranch One's planning and development created a vicious circle between locals and First Solar. For example, to meet DOD deadlines, First Solar used extended work shifts (e.g., eighteen hours) and overtime that further deepened the tension between WAV communities and the solar developer.

[62] On the toxicity angle see chapter 6 section on "Solar Developers and Local Communities in the WAV: An Unsustainable Relationship."

Antelope Valley Solar Ranch 1 (Solar Ranch One) Collapsing Consensus: Factionalism and the Loss of Legitimacy

In its design, Solar Ranch One was trumpeted as conforming to principles derived from LCA. The federal government had given its blessings to the project's life cycle metrics of GHG emission reduction, securing the project's financial viability founded on its quantified environmental sustainability. In addition, the world's foremost authority on PV LCAs, Vasilis Fthenakis, had not backed away from guaranteeing the safety of the First Solar CdTe modules: the DOE EA publication cited the researcher's LCA studies to justify the use of such modules in the design of Solar Ranch One.[63] As such, LCA would provide the epistemological grounding of a booming RE market in the desert Southwest US. "We were told in meetings by company representatives that LCA's guaranteed the transparency and sustainability of these projects," said one interviewee. Furthermore, the public hearing testimonies of West Antelope residents ostensibly suggested that utility-scale solar power had become a socially accepted technology. The actors representing the social effects of the project had seemingly expressed their assent.

<u>Setting:</u>

November 23, 2010, Meeting of the Los Angeles County Board of Supervisors, ninety miles away from the Solar Ranch One project site.

...

SUPERVISOR ANTONOVICH [LA County fifth district Supervisor]: AND DID YOU RECEIVE TESTIMONY FROM VARIOUS PARTIES DURING THE APPROVAL PROCESS?

KIM SOLAY [Solar Ranch One Project Planner, Department of Regional Planning]: YES.

SUPERVISOR ANTONOVICH: WHAT WAS THE ANTELOPE VALLEY ACRES TOWN COUNCIL'S POSITION ON THE PROJECT?

KIM SOLAY: IN 2009, THERE WAS UNANIMOUS SUPPORT BY THE ANTELOPE ACRES [Town Council]. AT THE HEARING, THERE WERE MEMBERS OF ANTELOPE ACRES THAT TESTIFIED. AND THERE WAS ONE THAT WAS NOT IN FAVOR OF THE PROJECT -- ACTUALLY THERE WERE TWO, ONE THAT WAS ON ANTELOPE ACRES AND ANOTHER THAT WASN'T. AND THERE WAS -- THE PRESIDENT REITERATED THAT THE BOARD DID HAVE A MAJORITY IN SUPPORT.

---

[63] DOE (2011).

SUPERVISOR ANTONOVICH: THE MAJORITY VOTED FOR IT. AND WAS THERE ANY OTHER SIGNIFICANT OPPOSITION TO THE PROJECT DURING THE PUBLIC REVIEW PROCESS?

KIM SOLAY: THERE WAS THE ONE MEMBER THAT I MENTIONED, NOT OF THE TOWN COUNCIL BUT...[one who] LIVES IN THE AREA, ANTELOPE ACRES, THAT HAD A LOT OF CONCERNS ABOUT THE PROJECT.

SUPERVISOR ANTONOVICH: ONE OUT OF ALL THOSE OTHERS WHO SUPPORTED IT?

KIM SOLAY: THAT'S CORRECT.

...

VICKIE NELSON [then President of the Antelope Acres Town Council]: HONORABLE CHAIRWOMAN AND SUPERVISORS. I AM VICKIE NELSON, PRESIDENT OF THE ANTELOPE ACRES TOWN COUNCIL. THE ANTELOPE ACRES TOWN COUNCIL VOTED TO SUPPORT A.V. SOLAR RANCH 1 PROJECT IN MARCH OF 2009. THIS ORIGINAL SUPPORT POSITION OF OUR TOWN COUNCIL HAS NOT CHANGED. THEY HAVE WORKED VERY CLOSELY WITH US FOR THE PAST TWO YEARS, AND WE ARE LOOKING FORWARD TO CONTINUING TO WORK WITH FIRST SOLAR AS THE PROJECT MOVES AHEAD. THANK YOU...[64]

Setting:

January 8, 2011, website of Friends of the Antelope Valley Open Space (FAVOS), a local organization aimed at "...ensur[ing] compromise between man and nature" (Figure 7.4).[65]

In point of fact, the WAV resident movement to push back Solar Ranch One only began in early June 2011, fueled by a local newspaper article that announced First Solar's intention—made public by the company's VP during a

---

[64] The Board of Supervisors (2010). Vickie Nelson was at the time serving as a director of the Antelope Valley Resource Conservation District Board, the parent agency to the DMCA, without identifying any conflict of interest.

[65] FAVOS. "AVSR1." Undated.
<http://avopenspace.org/html/av_ranch_1.shtml>
Accessed May 4, 2014.
Echoing FAVOS's website, Margaret Rhyne, a local conservation activist affiliated with FAVOS, and David Jefferies, who became active with rural Town Councils, argued in January 2011 that Solar Ranch One "had good communications with the communities, did an EIR and worked with their neighbors." Hedlund, Patric. "Neenach Area Files Two Appeals to L.A. County Solar Farm Permit." *The Mountain Enterprise*, January 7, 2011.

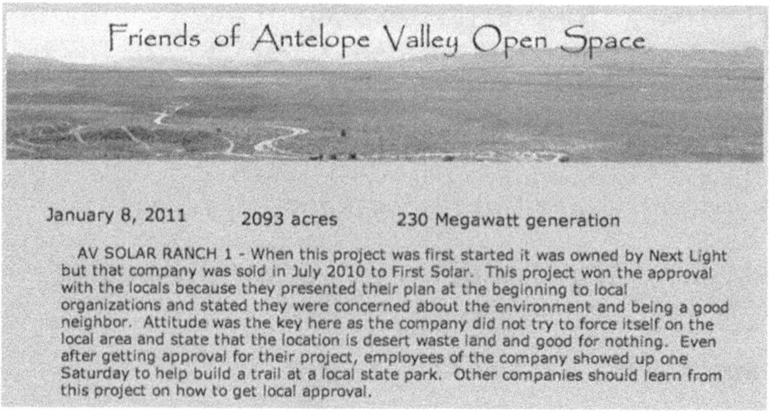

Figure 7.4 FAVOS' assessment of the Solar Ranch One project's early development stages. Courtesy of FAVOS.

"reception" event that had taken place a week prior to the article's publication—of "making a contribution of $140,000, to be placed in a non-profit trust for the Antelope Acres Town Council to fund projects." Written a month prior to the Solar Ranch One breaking ground scheduled in Fairmont, the report added that "similar contributions" were made to seven other local entities, including the Poppy Reserve/Mojave Desert Interpretive Association.[66] And finally, according to the same June 2011 piece, First Solar's disclosure that the company had bought mitigation land adjacent to the Poppy Reserve was met with "applause and cheers" amid Solar Ranch One's reception.[67]

Ultimately, continued conflict surrounding LA County's inaugural solar PV project was not simply a result of disregarding CEQA's intent; it was also caused by the incongruous nature of rural Town Councils: they could not insert themselves authoritatively into the RE project development process, except by brokering contentious financial arrangements between themselves and RE corporations. The Town Council structure initially gave the impression of marshaling engagement—in the form of corporate tax-deductible donations—but this actually excluded WAV residents from a legitimate structure for community participation in RE project development. First Solar promised to donate $140,000 to the Antelope

---

[66] According to another source First Solar gave $60, 000 to six such "community organizations"—i.e., $10,000 to each of them. Hedlund (January 27, 2012). Five months earlier, the Poppy Reserve's president was detailed to have asserted that Solar Ranch One "had worked with their neighbors."

[67] Anderson, Dennis. "Poppies and Panels." *Antelope Valley Press*, June 8, 2011.

Acres Town Council early on in Solar Ranch One's permitting process. This "offer" was given in a June 6, 2011, public reception meeting. Yet the Antelope Acres Town Council had not received approval by the IRS for its non-profit status until May 22, 2012. During the interim time, the Oso and Fairmont Town Councils were also after 501c3 applications in order to qualify for mitigation funds from Alpine's NRG Energy. When First Solar met with Antelope Acres Town Council members in November 2011, the company announced to WAV residents that it would not wait for them to acquire non-profit status and that there would be appropriations to whomever First Solar determined. Indeed, the developer decided in May 2012 to donate $140,000 to the Antelope Valley College, which is located thirty miles away from the Oso/Neenach communities. Only two WAV locals benefited from that scholarship. In the summer of 2013, the company still had "a chunk of money [another $140,000] that [they were] trying to divvy up amongst different groups" in the WAV.[68] If anything, after WAV citizens learned of First Solar and Antelope Acres' alleged deal, the clash and mistrust became greater than ever with regards to the role of rural communities in shaping the Valley's RE landscape.

Setting:

November 16, 2011, Antelope Acres Town Council Meeting, Westside Community Church, Antelope Acres, California, ten miles away from the Solar Ranch One project site. An Antelope Acres resident addresses the president of their Town Council.

"If you did get any money [from First Solar] you should not accept any funds; I think part of this money belongs to Fairmont…it would be wrong for the community of Antelope Acres to take money that belongs to Fairmont; you could take a proportion of the amount that everybody got but I think that if you're going to have executive meetings with FS regarding how the monies are going to be spent…you should…make them community meetings. And I think it's time that things weren't done behind closed doors…[69]

Touted by a trade news reporter as an "inspiring display of democracy in action," the above scene can best be described as a parody, reflecting the

---

[68] Participant observation, Oso Town Council meeting, June 6, 2013.

[69] Author's transcript. About a year later, on October 17, 2012, during Antelope Acres Town Council's regular meeting another political theater—in the form of a "coup"—took place. The meeting was called to order. Town Council members did the flag salute, then protesters got up and read a letter of complaint to the existing Town Council members announcing to them that they were fired.

reality of both rural community animosity and lost legitimacy on the part of RE project development in the WAV.[70] For all its unstructured mediation of RE mitigation arrangements—in fact due to it—the Town Council frame of governance in rural LA County fed factionalism between rural residents as well as an underlying mistrust of the RE project development processes: The Fairmont Town Council folks "are in it [RE project facilitation]," stated a September 2011 opinion piece in a local media outlet, "for the mitigation money and control of resources just like everybody else. Just like Oso."[71]

Interview material:

Oh, how can you trust the [LA] County and the green companies? Ours is a factionalized rural area…and this structure [i.e. the Town Councils] is what County and developers alike further extended to get your solar projects on the map. I have been personally following the advent of renewable energy in our valley since day one and I still can't make complete sense of it. But I can assure you that the view expressed on the FAVOS website [see Figure 7.4 above] is not shared by everyone in the group... Initially it seemed to be that Solar Ranch One was as [they] said, but later developments changed all that… It seemed they swore that their dust mitigation plan was the latest in science…and nobody looked at it carefully. Most of us were then focused on the wind turbines [fighting the "Blue Sky Wind Project" (225 MW) proposed by Next Era] and did not really know much about Solar Ranch One…

That's the impression [that Antelope Acres Town Council negotiated secretly]… There was a meeting—it was an executive council meeting yet everything was done above the board. Folks are… afraid to talk about that and I don't understand why… it was innocent. And it could have perhaps been done a little bit differently. And I [said], let's just come clean. Let's just tell them everything that happened. What we discussed there was that First Solar wanted to give the community some cash. So, they would say, for instance: What about a scholarship for somebody on the West Side? But nothing was ever set in stone… Eventually they gave…[$140,000] to the Antelope Valley College. They got tired of dealing with us—they had to dump this money. So, that sort of ended that.

[70] Trabish (November 29, 2011).
<http://www.greentechmedia.com/articles/read/element-power-first-solar-and-intense-local-politics>
Accessed May 8, 2014.
[71] Tuszynski (September 2011).

I thought the executive meeting [between Antelope Acres Town Council members and First Solar executives] was innocuous, but during the very next town council meeting somebody got wind of it...And... [a local resident] was screaming about it and then [the Antelope Acres Town Council president] requested that we don't reply...There is a public comment period. You have three minutes—but if you answer them or inquire from them they get three more minutes...So, you can just not say anything. So, this gave [the complaining resident] the opportunity to sit there and question the executive meeting with us sitting there and not answering. And of course they all knew this was going to happen. So, they all brought cameras and took pictures of us not answering the questions, so it even made it more suspicious than it should have been...

LA County and RE companies had generated a project that was contentious by design. At one level, the siting of solar PV projects in the WAV was predicated on a mechanism linking legitimacy and life cycle thinking. At another level, along with Solar Ranch One's contentious permitting and pre-construction phases, the question of legitimacy emerged from the LCA shadows and became a central issue in rural RE project development. As a semi-statutory act, approval of Solar Ranch One's CUP required the developer to adopt a public outreach plan to ensure the approval of local stakeholders. As far as neighboring communities were concerned, however, LA County's first utility-scale PV facility sought the approval—with the blessings of planning officials—of at least one individual: an unelected official from an entity not eligible to favor or disallow development. Recurring waves of mistrust in the WAV were stimulated by companies' promises to Town Council members—dissonant with the non-tax exempt status of rural Town Councils at the time—and the County's actions that excluded WAV residents from Solar Ranch One's permitting, while the County upheld the illegitimate conditions of mitigation arrangements between developers and Town Councils as well as between developers and non-local entities.

Recurring Waves of Factionalism and Mistrust: The Fairmont Butte Motor Sports Park and Solar Ranch One's Biological Mitigation

Interview material

You really cannot understand what happened with Solar Ranch One unless you are familiar with the racetrack story... NextLight [Solar Ranch One's first owner] had agreed to mitigate the impacts of their project by buying pieces of Fairmont Butte, which was supposed to be the site of a proposed racetrack.[72] Well, the racetrack drew huge public outcry because it was adjacent [one mile] to the Poppy Reserve. So what was going to be a racetrack became solar mitigation land... On the one hand, most of us felt there had been something going on in the background between the [LA] County folks, these different project [solar and racetrack] developers, and a couple of mitigation thirsty "conservation agencies" ... On the other hand, I think that a lot of people took a bit of sigh of relief thinking that solar wouldn't be nearly as bad as the racetrack would...

The Fairmont Butte mitigation was the ultimate act of First Solar's "we are free and clear" attitude. We wanted to have been involved in that mitigation—well, unfortunately the mitigation was already over by the time [November 2010] we got involved [via the creation of the Fairmont Town Council]. And nobody in the community had one word to say about it.

You bring up the racetrack. Oh my God. The racetrack thing got the whole community [of Antelope Acres] polarized. That guy [Thomas E. Malloy, the businessman who proposed the racetrack project] would come to the Antelope Acres Town Council meetings with his law team, and every single time the locals would change their minds about the project...The big thing, of course, that upset the community was the fact that they never intended it [the racetrack] to be open to the public at all.

In one of those meetings [in Antelope Acres,] Malloy's folks threw out a vacant threat that if he doesn't get his racetrack he's going to put a housing tract in. So, at that point, the Antelope Acres Town Council supported [the racetrack] because we didn't want the housing tract. But what really was the nail in the coffin was... [that] they could not guarantee that we would not be hearing the sound all the time...And the people who wrote the EIR came to our town council one night [circa December 2010] and then the next night they were going to Fairmont's [Town Council]. Yet, the Supervisor's deputy received a phone call before the Fairmont Town Council meeting that Malloy was going to abandon the project and that's when everybody found out that Solar Ranch One bought his land for mitigation purposes. And I think that this was in the back all the time because everybody kept saying, "why don't you just sell your property for mitigation purposes..."

[WAV locals who attempted to unite in a common conservation easement block] were purposefully thwarted by other not local, yet more powerful, conservancy folks and their minions...These folks are able to go ahead and work in secrecy and...make it look like they're doing conservation... They

---

[72] According to "Mitigation Measure 5.7-2: Off-site Mitigation for Loss of Habitat" included in Solar Ranch One's final EIR approved on December 7, (DOE, 2010): 26. The "racetrack" refers to the proposed, yet never realized, Fairmont Butte Motor Sports Park.

maintain that they don't need oversight; that they oversee themselves...Some of us figured that we would report the corruption and it would get fixed. Yet the people that we reported to...[in LA County and at the State level] are the people who appoint them, and chose them, and who are part of it.

Setting:

July 17, 2009, Governing Board of the Desert and Mountain Conservation Authority (DMCA), City of Santa Clarita, CA, fifty-three miles away from the Solar Ranch One project site. [A Joint Power Authority entity, the DMCA was created in 2006 through the Antelope Valley Resource Conservation District (RCD) and the Santa Monica Mountains Conservancy (SMMC)—the RCD operates as the budgetary agent of the DMCA. The DMCA was founded partly by Jim Dodson, current president of RCD, who has also held various positions with Sierra Club, California. The SMMC has a presence in the WAV, as it owns land at the top of the Poppy Reserve, approximately one mile southeast of the Solar Ranch One project site (Figure 7.5)]:

"Agenda Item VIII: Consideration of resolution authorizing a comment letter to Los Angeles County regarding the pre-draft Environmental Impact Report for the AV Solar Ranch One Project (Tract Map No. 071035) in the Fairmont Butte area, unincorporated Antelope Valley."[73]

Setting:

Tom E. Malloy, owner of Trench Shoring Company, addresses members of the Vintage Auto Racing Association (VARA), August 4, 2009, 556 Malloy Court, Corona, California, 130 miles away from the Solar Ranch One project site:

August 4, 2009
Dear auto racing fans, car enthusiasts and environmentalists:

---

[73] DMCA. "Memorandum, July 17, 2009." 2009. Courtesy of the DMCA. According to one informant, "the DMCA has failed to file the legally required annual financial reports to the State Controller's Office. The DMCA appears to have presented no financial reports to the AVRCD as required by its founding document, the Joint Powers Agreement. The DMCA cancels almost every one of its regularly scheduled meetings, fails to notice or hold public meetings, and appears to fail to hold the minimum required annual meetings per its founding document, the Joint Powers Agreement. The DMCA Minutes read, 'So and so asked a questions, so and so made a comment,' without giving any insight into what was discussed. Yet the California Dept of Fish and Game/Wildlife has held that the DMCA is exempt from its Authorization process for conservancies, which seeks to ensure the integrity and legal operation of entities holding mitigation lands."

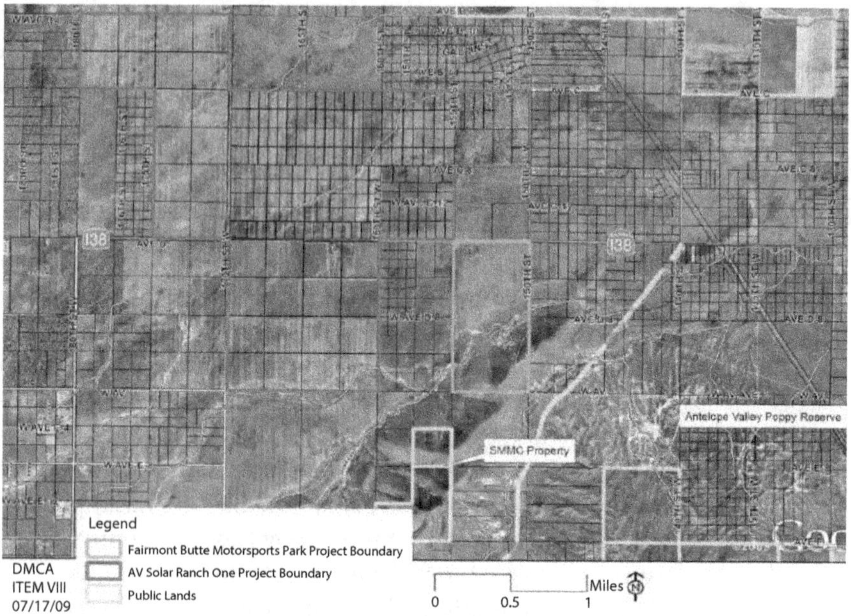

Figure 7.5 The Solar Ranch One and racetrack project sites. North of the proposed racetrack's boundaries lies the SMMC property, to the West of the Antelope Valley Poppy Reserve. The Fairmont-Antelope Butte SEA is located approximately 900 feet southeast of the Solar Ranch One project site. Courtesy of the DMCA.

For the past seven years I have been attempting to develop the "Fairmont Butte Motorsports Park" on 140 acres of my 320-acre property... The Draft Environmental Impact Report (DEIR) has been distributed and the County is receiving opposition comments from a small but vocal group associated with the nearby Poppy Reserve. Despite the facts clearly stated in the report, this group has distributed incorrect information about the project...The proposed zone change to Commercial-Recreational is consistent with the recreational uses of the Poppy Reserve. A small amount of development [what would have been the southern part of the project] is proposed within the north end of the Sensitive Ecological Area 57 (SEA). The majority (84%) of the SEA will be preserved in perpetuity through a conservation easement...The track will operate as a private facility not generally open to the public...[74]

[74] Vara.freeforums.org, August 13, 2009.
<http://vara.freeforums.org/read-this-now-and-send-your-comments-this-is-a-must-do-t923.html>
Accessed May 10, 2014.

The Fairmont Butte Motor Sports Park (racetrack) DEIR mentioned in Malloy's letter was filed with LA County's DRP not long before the DMCA leadership met in Santa Clarita to weigh their judgment about Solar Ranch One's layout and the possible construction of a solar facility in the WAV. On September 2, 2009, comments about the racetrack's DEIR were collected during an RPC public hearing at which the Commission recommended the project—several of the racetrack's adherents at that hearing were VARA members recruited by Malloy. Around that time, Mike Powell, a local volunteer at the Poppy Reserve, spoke for many in the Valley when he conveyed with a sense of urgency to a *Los Angeles Times* reporter that the racetrack "represents the start of a significant threat to the kind of lifestyle and environment that many people moved out here for."[75]

Yet DMCA chairperson Jim Dodson wrote to regional planning in conditional support of the racetrack on the same day that the DMCA governing board convened to compose its comment letter to LA County regarding the Solar Ranch One DEIR. Though noting that the project would entail "unavoidable" impacts, Dodson stated that "DMCA['s] "suggestions…include a modification" of the racetrack's design—namely moving the project 700 feet to the north. Later in his letter he explained that the final EIR "must include" a mitigation measure that would require the dedication of the property located to the south of the new project design's disturbance line. "The DMCA would accept that dedication," Dodson asserted.[76] Likewise, addressing the Solar Ranch One DEIR, the DMCA authorities did not oppose the RE project; rather, the conservancy supported NextLight's application for a CUP while advocating for "approximately 650 acres of permanent protection" to mitigate the facility's construction.[77] Thus even as he was diagnosing the "insufficiencies" in the two adjacent projects' planning documents, Dodson was in reality coveting development—particularly the concept of land mitigation.

---

75 Simmons, Ann. M. "Wildflower reserve concerned about proposed racetrack." *The Los Angeles Times*, July 27, 2009.
<http://articles.latimes.com/2009/jul/27/local/me-poppyfields27>
Accessed May 10, 2014.

76 DMCA. "Draft Environmental Impact Report Comments for the Fairmont Butte Motorsports Park Project, Parcel Map 26805 (SCH No. 2005031170)." July 17, 2009.
<http://smmc.ca.gov/pdf/attachment2537_Attachment%201.pdf>
Accessed May 18, 2014.

77 The Board of Supervisors (2010).
Edmiston, Joseph T. "Memorandum." April 6, 2011.
<http://www.mrca.ca.gov/pdf/attachment2084_Staff%20Report.pdf>
Accessed May 20, 2014.

The idea of using a government regulated structural mechanism, like the Joint Power Authority, to acquire and maintain mitigation lands belongs to SMMC's founding executive director Joseph (Joe) Edmiston. Termed by *Los Angeles Magazine* as the "King of the Hills" and the "most powerful unelected official" in the state of California, Edmiston is believed to be a man for whom "there is hardly a large, privately held of land in Southern California that he doesn't have his eye on."[78] If the Joint Power Authority was to be the mechanism to sustain a seemingly freewheeling land management state agency with an unelected and unaccountable board, then the politics of Solar Ranch One's mitigation property claim was barely challenging for Edmiston, who had won longer and more complex clashes over California land.[79] From the time the DMCA submitted its Solar Ranch One comment letter, the agency held private meetings with NextLight representatives until both sides decided to enter into "an agreement in which the DMCA would be the recipient agency for approximately 450 acres of additional open space to be acquired" by Solar Ranch One's owner company.[80] Far from making conservation and development antithetical, the politics of solar engineering in the WAV included battles over what role Southern California land trusts may play in managing mitigation lands for RE projects.

While Solar Ranch One managers were meeting with Edmiston and Dodson's DMCA, they also viewed local factionalism between WAV conservation-activists as being especially beneficial to the company's cause. Although she expressly admitted in personal communication with me that she made no pretense of seeking exclusivity in negotiating with planning or RE corporate entities, AVC president Wendy Reed's efforts to "engage the Valley's population in a public process" were met with reluctance and confusion by fellow WAV conservationists. Quickly outmaneuvered

---

[78] Macadams, Lewis. "King of the hills: environmentalist Joe Edmiston might be the most powerful unelected official in Los Angeles." *Los Angeles Magazine*, July 1, 2005, 42, 44, and 42, respectively.

[79] In June 2004 the State of California's chief auditor Samuel E. Hull accused Edmiston's conservancy for not "adequately manag[ing], control[ing], or oversee[ing] $115 million in bond funds" and for failing to provide "financial separation between the conservancy and the authority." Ibid., 54.

[80] Edmiston (2011). Between September 2009 and December 2010 LA County consistently postponed public hearings on the racetrack project. The deal between Solar Ranch One and the DMCA—which besides donating mitigation lands required that the solar project developers would provide the DMCA funds for administrating and restoring the property—was executed on September 14, 2010. The minutes from the December 15, 2010 meeting of the LA County RPC show that on that day the racetrack was "recommended to be taken off [the] calendar."

by the FAVOS group and affiliated members of the Transition Habitat Conservancy located in Pinon Hills, California, the AVC was "blacklisted" from partaking in a common conservationist-minded local activism forefront with regards to RE projects in WAV:[81]

> When the first wave of projects was coming… I said 'we need a public process.' I will get the grant funding through the conservancy [AVC] because we are already a non-profit organization... And we will hold meetings. We will hold them in large places, like the library, like the cultural center in Palmdale…we will go around and engage the community. It was my great desire to bring everyone; Roosevelt Town Council, Little Rock Town Council, Lake Los Angeles Town Council—not just Antelope Acres Town Council—on the table. The FAVOS and WAVCC asked, "Why should the Antelope Valley Conservancy be in charge?" and was told that the AVC would not be in charge, merely a sponsor of the meetings, and we will all be a part of that discussion. [But, alas,] an alliance was built and AVC was excluded—[these folks eventually] started working with the DMCA and the Transition Habitat Conservancy…[82]

Cheered and applauded, as it were, by attendants during the Solar Ranch One reception in the summer of 2011, the biological mitigation agreement between LA County's first utility-scale solar PV project and DMCA—supposedly motivated by legal commitment to promoting environmental sustainability—naturalized backroom dealing as well as the evasion of due diligence and community participation in RE project mitigation procedures. As such, it bred and sustained factionalism between local residents, together with mistrust against the network of the developer, LA County and other local government entities that—under the aegis of DRP—had been or would be involved with Solar Ranch One during its planning and construction phases, respectively.[83]

LA County's decision making regarding the project's biological mitigation occurred similarly to the planning and permitting of Solar Ranch One: the

---

[81] For example, the FAVOS website does not even mention the AVC and its activities in the Valley—although the AVC is the only legally incorporated local land trust and has held Authorization from the California Department of Fish and Wildlife since 2007.

[82] Wendy Reed, personal communication.

[83] The "SMMC, through DMCA, is uniquely suited," argued LA County DRP's director Richard J. Bruckner, in an August 2011 letter directed to AVC personnel, "to manage the Solar Ranch One mitigation lands given the mission and experience of its member agencies." Reed, Wendy. "Re: AV Solar Ranch I, First Solar." August 25, 2011. <http://avconservancy.org/Letters/LTR-LACountyDeptRegPlanning-AV_Solar_2.pdf> Accessed May 22, 2014.

County swiftly and conveniently relied on Antelope Acres Town Council's advisory "authority," and on the dearth of official and endorsed structures that incontestably operationalize public engagement in RE project development. This is hardly surprising, as local division along with conflicts of interest in designating RE project mitigation overseeing entities with the Antelope Acres Town Council and California State Parks personnel in fact facilitated and expedited Solar Ranch One's mitigation agreement.[84] Factionalism and conflicts of interest rendered the proposed racetrack property vulnerable to actors like DMCA executive director Joe Edmiston who have reportedly made themselves "invaluable to a powerful network of county and local officials who depend on…[them] to deliver" land in Southern California.[85]

How accurate is the interpretation that Antelope Acres residents chose DMCA as the overseeing authority for the Solar Ranch One mitigation? as outlined in the trade news article's excerpt in the beginning of this section? Clearly, on the one hand, Solar Ranch One's mitigation agreement—like the project's planning and permitting processes—involved unauthorized entities, and their representatives became entangled in professional and local alliances. On the other hand, both in practice and in theory, rural Town Councils do not represent their constituencies nor are they entitled to empower semi-autonomous million-dollar organizations like DMCA to act on behalf of local residents. That interpretation implies, echoing the developers and LA County's rhetorical strategy, that WAV residents actually had agency in organizing and actuating their participation in RE project development: it would be misleading, however, to contend that such agency was either carefully planned or meaningful.

In this fashion, rural resident factionalism and mistrust of solar PV projects advanced concurrently. The simultaneous growth of factionalism and mistrust undermined developers' *ad hoc* attempts to foster active community engagement in the WAV, and partly determined the fate of LA County's inaugural wind project proposals as well.

Sowing seeds of Mistrust: From Non-Transparent Permitting and Mitigation to Boomerang Construction Hurdles:

---

[84] Reed, Wendy. "Re: AV Solar Ranch I, First Solar." July 6, 2011. <http://avconservancy.org/Letters/LTR-LACountyDeptRegPlanning-AV_Solar.pdf> Accessed May 22, 2014.
The conflicts of interest regarded one Antelope Acres Town Council member that worked with Jim Dodson's RCD and one FAVOS member that is allegedly connected to the State Park system.

[85] Macadams (2005): 46.

Setting:

Wall Street speculators commenting on Solar Ranch One's financial complications, February 10, 2012.

"Shares in First Solar fell nearly 9% Friday after the solar panel maker and project developer reported a delay in funding for a massive solar farm the company is building near Los Angeles....If the construction permit [suspension at the Solar Ranch One project site] and subsequent funding don't come through by Feb. 24 [2012], the company may be forced to cancel a deal it has struck with the utility Exelon to buy the solar farm, known as Antelope Valley Solar Ranch."[86]

"We believe Exelon will NOT exit Solar Ranch One over minor issues like a construction permit. Solar Ranch One is a large, viable, prestigious, and controversial (due to local resistance, compensation and development cost issues) solar project. *We expect it to proceed*."[87]

Setting:

Letter entitled "First Solar versus WAV" posted on the Oso Town Council's website, dated October 8, 2012.

First Solar has tainted the relationship between locals and themselves and any other energy company with plans for our valley. Without local approval the County will need to steamroll over the objections of the community if they want to see our valley become a showpiece for renewable energy. Or, if the County takes the residents into consideration, maybe First Solar will become the solitary energy company in the valley because they'd poisoned the situation for everyone else... Fool us once, shame on you. Fool us twice, shame on us.[88]

By highlighting their role in securing the construction of Solar Ranch One and adjacent Alpine solar projects and thereby bringing semi-hallowed jobs for federally supported PV projects, LA County and First

---

[86] Hargreaves, Steve. "First Solar shares plunge after project delay." Money.cnn.com, February 10, 2012.
<http://money.cnn.com/2012/02/10/markets/first_solar/index.htm?iid=EL>
Accessed May 23, 2014.

[87] Unknown. "Auriga Maintains a 'Buy' on First Solar (FSLR); The Antelope Raises its Head Again." Streetinsider.com, February 10, 2012.
<http://www.streetinsider.com/Analyst+Comments/Auriga+Maintains+a+Buy+on+First+Solar+(FSLR)%3B+The+Antelope+Raises+its+Head+Again/7159467.html>
Accessed May 24, 2014. Emphasis added.

[88] Manuscript, courtesy of Jan Hellsund.

Solar executives sought to counter the confusion of combined permitting uncertainty, financial hurdles, and local resident outcry.[89] On their part, by envisioning Solar Ranch One as an albeit temporary solution to the Valley's hemorrhaging economy and by propagating that vision to their constituencies, Town Council leaders may have facilitated the manipulative collusion between County and RE project developers. Oso Town Council members, for instance, went so far as to buy First Solar stock and attended the company's annual shareholder meeting in May 2012, reporting that "the County was more responsible for impediments to continued construction [at Solar Ranch One] than the company."[90] One month later, still assuming their landscaping objectives would be met at the periphery of Solar Ranch One, the same Town Council leaders together with twenty-six laid-off solar workers attended an LA County Board of Supervisors meeting, attributing the certification standard dispute to Michael Antonovich. The dispute was readily resolved three days after WAV residents made an appearance at the Board of Supervisors meeting. In contrast, the requests of residents for landscaping at Solar Ranch One were dealt with in a much less synergistic fashion by First Solar and Antonovich, although they triumphantly congratulated each other on the fruits of a "very collaborative process."[91] The gap between urban and rural politics in LA County hence fostered community manipulation while it literally shaped the RE landscape in the WAV. This gap also disconnected the use of life cycle metrics from a participatory framework of community engagement that might have added social grounding to those metrics.

---

[89] Although First Solar was the developer in Alpine, that significantly smaller project was not supported by a federal loan guarantee.

[90] Trabish, Herman K. "Update: First Solar 'Furloughs' Half its Solar Ranch One Workforce." *Greentech media*, May 30, 2012.
<https://www.greentechmedia.com/articles/read/First-Solar-Furloughs-Half-its-Solar Ranch One-Workforce-1100MW-in-Jeopardy>
Accessed May 28, 2014.
Describing how LA County officials, instead of facilitating collaboration between developers and WAV groups, in fact nullified the effectiveness of rural Town Councils, one interviewee noted that: [the meetings between planners, First Solar personnel and Town Council representatives were "interesting [in that] the planning and zoning administrators... [summoned] with the Solar Ranch One people and were sitting there…It was basically obstructing. It was a way of providing so much inane information…[that] did not get down to all the real points of concern that… [locals] were there to accomplish… The County Liaison was right there, more or less facilitating what was going on…"

[91] Alan Bernheimer. "LA County Approves First Solar to Start Module Installation at AV Solar Ranch One." Investor.firstsolar.com, June 22, 2012.
<http://investor.firstsolar.com/releasedetail.cfm?ReleaseID=685651>
Accessed June 5, 2014.

NRG Energy Alpine Solar Project

NRG Energy-owned Alpine Solar is Solar Ranch's non-federally supported sister project. Conjointly built by First Solar in the same region and featuring the same Series 3 PV thin-film CdTe technology, the Alpine facility was granted a CUP exactly two weeks after Solar Ranch One's permits were authorized by the LA County Board of Supervisors, in early December 2010. The same argument—shared by the Fairmont and Oso Town Councils in the summer of 2011—that "no member of NRG ever met with residents" was conveyed at the Alpine Solar project's controversial debut in the Valley.[92] As it was constructed with private rather than government funding, however, the Alpine Solar facility did not embody the "green jobs" rhetoric to the degree that Solar Ranch One did. The laborer salary at the Alpine Solar site was $16 per hour less than the $42 per hour "prevailing wage" offered at neighboring First Solar. What is more, Alpine's adoption of a Mitigated Negative Declaration implied that public input was redundant.[93] This meant the process remained fundamentally unstructured: the arrangements between rural Town Council leaders and NRG engineers existed yet were not held accountable.

The resolution of Alpine's mitigation transactions between NRG executives and the project's adjoining communities was equally murky. In early January 2011, the same actors who proclaimed a short-lived support for Solar Ranch appealed Alpine's negative declaration CPU. In downtown LA, the DRP was able to orchestrate the rejection of the local environmental organization coalition group's appeal, while a second appeal filed by one of Fairmont's residents was privately settled in the Valley before reaching the County's Planning Commission meeting on March 30, 2011.

In settling the latter appeal, Fairmont Town Council leaders signed a confidentiality agreement with NRG engineers. Based on this document, Fairmont was not allowed to acknowledge, notify, or advertise the specifics of the mitigation until on-site construction was launched. Subsequent reporting indicated that the deal included $300,000 paid in several installments in cash to Fairmont Town Council over the lifetime of the project, in addition to a number of acres preserved for project mitigation. The deal specifically required that the Fairmont Town Council acquire tax-exempt status under section 501(c)(3) of the Internal Revenue Service

---

[92] Fairmont Town Council. "Reasons for Appeal." January 4, 2011.

[93] This is a permitting status endorsed by County planning authorities which allowed the developer to avoid conducting an EIR as well as the corresponding public hearing requirement mandated by CEQA.

(IRS).[94] It then transcended Solar Ranch One's legacy in that it became grounded in negotiations between RE project owners and WAV locality representatives *before* it was even fully permitted. Alpine's permitting process showcased an "everything goes" politics of RE project mitigation, characterized by a combination of precariousness and unaccountability.

Setting

Oso Town Council president Richard Skaggs writes to First Solar's Jack Pigott on July 9, 2011, Neenach, California, eight miles away from the Solar Ranch One project site.

"Despite all the rancor I hope you've been getting the impression that the people here in West Antelope Valley are actually mostly in favor of sharing the area with solar energy companies. They might argue over the details but they're generally on board...One exciting thing the Oso Town Council... is considering is incorporation as a city, which would allow us to control our own futures and not have our lives dictated by the County."[95]

Setting

Fairmont Town Council President David Kerr recalls his experience of dealing with mechanical engineer and NRG Alpine LLC project manager Keith Latham during private negotiations between the solar company and Kerr's Town Council, with regards to the appeal that Fairmont filed against Alpine. Fairmont, California, two miles away from the Solar Ranch One project site.

[Latham] is a pleasure to work with...[;] most unlike the used car dealer types from the other solar company [First Solar] in the area...[I hope] Keith will do other projects here in the future.[96]

---

[94] Interview data. As it was reported in the local press, NRG executives promised that the release to the public of the terms included in the contracts between NRG and Fairmont/Oso Town Councils would take place in June 2011. This never happened, while the first payments to the Oso Town Council were made in May 2013. Hedlund, Patric. "Valley Voices. News from the Western Antelope Valley." *The Mountain Enterprise*, December 14, 2012; interview data.

[95] Manuscript, courtesy of Richard Skaggs.

[96] Trabish, Herman K. "How does a solar power plant developer win over the community?" *Greentech media*, October 11, 2011. <http://www.greentechmedia.com/articles/read/how-does-a-solar-power-plant-developer-win-over-the-community> Accessed May 22, 2014.

Interview material:

NRG was open about what they were doing, and that was the difference; they worked with us. They were very open about what they were doing, whereas Solar Ranch One was very closed...We had some input in Alpine, while Solar Ranch One folks wouldn't have anything to do with us. ...It got to the point where NRG even gave away chairs to our town council so they could have something to sit on. In other words, NRG engineers worked with the community and Solar Ranch One did not... It was a success; that was the main difference.

They [NRG personnel] listened to us, if that makes any sense. They listened to us and did everything they could to...it's the wrong word...[,] but appease us is the best word I can think of right now... [T]hey not only went to [mitigate with] the county, but they went to the individual people themselves.

As tempting as it is to interpret Alpine's mitigation settlements as manifestations of community engagement, I explain in this section that the project's permitting resolution reveals an environmental and social blind spot when it comes to transparency and procedural justice. Alpine's fiasco was transformed into a "success" by a coalescence of the Town Council structure, and LA County's allowance for illegitimate mitigation arrangements between developers and Town Council representatives.

Not long after WAV rural Town Councils were formed or revived in order to dispute the racetrack, they got wind that money could be made by patronizing the RE companies that had been flocking to their area. Cloaked in seclusion and behind-the-scenes deals, Fairmont Town Council's arrangements with NRG—later duplicated by Oso—subverted individual residents' demands to become part of Alpine's design process throughout the EIR. Yet, Fairmont Town Council later changed its stance. Allowing a negative declaration CUP for Alpine "sets a terrible precedent for the standard of care expected in the design and construction" of RE engineering projects, stated a 2011 Fairmont Town Council letter that contained the rationale behind filing an appeal with the County planning authorities.[97] An EIR carried out in accordance with the intentions of CEQA may have preserved a consolidation of local knowledge and procedural justice in the WAV. Thus, Fairmont's flip-flop on its decision to challenge

Likewise, on March 17 2011, a Fairmont Town Council member was surmising that "[t]his very nice gentleman...from NRG...they've been working really hard with us. They are an example of what the rest of... [the developers] should be doing," author's transcript.

[97] Fairmont Town Council (2011).

NRG demonstrates that contentiousness was a persistent characteristic of the design of RE projects in the WAV.

Interview material:

[NRG engineer and Alpine Solar project manager Keith Latham] is quoted to have said he was "working" with us in Antelope Acres,[98] but when I saw him at that meeting with the newly formed Fairmont town council, I told everyone he was a liar. He kept saying he mailed us this and he did that… and he never actually did.

The only project we, here in Oso, caught in time was Alpine…We actually met…with the engineers from NRG. They were offering something like $10,000 for us to let them move with the project—they were offering a very low amount…[;] they didn't say "we'll give you this to go away," but... So, then we walked out—walking back to go to the bathroom and one of them says "c'mon let's talk a bit more." They were very business like, so we decided we are going to continue this…We were two people left and we were running down to the last day for their final approval. So, we either had to show up there with approval or say no to them [and appeal the project]… Eventually we did it mostly by emails and phone calls…in the last two days we went back and forth a hundred times and all I was able to do was to say "we got to get basically the same contract as Fairmont." And we added a clause there that says if some other town council or organization has a better contract than ours, then ours will improve likewise as well…

So the ensuing fiasco that came about over Alpine, was that NRG petitioned the county en lieu of doing an EIR, to do a mitigated negative declaration. Well, that project is only five miles away from Solar Ranch One... The same lizards go back and forth. The same burrowing owls are on the property…

What's divided the Valley and created mistrust among rural folks and energy companies more than anything else is the finances going on through handing negotiating power to these unelected entities [Town Councils]. People out in Oso and Fairmont can immediately smell money—that's when Alpine was trying to get their permits and they were in a big hurry. A group…appealed NRG's project—[they were]… however, contested. But those two town council groups made secret contracts with NRG… They supported the project at the hearings down in LA; whereas [environmentalists] were legitimately trying to stop it…

[98] Trabish (October 11, 2011).

During that meeting at the Grace Chapel [the church out on California State Route138] the LA County folks admitted that the County was fooled by the negative mitigation of NRG—and never again are they going to approve a project without an EIR. And that was a flat out statement that they made at that time.

County facilitation of the Alpine permitting phase and the circumvention of State planning policies together with the California Environmental Quality Act was equivalent to harmonizing with private, "business-like" approaches to project mitigation negotiations, congenial to engineering corporations. Five miles down the road from Alpine, Nextlight followed an EIR procedure that was publicly challenged and took two years. NRG Energy, on the contrary settled mitigation measures with WAV representatives over email and a few privately held communications. In that regard, precisely because of LA County's commitment to the unauthorized structure of Town Councils, in February 2011, NRG engineers ceased to conduct meetings with representatives from the local environmental organization FAVOS—where the overseeing of the environmental and landscape substance of Alpine's mitigation by a legitimate conservation authority was at stake.

The loss of trust and legitimacy in RE project development in the WAV began before the arrival in 2011 of First Solar and NRG Energy in Fairmont, and it was a by-product of REAT's engineering of solar and wind energy "competitive zones." However, it was also deeply rooted in the loss of legality for democratic institutions in the WAV in the early 1990s, and as Supervisor Antonovich reestablished Town Councils and then oxymoronically merged them with rural LA County residents' imperatives to "control [their] own futures," questions of trust and legitimacy returned. Rural residents' claims to procedural authority were impaired even as they upheld the very foundation upon which rural LA County non-legitimacy governance rests. The opacity and contentiousness inherent in the development of California's inaugural utility-scale solar PV projects corresponds with rural Town Council unaccountability, nurtured and sustained by LA County. Having examined the politics and procedural justice pathologies of engineering PV projects in the WAV, the next section investigates why utility-scale wind projects were not equal to their solar counterparts in planning or execution, and thus did not proceed.

Blue Sky Thinking: Siting Utility-scale Wind projects in Rural LA County

As a result of Solar Ranch One's permitting hoax and Alpine's covert mitigation legacy, by the time NextEra Energy Resources (based in

Juno Beach, Florida) and Element Power US (Portland, Oregon) hosted independent scoping meetings in late 2011 for their respective proposed utility-scale wind projects in the Valley, local residents had thoroughly challenged the legitimacy of RE project development in the WAV. Moreover, mistrust and division had grown substantially between LA County, RE developers and rural communities. This section details an uneven record: while a low level of legitimation in permitting solar PV facilities of over 100 MW in LA County encouraged and expedited the proliferation of RE in California, the same attenuated legitimacy might have simultaneously facilitated the delay or cancelation of wind projects south of the Kern County segment of Antelope Valley. This contrast also provides insights into the dynamically changing attitudes and experiences of local residents as they learn more about the sustainability of RE projects. This comparison illustrates that local groups are indeed learning about RE and applying their new knowledge to the politics of permitting. Most notably, this section explains that RE projects should not be treated as having obtained stable communal consent once and for all.

Compelled by the high stakes of the utility-scale solar PV projects' permitting *process*, County officials justified their judgment to block the meteorological towers—required for proposed projects' wind energy analysis—by making assumptions about wind *technology*. "[Wind] turbines," in the words of Edel Vizcarra, Planning and Public Works Deputy under LA County supervisor Mike Antonovich, "pose[d] another set of questions." In the aftermath of a unanimous decision in January 2012 by the County Board of Supervisors to freeze NextEra and Element's plans to site meteorological towers on their adjacent properties located near Solar Ranch One,[99] thus blocking their attempts to secure financing for their facilities, Vizcarra added that "typically you're going to accept a solar project over a wind project where you don't have the same impacts to the wildlife."[100] In his statement, Vizcarra sketched the Valley's biological importance along with its relationship to habitat loss and wildlife corridor connectedness as the fundamental difference between utility-scale solar and wind facilities—which makes it harder for companies to secure planners and communities' approval for projects featuring last-generation wind turbines. Nonetheless, wind developers perceived "Blue Sky" and "Wildflower" (Table 7.4) as sustainable and transparent schemes, much like the participatory design

[99] The Board's decision overruled DRP's determination to award a CUP to both NextEra and Element.

[100] Currier, Craig. "The question: Turbine or not turbine?" *Antelope Valley Press*, Wednesday January 25, 2012.

Table 7.4 California's proposed inaugural utility-scale wind projects.

Project name **Area (acres)** **MW** **Electricity buyer[1]**	**Technology**
Wildflower 3708[2] Private 259 SCE	The proponents of the Wildflower project proposed four alternative design options: First, a hybrid design that combined wind and solar energy. In the northern section of the project site (Healy Ranch)—where the land was assumed to be 2% or less grading—the Wildflower would have featured 100MW single access tracking solar PV; the rest of the site would have been covered by 50, 3MW, 491-foot turbines (112-meter blades with a 336-foot rotor swept area). The second and third options would have included more, yet progressively smaller sized (2.3MW or 1.5MW), wind turbines. The fourth option discussed by Element Power included solar PV only.[3]
Blue Sky 7500 Private 225 SCE	Blue Sky's NOP illustrated three different technology scenarios for the facility: The first scenario proposed approximately 90 GE XL, 2.75 MW turbines; the second, approximately 90 Siemens SWT 2.3-93 101 2.3 MW turbines; and the third, the same number [90] of GE XL 2.5 MW turbines.[4]

1 Electricity produced in both projects is sent to Southern California's Whirlwind substation that is located about three miles away from Solar Ranch One.

2 Although Element Power has purchased close to 4,100 acres in the area of the so-called Healy Ranch (adjacent to the Fairmont Butte and the Poppy Reserve), the project was designed to cover 1,000 acres or less. Wildflower's border on the left side touched the Solar Ranch One project site. According to interview data, Element Power had purchased leases on the lands, not ownership.

3 Interview data.

4 LADRP. "Notice of Preparation, Blue Sky Wind Project." September 20, 2011. <http://planning.lacounty.gov/assets/upl/case/r2011-00408_nop.pdf> Accessed May 20, 2014.

Figure 7.6 Breeding mistrust and contention in RE project development. NextEra's Met Tower Notice of Public Hearing was, in fact, not viewable from a public road, placed as it was beyond warning signs reading "No Trespassing" and "Do Not Enter." Photo, Courtesy of John Calvert, photo editing, Courtesy of Margaret Rhyne.[101]

methods they reportedly planned to employ in the Valley. Solar Ranch One marked an assault on transparent siting and community involvement; by contrast, Wildflower's project manager declared in an early 2011 Town Council meeting in Fairmont that "all that I can commit to is that my company [Element Power] is trying to approach this sincerely and differently [from First Solar]."[102]

It may not be enough to confirm LA County planning officials' convictions that the fate of what was envisioned to be the first wind projects in LA County was actually dependent on generators' impacts on environmental sustainability. Nor can we assume that the most recent wind facility communication advances embody a reconceptualization of RE project development in the WAV as an open and participatory procedure rather than as an opaque and closed one. In fact, the similarities between utility solar PV and wind facility permitting pathologies in the WAV have proven to be stronger than developers and planning authorities alike had conjectured. The virtual inaccessibility of Next Era's Notice of Public Hearing, representing the dual failure of planners and developers to provide a proper facility intent notification to long-term residents residing adjacent to the Blue Sky and Wildflower project sites, just like in Solar Ranch One, was an additional symptom of a legitimacy crisis in utility-scale RE project siting in LA County (Figure 7.6). Similar to LA County's decision to exclude the Fairmont and Oso communities from Solar Ranch One's public

---

[101] NexEra (and, for that matter, Element Power's) CUP request for siting a meteorological tower in the company's project site was exempted from CEQA. The CUP, however, did require a "Public Hearing" to be held on DRP's premises in Los Angeles because the tower would have been higher than 85 feet, placed on a parcel greater than 2 acres. The appeal of NextEra's tower was filed by a local resident, John Calvert, who after being informed about Blue Sky by a neighbor, was able to trace the "Notice of Public Hearing" featured above in Figure 7.6.
[102] March 17, 2011, Fairmont Town Council meeting, author's transcript.

consultation, Next Era's first map of Blue Sky as well as Element Power's EIR did not show any residents living in the Lakes Communities of the West Antelope Valley.[103] This omission further excluded the participatory experience of locals in RE project planning. And bitterness and mistrust colored the relationship between wind companies and County officials, as well as between developers and long-term Valley residents. In sum, the failure to successfully begin utility-scale wind projects in the Valley was connected to the loss of procedural justice during the earlier politics of siting solar PV projects in Los Angeles County.

In contrast to the economic and political promptness in authorizing both Solar Ranch and Alpine Solar, this section argues that NextEra and Element Power's ambitions to kick off utility-scale wind project development in LA County were restrained by a confluence of circumstances. In particular, it is hardly accidental that the conditions that led to Blue Sky and Wildflower's cancelation converged just as the connections between solar PV facility development, unfair permitting, and factionalism became more obvious in rural LA County. These circumstances converted the notion of RE engineering in the WAV from something inevitable to something that could be fought against—even halted. It is, then, the contextualization of RE projects that is precisely the essential implication of the question of why it is that solar *seems* to have had a more favorable meaning compared to wind with the WAV.

First, the Blue Sky and Wildflower project sites were immediately adjacent to the Valley's treasured Poppy Reserve, a fact that was not lost on long-term residents, who promptly identified their rural identity with the physical boundaries of their local national park; they were also aware of the on-site presence of endangered fauna and flora species because of the park. A second circumstance involves suspected manipulation: supervisor Mike Antonovich firmly endorsed LA County's solar PV projects. The supervisor's ruling, for instance, had been instrumental in overriding US army contractor Northrop Grumman's objections to AVSR1, in spite of the firm's 2010 manufacturing work support for the Department of Defense that employed several thousand people[104] in Antonovich's supervisorial district. This political manipulation was becoming clearer to residents. Third, a small group of locals formed a new anti-wind action group called Concerned Citizens

---

[103] "The first map that we saw from NextEra showed no residents here [in the Lakes Communities]. And there are something on the order of thirty... It was like we did not exist here in this canyon." Interview data. See also Zahnter (2012).

[104] In all LA County, Northrop Grumman employed 15,000 people. The Board of Supervisors (2010).

of the West Antelope Valley (CCWAV) while maintaining a cautious distance from rural Antelope Valley Town Council politics of infighting. In a related fashion, it might have been that the local area is associated with solar energy—portrayed as "clean" and "new"—whereas there has been many years of experience with bird kills in wind farms elsewhere in California that WAV locals were familiar with. As a fourth and final circumstance, the wind industry was not especially interested in single-mindedly advocating for wind facilities under difficult permitting politics. The industry preferred to successfully commission projects already approved elsewhere.[105] The fact that wind developers sought to endow their projects with an aura of environmental *and* social sustainability (unlike the solar industry) in a time of acute uncertainty for the industry is worth investigating. Moreover, the Fairmont area had a particular chronology: solar developers arrived first—what if it had been wind developers coming first?

Attitudes one is Bound to Mistrust: NextEra Energy and Element Power's Attempts to Reset RE Industry Local Engagement Standards in the WAV

Setting

January 20, 2011, and April 21, 2011, Wee Vill Market, Fairmont, California, two miles away from the Solar Ranch One project site.

[Blue Sky's project manager] Cliff Graham of NextEra shows photos of the 472-foot next-generation wind turbines his company plans to build. But Graham has assessed the situation in the room accurately, and makes a shrewd move. Before he goes into any detail, he woos the crowd by announcing he's done four deployments in Iraq as a Green Hornet supporting Special Operations—a master move on the public presentation strategy board that trumps the thick attitude from several of the more confrontational fellows in the crowd. Part of the "guy culture" in this wind-whipped valley is wrapped in tales of military or intelligence service duty....[So] Graham has another round in his chamber. He lets it casually drop that he has been a pilot since age 12. A hush of respect falls over the meeting...[106]

....

[105] The Production Tax Credit (PTC) threatened to expire in late 2011 and early 2012.

[106] Hedlund, Patric. "Neenach-Fairmont Area Meet Energy Developers-Part Three." *The Mountain Enterprise*, February 11, 2011.

"One of the hardest things we are going to have to do is show people what is different about us," [Nat] Parker [Wildflower's project manager] said, making a focused pitch for community support. "We've done a year of raptor study in helicopters and done burrowing owl studies, figuring out how we can avoid impact to bats and other species."[107] "We have looked at the biotic resources, the unique vegetation communities, the wildlife movement behaviors we've observed for a year, and the things that are important to the people in the community," [Element Power's Nat] Parker said, "and we have attempted to incorporate mitigation into the very design of the project..."[108] "[In other words, we] have not baked the project and said 'here it is.' We want your input," [he added].[109]

Setting

November 16, 2011, Antelope Acres Town Council Meeting, Westside Community Church, Antelope Acres, California, ten miles away from the Solar Ranch One project site.

[Siting the Wildflower wind project] is a big issue[,] [Nat Parker asserted while addressing a crowded room in Antelope Acres]; [this is] complicated stuff... But we need to have a dialogue where we put the facts on the table[;] and I know that people here are emotional. But let's just talk about the facts...[110] Let's be scientific, [Parker concluded], we invite your input [on Element Power's environmental studies performed for Wildflower].[111]

Interview material

We were all suspicious of the town councils—their secret dealings with solar companies and their meaningless yet continuous fighting over jurisdictional border zones—so we formed a group of our own. Hence even

---

[107] Hedlund (April 29, 2011).
[108] Trabish, Herman K. "The biggest US wind-solar hybrid project." *Greentech media*, October 18, 2011.
<http://www.greentechmedia.com/articles/read/element-power-announces-specifics-for-the-nations-biggest-pre-planned-wind->
Accessed May 2, 2014.
[109] Hedlund (April 29, 2011).
[110] November 16, 2011, Antelope Acres Town Council meeting, author's transcript.
[111] Trabish (November 29, 2011).
<http://www.greentechmedia.com/articles/read/element-power-first-solar-and-intense-local-politics>
Accessed April 28, 2014.

before the January 2012 hearings [that determined the fate of NextEra and Element Power's Met Towers] we started following the wind companies in every single town council meeting. If Element was going to show up at Fairmont, we were there. If Element was going to be at Antelope Acres, we were there... Nat Parker just wept—"oh, you again."

To some extent, the interactions between groups of long-term WAV residents and proposed wind project managers Cliff Graham and Nat Parker showed a curious selectivity in the approach of RE proponents in the Valley, as well as the dysfunction of locals' experiences with corporate representatives. Graham's "master move[s]" in project communication strategy epitomizes the relative sophistication of an industry gradually scaling-up community engagement in a wind project's hierarchy of priorities. Because of their more elaborate communication codes, NextEra and Element Power's representatives in the WAV were able to maneuver themselves and their projects into positions of authority. Instead of employing, as did First Solar, a non-local, fresh-out-of-college community liaison whom interviewees described as a "chameleon," Element Power assigned community affairs to a local management expert who was also former chief financial officer with the AVC—a conservancy located in the WAV. Whereas NextLight—and later First Solar—had assigned the management of communal conflicts at Solar Ranch One to the unforthcoming and arrogant project developer Jack Pigott, Element Power reached out to former Oregon Sierra Club manager Nat Parker. Besides being the sole RE developer to locate offices in the WAV in 2010, Parker legitimated his communication not only with the argument that Element would provide "green energy," but also with his personal position:[112]

> [Element Power is] not going to be able to be here if you don't want us here. And the reason that I know that is because I don't come from a development background... So, I have been in this room before. I have helped lead citizen groups like your own to get together, to get organized, to become empowered, and to stop projects...

Community empowerment was supposedly one of the defining corporate features of Element Power. Formed in 2009 by engineers and executives who had previously been involved with top international corporations like Iberdrola Renovables, Horizon Wind Energy and EDP Renewables,

---

[112] Nat Parker speaking at the Fairmont Town Council on March 17, 2011, author's transcript. Nat Parker staffed his Antelope Acres office by hiring Antelope Valley Conservancy's board member Sean Ponso.

and backed by financiers who used to run Goldman Sachs' Clean Energy Investment Fund, the Element Power US team was built on its founders' assumption that "the bigger the company, the worse a job it was doing at working with people locally."[113] Like former Iberdrola and Horizon executives, who founded Element to catalyze a structural transformation in the utility wind business, the Wildflower project manager's declaration of his environmental organizing background presumably meant that he condemned an approach to RE development that did not examine social sustainability issues. Thus, on the one hand Parker publicly and openly invited Wildflower's opponents to "put facts on the table" in order to demonstrate not only a different communication perspective—that solar project executives apparently lacked—but also his company's intention to integrate impacts mitigation into Wildflower's design.[114] The facility's four alternative design proposals (Table 7.4) and Parker's own claims to getting people "on board," for instance, sketches a possible basis for infusing local concerns into the engineering of Wildflower.[115] On the other hand, from the perspective of Wildflower's advocates in the WAV, an ideal decision-making process ought to be kept unemotional.

When NextEra's Cliff Graham sauntered into Town Council meetings and pressed the buttons of agitated rural citizens, for example, the RE developer's professional identity was no longer unresponsive nor pompous. Furthermore, Element Power's $100,000 contribution to a student program at a WAV historic park six months after the Board of Supervisors denied the Met Tower permits also demonstrates that wind companies were anticipating the controversial nature of RE development, as well as the emotionally laden perspectives of WAV communities.[116]

Yet where does this leave wind companies' approach as means to integrate social and environmental sustainability in RE project development? With solar PV projects' problematic siting legacy and the ineffectual Town Council structure that provided the common setting for RE facilities in the Valley, LA County's proposed wind projects served as channels of animosity and mistrust. For instance, the community investment package that

---

113 The names of the individuals and the Fund are: Ty Daul, Raimond Grube, Chris Taylor, and Hudson Clean Energy Partners, a Teaneck, N.J.-based private equity, respectively. Nat Parker speaking at the Fairmont Town Council on March 17, 2011, author's transcript.

114 The names of the two proposed facilities are, for example, illustrative of wind developers' more sophisticated communication strategies.

115 Breault, Rich. "Solar, wind: Learn plan, enjoy early Halloween." *Antelope Valley Press*, October 25, 2011.

116 Currier, Craig. "Developer again seeks energy plant near poppy reserve." *Antelope Valley Press*, August 23, 2012.

Element Power brokered with the Poppy Reserve was quickly branded by locals as a buy-off deal, as were Wildflower project manager's assertions that RE development in the WAV be seen as a "wholesale equation."[117] Likewise, Element Power's undertaking to establish a community investment group—a 501c3—with member representatives chosen by the Valley's Town Councils to administer an annual budget for investment in environmental education, habitat renewal, etc., was locally perceived as "doing masterfully something in lawyer terms[;]...tak[ing] all the local mitigation money and [making] it dependent on LA County..."[118] By the end of 2011, legitimacy was undercut within the broad network of RE project development in the WAV, dramatically limiting NextEra and Element Power's efforts to build trust in Blue Sky and Wildflower, respectively, even as these facilities represented a strategically improved approach for reaching out to local communities.

Interview material

The companies thought that they would outdo us financially, but they did not realize that after all this work, organizing, and relentless outreach to the local community we still had another 800 bucks for another appeal.

Our group definitely made an intervention in the decision about the wind projects. For example, I have to give one guy out there an awful lot of credit with how good of a job he did at presenting his arguments to the County folks. Personally, I got some equipment and I started shooting tricolored blackbirds that are going to go extinct in our area because of the [RE] projects. I took photos of rare hawks, bald eagles, and burrowing owls... By all means these photos did help to stop the wind turbines... Some were presented to the board of supervisors [in January 2012] and Antonovich asked if he could have them...

Our anti-wind activism was a real team effort. We've formed a pretty cohesive group—no town councils, no backstabbing, no big money deals—and it all clicked. We were amazed at what we accomplished.

Prior to January's meeting, after we had submitted the second appeal [against the Met Towers] we made an appointment and about nine of us

---

[117] "But the real question is," said Nat Parker on March 17, 2011, while addressing a Fairmont Town Council meeting whether "there [is] any net benefit to your community that you can extract out of this admittedly painful process..." "Meaning, what's our price?" scoffed one local resident. Author's transcript.

[118] This idea of the community investment group was based on a similar dotcom mitigation mechanism employed in the case of the Santa Clarita landfill. Interview data.

went to Antonovich's office in downtown LA ...We actually sat in his office and each of us in turn said what we had to say to the supervisor. After that meeting we were beginning to think already that maybe the supervisor wasn't all that enthusiastic about these massive towers. So all of the work we did simply made it easier for him to halt the Met Tower...

Conclusion

Wrong Time, Wrong Place

"One big thanks to Mike Antonovich...for voting in our favor, against the Wind Turbine Projects on Portal Ridge a year ago," reads a letter written by a resident of the Lakes Communities in the aftermath of a fire that burnt more than 30,000 acres in the Western parts of the Antelope Valley during the early summer of 2013. "If 500-foot tall Wind turbines had been allowed to be built on our mountain, our Kings Canyon community of 35 homes would be toast," they added, explaining that the realization of wind projects would have rendered impossible the effective retardant drops at 30-40ft from the ground level. When Antonovich, one of LA County's five supervisor-kings, read his statement in favor of CCWAV during the January 2012 Board of Supervisors meeting, some WAV locals sat back in disbelief, others burst into tears. Echoing LA County planners' insistence about industrial wind turbines' detrimental effects to environmental sustainability, a subsequent note from the supervisor's district liaison to CCWAV members argued: "we must stay vigilant in protecting the valley."[119] In a highly factionalized environment such as the WAV, and by buttressing the resolve of a local minority group against RE development, LA County politicians carried their flag as protectors of the Antelope Valley and stood out as conciliatory in their publicized endeavor to bring conflicting parties together.

On the one hand, Antonovich's ruling to at least temporarily inhibit the construction of utility-scale wind projects in LA County reflected the social and environmental sustainability-related considerations of Lake Communities' rural residents who, ironically, one-and-half years prior had suffered through a devastating fire in the WAV. They had foreseen that siting 200- to 500-foot engineered structures in the vicinity of Portal Ridge

[119] People in the Valley refer to the five supervisors as kings because they have an unusual amount of power in an elected office. Manuscript provided to the author.

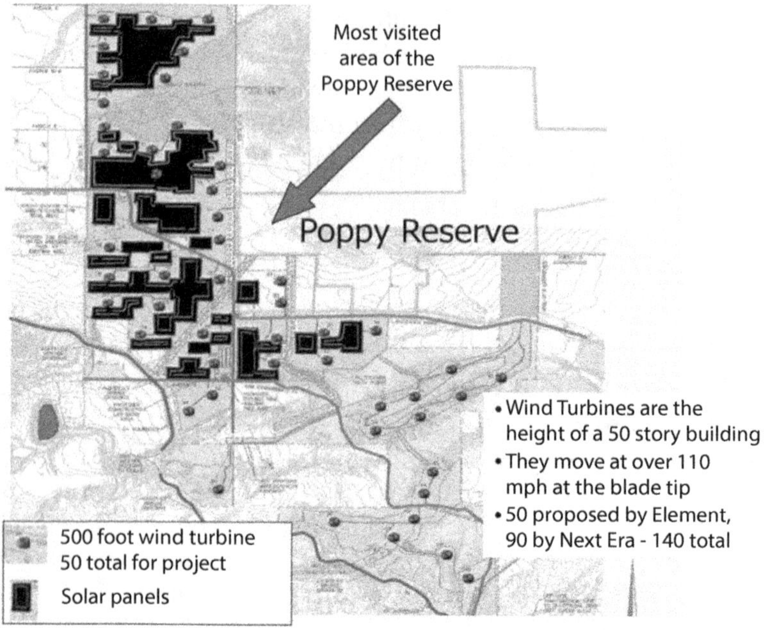

Figure 7.7 An unsustainable idea: Element Power's design proposed the erection of fifty 500-foot-high industrial turbines on the western and southern borders of the Valley's Poppy Reserve. Image Courtesy of Margaret Rhyne.

would have interfered with the protection of their homes and properties.[120] On the other hand, the sustainability of Wildflower as a concept in and of itself strongly depended on the facility's proposed location (Figure 7.7). That is, for all its rhetoric about "the Right People, Right Place, Right Project," what may have symbolized the kiss of death for Element's initial venture in the Valley was the proposal which suggested that industrial wind turbines and poppies—considered sacrosanct—would coexist as mirrors of one another in both relational and physical proximity.[121]

"Is the reverence and legacy of dedication that local citizens have for the Poppy Reserve," inquired a December 2011 letter submitted to LA County by the Park's president, "compatible with the mechanized structures of a massive industrial energy plant?" The letter encouraged locals to consider the effects of Wildflower's potential construction on rural residents'

---

[120] A letter written by Susan Zahnter dated January 5, 2012, wrote: "How will increased access eliminate the need for further analysis if firefighters will not fight fires in areas with electrical hazard, and who rely upon firefighting aircraft for support?" See Zahnter (2012).
[121] Element Power. Brochure." Undated. Courtesy of Element Power.

identity, the types of effects that developers dismissed as NIMBYism. The author inferred that Element's hybrid design would result in "this source of so much local pride [to be] dwarfed by manmade vistas of giant steel towers set in a sea of glass."[122] In fighting Element Power, the Poppy Reserve was considered to be simultaneously the aesthetic dimension and the environmental equivalent of a rural "historic experience." In this context, there was only room for wildflowers in the WAV.

Adjacent to an existing SEA and contained within a proposed one, Next Era's wind facility was similarly viewed as a manageable environmental externality problem. Moreover, if its project site had been located north of Lake Hughes and Lake Elizabeth, for example, Blue Sky would have been part of an area classified by the Audubon Society as an Important Bird Area (IBA), providing key habitat for species such as the Mountain Plover and the Tricolored Blackbird. Besides, the site itself—a wildlife corridor—bridges the Desert Pines Wildlife Sanctuary with Ripley's Woodland State Park as well as with WAV's emblematic Poppy Reserve. For most rural WAV residents who, a few months before NextEra and Element's scoping meetings, had been presented with two solar facilities whose design manifested the evasion of public input, the Valley's unique habitat offered a way to reclaim their identity as effective members of that ecology.

By mid-2011, the lands surrounding the Reserve were viewed by WAV residents as a means for rural emancipation. They thus offered locals the opportunity to define their lifestyles in ways that Solar Ranch One and Alpine were seen as categorically violating. If the planned solar PV facilities had deprived rural residents of their chance to integrate their voices into RE project development, the permitting fight against utility-scale wind projects offered them both an opportunity to defend the ecological richness of their area and a chance to restore a vision of preservation in the WAV: "Just as the City of San Francisco could have solved its water problems by…[not damming] Hetch Hetchy Valley, Los Angeles County does not need to destroy Portal Ridge and the unique biological area of the Western Antelope Valley to help California meet its goals for renewable energy generation."[123]

To some degree, the formation of the western Valley's first anti-wind coalition reflected disgruntlement over Town Council local politics, in the same way that claiming the Poppy Reserve as a source of local pride and identity had partly been a reaction to the impression of a legacy of

---

[122] Rhyne (Dec. 31, 2011).
[123] Rhyne (October 6, 2011).

illegitimacy that big solar had left with WAV communities. "Don't let the council do your thinking or make decisions for you," prompted one CCWAV member in a letter to an Oso Town Council official. "Speak up!" Meanwhile, wind activists attempted to restore a sense of political representativeness in the Valley by organizing themselves outside the Town Council networks, reaching out "door to door" to other WAV groups. Yet the factionalism that plagued rural Town Councils made envisioning a politically harmonious alternative to the rural councils bitterly unrealistic. As some Fairmont residents put it, in their practices, CCWAV members were "obnoxious, browbeating and flat intimidat[ing]; they were, in short, nothing but "NIMBYs."[124]

In all respects, CCWAV's vision reinforced the lack of a formal structure to assure that rural residents' concerns would be conveyed to the County, and lacked a format that was representative. And like the critique raised against Solar Ranch One, CCWAV members made the argument that the operation of RE facilities would have interfered with their rural lifestyles and unobstructed vistas—indeed the very reasons most locals had moved to the WAV in the first place. The individuals who formed CCWAV did not seem to have been interested in the money that some of their neighbors were either making, or trying to make by dealing with NRG Energy and First Solar. For them, it appears, "[m]oney [could not] buy peace of mind."[125] A local who later co-formed CCWAV explained to NextEra's lawyers, when they approached him at the end of the first public hearing on the company's proposed Met Tower CUP on March 15, 2011, that he did not expect something "out of the deal"; his position, instead, was that he desired to "trip [NextEra] up at every opportunity. To make sure [engineers and corporate executives] followed the rules."[126] A similar attitude characterized the group's reaction to MidAmerican's mitigation arrangement with the Transitional Habitat Conservancy over 2,000 acres of hawk

---

124 Manuscript, courtesy of Judy Watson. See also comments on the online forum avhidesert, January 25, 2012.
<http://www.avhidesert.com/forum/showthread.php?tid=2181>
Accessed June 2, 2014.

125 Manuscript, courtesy of Judy Watson.

126 Calvert, John. "Email from John Calvert." AVHIDESERT, March 18, 2011.
<http://www.avhidesert.com/forum/printthread.php?tid=1663>
Accessed June 2, 2014.
According to John Calvert, the NextEra lawyer "indicated that NextEra could go easy on [him] or go hard if [he] did not cooperate. That if [he] did not file the appeal things would be a whole lot better for [him] and [his] family. [The NextEra lawyer] wanted to know what [he] wanted out of the deal..."

habitat: "The money to buy the land had actually come from one of these green energy companies," explained one rural resident as they managed a bittersweet smile. The oxymoronic feeling that was conveyed to CCWAVers as they witnessed the conversion of the area that would have been "Blue Sky" into conservation land was that "we just sold our soul to the devil."[127]

CCWAV members were adamant in their determination not to settle for wind project mitigation arrangements. Their efforts were significant in that the Blue Sky and Wildflower projects became challengeable. Over the period of the Met Tower permitting process, long-term WAV residents developed a political view and a practical strategy to confront the wind companies.

For instance, while Cliff Graham had struck a public relations victory pose in Fairmont, NextEra's record did not yet portray an image of earnestly appreciating desert residents' concerns. As CCWAV members argued, quite the contrary was true in light of a lawsuit that several conservation groups, including the Sierra Club, had raised against the developer's North Sky River Project in adjacent Kern County. Besides information regarding NextEra and Element Power's controversial activities, concepts contained in interminably long EIRs or project brochures were extracted from their internalized, taken-for-granted, rhetoric and made sensitive to the interests of WAV locals. In this context, facility opponents stirred up a challenge on developers' own turf.

Commenting on a visual simulation adapted from one of Element Power's color brochures, a CCWAV member compared the plants that would have supposedly grown underneath Wildflower's panels—in both shade and full sun—with the bare ground and non-native weeds that "decorated" solar module structures in nearby Lancaster (Figure 7.8). "What are [Wildflower's] plants," the member inquired, surmising that Element Power had yet to disclose that information. "Will they spread to the Poppy Reserve, crowding out native wildflowers? How will they impact existing wildlife? How will planting and growth be managed?" Thus rural anti-RE politics in the WAV, vague and unsubstantiated as they had been in the case of solar PV facilities, became, in a sense, "scientized": "[According to] biologist, environmental consultant, and expert in windfarm avian mortality..." wrote an activist, quoting a 2008 *Environmental & Energy Law & Policy* article, "guywire-free met towers" should not be used at all in the Valley.[128]

---

[127] Interview data. The mitigation regarded the "Antelope Valley Solar Projects."

[128] Zahnter, Susan. "Re: Met Tower Appeals, NextEra Blue Sky Windfarm, Project No. 201001402-(5), CUP No.201000136; Wildflower Green Energy Farm (County Project No.

Figure 7.8 Challenging developers on their own turf. Comparing Element's visual simulations of Wildflower with existing solar PV structures in Lancaster, California. Image courtesy of Susan Zahnter.

To some degree, the allegation raised against CCWAV by Fairmont folks, namely that the group lacked the "evidence that the [meteorological] towers were supposed to have provided" resonated with the highly factionalized environment under Antonovich's rural Town Council structure. Notably, such insistence upon getting access to information—in this case a quantified study of the area's wind potential—reflected locals' deep sense of exclusion from the trust and justice that a more participatory and legitimate solar PV permitting process would have ensured. The allegation also echoes the opportunism—a result of an "anything goes" framework for mitigation arrangements between Town Councils and RE companies—that had become part of the Valley's economic reality.

Perceptions of Community Engagement

RE companies learned the hard way that it is essential to incorporate community engagement in every single phase of a project's life cycle. As Joyce McLaren, currently an energy analyst with NREL has shown, the early 1990s witnessed a large number of government-funded, top-down-implemented, wind projects in the UK that encountered significant pushback both from communities and from local governments. Meanwhile, local

R2010-00256-(5),Conditional Use Permit No. 201000121, Environmental Review No. 201000063." January 23, 2012, courtesy of Susan Zahnter.

planning offices lacked experience entirely with siting wind generators, and opposition groups gained significant momentum. McLaren points out that many of the aspects that are currently considered part and parcel of "community engagement" in RE practice—notably participatory design methods—were in fact the criteria that determined whether a facility would make it through or not.[129]

By the late 1990s, the National Wind Coordinating Committee (NWCC), on the other side of the Atlantic, had begun documenting the procedures—purportedly empowering—through which RE companies had sought to structure wind project development in the US. In a 1997 paper entitled "Siting issues for wind power plants," the NWCC argued that local community involvement was, by design, an essential constituent of engineering RE projects. "[W]ithout public involvement," the article alleged, "a much greater likelihood exists for later opposition and costly litigation." Although there was no one clear approach to engage the public in wind siting, the authors admitted, the wind industry apparently already had extensive expertise with community involvement: the industry's imperative was to forge and sustain a participatory structure involving engineers and permitting agencies.[130]

Throughout the wind industry's later years, the precepts for a more fair and inclusive permitting framework were embraced by professional organizations, national governments and project developers alike. This framework was to be integrated with academic research that verifies the critical nature of early local involvement for successfully introducing RE engineering projects. In early 2011, the Canadian Wind Energy Association (CanWEA) released its *Best Practices for Community Engagement and Public Consultation*, in order to bolster developers' efforts "in continuously improving their work with local communities while ensuring that they meet *and strive to exceed* provincial requirements for public consultation."[131] In the same year, Australia's Clean Energy Council created a new position for a Community Engagement Manager in collaboration with the national wind energy indus-

[129] McLaren Loring, Joyce. "Wind energy planning in England, Wales, and Denmark: Factors influencing project success." *Energy Policy* 35, no. 4 (2007): 2648–2660.
[130] NWCC. "Siting Issues for Wind Power Plants." *Wind Energy Series*, no. 3 (1997). <http://nationalwind.org/wp-content/uploads/assets/archive/Issue_Paper_3.pdf> Accessed May 22, 2014.
[131] CANWEA. "Best Practices for Community Engagement and Public Consultation." December 2013.
<http://canwea.ca/communities/best-practices/>
Accessed June 2, 2014.
Emphasis added (earlier edition 2008).

try to address activism in opposition to rural wind facilities.[132] And a year later, in 2012, the British Department of Energy and Climate Change invited companies, utilities and other organizations to examine how the question of community engagement is dealt with in the UK and elsewhere.[133] In the US, NREL scientists reporting on behalf of the International Energy Agency (IEA-Wind) further contend—drawing on domestic case studies and academic literature—that efforts to engage communities in wind project development are shaped equally by responding to the political struggle between federal/state policy and local permitting tradition, and by addressing perceived notions of trust and procedural injustice.[134] Furthermore, community engagement has presented the industry with a market opportunity. For instance, under the banner of community engagement, US RE companies Own Energy and National Wind have created business models through which they are able to spread the economic benefits of wind facilities more widely—including not only to landowners that maintain generators on their properties, but also the local communities at large.[135] Even though the risk and requisite liability issues are significant, WAV residents reported that consideration of community ownership of RE projects, in combination with

---

[132] Unknown. "Engaging the wind community." ecogeneration.com, September/October 2011.
<http://ecogeneration.com.au/news/engaging_the_wind_community/063231/>
Accessed June 4, 2014.

[133] Department of Energy and Climate Change. "Onshore wind: call for evidence." June 2013.
<https://www.gov.uk/government/uploads/system/uploads/attachment_data/file/66577/6437-onshore-wind-call-for-evidence-document-part-a-com.pdf>
Accessed June 5, 2014.

[134] Lantz, Eric, and Larry Flowers. "Social Acceptance of Wind Energy Projects, 'Winning Hearts and Minds' State-of-the-art report. Country report of: United States." The International Energy Agency, 2009.
<http://www.socialacceptance.ch/images/State-of-the-Art_Acceptance_Wind_Energy_USA.pdf>
Accessed June 5, 2014.

[135] Small and Community Wind Conference, September 15-17, 2011, Des Moines, IA, participant observation. Interestingly, Own Energy's site features McLaren's work as an example of the role of public participation in ensuring success of wind project development.
Unknown. "Wind Energy planning in England, Wales and Denmark: Factors influencing Project Success (McLaren 2007)." Undated.
<http://www.ownenergy.net/knowledge-center/research/wind-energy-planning-england-wales-and-denmark-factors-influencing-project>
Accessed May 23, 2014.

a significantly smaller scale of development, would have offered engineers and localities new possibilities for navigating public engagement:

> The more opposition…[corporate executives] saw, they said "oh, we'll give you money." Yet they don't understand that after Solar Ranch One that's not going to look good… Had community ownership ever been considered by these companies, the relationship between "us" and "them" would have been a totally different one…[But] for the time being, someone who has power and answers must come to our meetings. [Project community liaisons] don't have any prerogative and at times they're flat-out lying ...One of them recently told us "you're two years too late—we've already got our permits." Most times, yet, all they can say is that they want to take [our concern] back and go to the big guys and technical guns...Besides, all company representatives, liaisons or executives, who supposedly do "community engagement"… their job is not to say "what would you like to see in a power plant"; they already have a project designed and submitted to the County.[136]

At one level, the recent calls for increased local involvement in RE facility building convey the impression of relevancy and genuineness.[137] That is, in the words of UK-based green energy consultants, a "new model" for engaging communities in RE project development is underway, one which they identify as reaching beyond the institution of payment schemes.[138]

---

[136] Interview data.

[137] Community engagement in wind energy, for example, has had actual effects in the very design of wind facilities in the US. According to NREL's Eric Lantz, the laboratory has compiled data on the distance between turbines and homes: "5-8 years ago developers… might [have] place[d] turbines 1000-1200 feet away from homes, and what we are hearing today is that its 500-700 meters, which is something more on the order of 1500-2000 feet away from homes…[S]o I think maybe the industry is responding [to social acceptance issues] in some ways without the policy requirements circulating." Personal communication.

[138] Hiles, Cheryl. "Regen SW response to the Community Energy Strategy call for evidence." Undated.
<http://regensw.s3.amazonaws.com/regen_sw_response_to_community_energy_strategy_call_for_evidence_cf5b4f70f1c7c79b.pdf>
Accessed June 4, 2014.
As the WAV case studies indicate, a common way developers have thought about "community engagement" is by making payments to individuals or local constituencies. These payments, which used to be called "good neighbor" payments, now typically occur in a gradient manner and include people who will be impacted—even visually—by RE. According to NREL's Suzanne Tegen, such payments "have been going on for the last ten years or [so] and I don't know if anybody is tracking that kind of thing, but it does make a difference…[W]e can't tell exactly what the success rate of projects is due to [payments]

What is more, a particularly effective way to engage local constituencies, aside from creating community liaison committees, is to ensure significant commitment from senior management. At another level, as the previous section illustrated, the perceived rupture between executives' affirmations of commitment to local community and development realities suggests that, in the case of the WAV, "engagement" may have depended on a more sophisticated understanding of corporate public relation strategies. In addition, the alienation of project liaisons from the local Valley community had been driven primarily by opaqueness and secondarily by the example set by senior managers themselves.[139]

In other official contexts of "community engagement" discourse, there are parallels with California's inaugural utility-scale PV projects, too: "Public meetings [regarding wind projects] are rarely of value," British communications practitioners maintain, as they favor up-to-date web pages and other media. Likewise, in conducting research for RE developers like Shell Wind, corporate social responsibility consultancy BSR suggests a "stakeholder engagement plan" as a way to broaden communications with residents; but whether BSR and Shell Wind considered the design of the project to be such an integral part of that plan is questionable.

NextEra's strategic positioning in the Valley and Element Power's declarations advocating "mitigation by design," coincided with the publication of North American wind industry professional organization siting guides. The US handbook—certainly less comprehensive than its Canadian equivalent—mentions the importance of early local involvement, and does not elaborate on the role of corporate spokespersons in conducting public meetings.[140] How do these documents handle RE engineers' models of

but we do know that in parallel to that the developers have been much better about working with communities and stakeholders up front for the past few years."

[139] "Of course the attitude was 'we've got our permits, there's nothing you can do about it.' Which is a bad attitude…They could have done a lot more to smooth things over, but they didn't… We would ask them simple things like 'how many veterans did you hire?' They wouldn't answer us. Oh, we have that information. Okay, well, how many. But they wouldn't answer us. We'd ask 'What are you going to put on the ground to stop the sand from blowing.' They wouldn't tell us." Interview data.

[140] Tetra Tech EC, Inc. "Wind Energy Siting Handbook." American Wind Energy Association, 2008.
<http://awea.files.cms-plus.com/AWEA_Siting_Handbook_Feb2008.pdf>
Accessed June 6, 2014.
CANWEA. "Best Practices for Community Engagement and Public Consultation." December 2013.
<http://canwea.ca/communities/best-practices/>
Accessed June 2, 2014.

engagement? "Don't feel obliged to provide an immediate answer to every question," asserted CanWEA's siting guide in 2011. Complying with this suggestion, "We don't give straight answers. Our company does not operate that way," affirmed one community liaison in a March 2013 Town Council meeting in Antelope Acres.

Despite their evolving intentions and ability to work with communities and project stakeholders, some RE project developer strategies can go both ways: On the one hand, in combination with the arbitrary Town Council structure, community engagement in the form of payment schemes offered developers a compelling venue for realizing RE projects in the WAV. On the other hand, for anti-renewables activists, the same notion of engagement served as a target to challenge companies' motives and instigate mistrust.[141]

[141] "I actually brokered a community investment package with [one individual]" said one developer in personal communication, "and we had looked at the issue funds, education program, and funding… And [local and state organizations] very much behind it, but again, the locals thought, or just branded it as a buy-off kind of deal." Interview data.

8

Epilogue

With his confident voice and authoritative presence, Hank Hatch, former chief engineer in the US Army Corps of Engineers and key figure in promoting "sustainability" into the engineering profession, had the last word during a 2010 conference on "Engineering towards a more just and sustainable world." Trained at West Point, Hatch was asked to provocatively "connect the dots" in less than fifteen minutes. In complying, Hatch argued emphatically that a reason so many civil engineers have supported the American Society for Civil Engineering's *Vision 2025,* which prompts professionals to "*control their own destiny* rather than letting events control it for them," is "[Vision 2025's] relation to sustainability."[1] "[Whether] you like the term [or not]," he added, American engineers "have accepted a case or cases, depending on their particular condition of servitude, why

[1] ASCE Steering Committee to Plan a Summit on the Future of the Civil Engineering Profession in 2015. *The Vision for Civil Engineering* in 2025. Reston, VA: ASCE, 2007: 5. Unless otherwise noted, all quotations in this epilogue are adapted from the author's transcript of Hank Hatch's March 6, 2010 presentation during the NAE and APPE conference in Cincinnati, Ohio.

sustainability makes sense...They have basically bought it." It was the "condition of servitude" that made "social justice"—as opposed to "sustainability" or "social responsibility"—such a contested topic during Hatch's presentation. The idea of linking the concepts of sustainability and social justice was still quite novel: it was only in 2008 that the prestigious National Academy of Engineering had held a conference on "Engineering, Social Justice and Sustainable Community Development."[2]

During the questions and answers following Hatch's talk, Harvard University's Sheila Jasanoff and Smith College's Donna Riley (currently with Purdue University) sat across the table from Hatch and engineering ethicist Michael Davis: they were literally divided from their male colleagues by the podium that stood between them. In response to Jasanoff and Riley's contention that engineers' concept of servitude must address questions of power, such as "Who are we serving as clients?", Hank Hatch responded that "You are not yet engaged meaningfully" with translating social justice in a way that engineers can both understand and identify with. Hatch also claimed that because social justice defies a precise definition, "[social justice advocates'] philosophical discussions must fit into a realistic context, [one] in which the engineer is working." He surmised that there would be no warm welcome for philosophical arguments in organizations like the American Council of Engineering Companies (ACEC). "I know," he concluded with a tone full of conviction, "I did that with sustainability in 1990. There weren't no standing ovations."

Extra-engineering values, it seems, are unlikely to have a significant impact on engineering affairs. And yet, as Hatch understands probably better than any other active industry or academic engineer in the US, making an extra-engineering idea attractive for engineers and non-engineers is far from being an unusual extra-engineering practice.[3] Indeed, such

---

[2] Rachel Hollander, who directs the National Academy of Engineering (NAE) Center for Engineering, Ethics, and Society, summarized the results of the 2008 NAE Conference on "Engineering, social justice and sustainable community development" as follows: "It turned out that the question of engineering and social justice was a hotly contested topic..., while humanitarianism and engineering or engineering and social responsibility was not" (emphasis added).
Online Ethics Center. "Results from APPE Mini-Conference: Engineering Towards a More Just and Sustainable World." National Academy of Engineering, 2010. http://www.onlineethics.org/Topics/Enviro/EnviroResources/APPE2010/23243.aspx. Accessed December 10, 2013.

[3] "I lived through the dark ages of gaining engineers' awareness, understanding and finally full acceptance and application of the principles of sustainability," admitted Hatch in that meeting. "It has taken over two decades."

appropriation of "non-engineering" values has historically been the way engineers successfully redefine themselves and the work they do to claim that they can "control their destiny." Certainly, fuzzy verbiage or uncomfortable ideological questions stand in the way, as well as resistance from professional societies and corporate actors. Such process of cultural value assimilation, this book argues, is by design philosophical: in his introduction to a 2010 piece entitled "Philosophy for a Sustainable World," the founder of engineering NGO "Engineers for a Sustainable World" Alexander Dale expressed a preference for "quantifying impacts through life-cycle assessment [LCA] rather than just talking about them in a qualitative sense." Yet he also asserted: "we need to look at areas including and beyond engineering to come up with cohesive and workable solutions."[4]

To meet the challenge of environmental sustainability that emerged in the late 1980s and 1990s, an elite group of American engineers led by Hank Hatch, together with senior US and international environmental consulting professionals, revisited the professional adage of "social responsibility." They redefined engineering professional identity by appropriating the concept of sustainable development. In doing so, advocates of sustainability engineering invoked arguments that were philosophical, social, and political in character. For example, engineering projects and interventions were seen as parts of systems whose boundaries were to be construed such that lifecycle impacts and costs could be incorporated in the analysis of alternative, more "sustainable" solutions. Engineers were to seek the financial, institutional and public support necessary for the internalization of environmental costs—for example, energy pricing. Engineers were to commit to a professional role as "facilitator of sustainable development," explaining "cleaner technologies" to educate an eager and receptive public. Further, historically utilitarian and homocentric engineering codes of ethics were to be rewritten to incorporate principles of sustainable development, while using "development" to legitimize the primacy of the market economy.

Part I of this book explored the disparate meanings of sustainability by focusing on social theories of technology as developed by engineers. My working hypothesis was: in conditions that engineers perceive as a technology crisis, where professional/self-interest and public interest appear incommensurable, they will contextualize themselves and their work.

---

[4] Dale, Alexander. "Philosophy for a Sustainable World." www.eswusa.org, January 13, 2012.
<http://www.eswusa.org/drupal/content/2012/01/philosophy-sustainable-world>
Accessed December 26, 2013.

Thomas Kuhn's work relates to my hypothesis. He was referring to *internal* crises: specifically, those involving elite communities in which practitioners confront the deadlock of the scientific paradigm driving their research.[5] In contrast, I showed that the sustainable technology crisis (1989–2003) and the requisite engineering contextualizations arose primarily from "external" (i.e., non-engineering) influences.

In the 1990s, engineering contextualizations of sustainability took place under the guidance of a broad consensus that environmental protection is inherently linked with technological innovation and market capitalism; as such "sustainable development" was viewed by technological change advocates of sustainability as the model *par excellence* for engineering progress. Moreover, the normative portrayal of the engineering expert as "global steward" was bolstered by the expansion of LCAs. Although before 1989 LCA-type of studies were popular only among a handful of European and North American practitioners, in the 1990s—particularly in the US—environmental organizations expressed concern about LCA studies produced by corporate engineers that were being used to buttress the *status quo* in product manufacturing or to counter environmental claims made by competitors and organizations. The result was that debates around the potential of LCA studies and methodologies to embody sustainability contributed to redefining engineering identity politics.

Part I of this book also detailed that life cycle thinking provided the elusive wholeness in sustainability engineering. LCA's key conceptual attribute—holism—has a traceable engineering history in "systems analysis." Nevertheless, this book argued that the historical development, current popularity and future role of LCA as regards the sustainability engineering community cannot be appreciated in teleological terms: that is, LCA inherently operationalizes sustainability. Rather, LCA involves many socio-political decisions and value choices. I showed that while non-engineering organizations and disciplines, but also subaltern engineering groups, have articulated sustainability more broadly, an input-output, life-cycle inventory approach dominates the operationalization of sustainability in engineering circles.

---

[5] "I think, particularly in periods of acknowledged crisis, that scientists have turned to philosophical analysis as a device for unlocking the riddles of their fields. Some have not generally needed or wanted to be philosophers… But that is not to say that *the search for assumptions cannot be an effective way to weaken the grip of a tradition upon the mind and to suggest the basis for a new one.* Thomas Kuhn. *The Structure of the Scientific Revolutions.* Chicago: University of Chicago Press, 1996: 88, emphasis added.

In the aftermath of what I described in this book as the "sustainable technology crisis" in engineering (1989-2003), practitioners perceive life-cycle-based sustainability assessment of products, systems, and operations as an extension of LCA.[6] One of the fathers of LCA considers it "almost the only possible methodology" for sustainability.[7] The view that LCA is "the only possible" technique for operationalizing sustainability stems from the dominant professional culture of engineering, which frames "social" issues—and particularly political questions like social justice—as inherently extra-engineering concerns. The case studies describing resistance to sustainability by engineers illustrate that depoliticization, an important ideological element of technological change, has reinforced the assumption that engineering is about the technical, not the social or political. The ideology of technological change thus contributed greatly to shaping sustainability engineering identities and what practitioners consider as "sustainability engineering work." Because mathematics is perceived to be the foundation of engineering knowledge, LCA, with its emphasis on quantifiable impact categories, resonates with the dominant image of an objective engineering design methodology—free of social and cultural "irrelevancies." In the words of Garvin Heath, a renowned LCA and renewable technology specialist at the National Renewable Energy Laboratory, the convergence between sustainability and LCA is driven "mainly by the desire for numbers. The desire for quantification. Sustainability has been incredibly challenging to... operationally define, and LCA provides a... holistic framework" to do so.[8] Other authors have reached similar conclusions in comparing detailed statements of suggested "sustainability principles" with engineering curricula and sustainability-related course descriptions.[9]

In today's professional engineering environment, the tendency remains to grant LCA transformative dynamism: In 2012 ASCE stated

---

[6] See, for example, Keoleian, Gregory A., and V. David Spitzley. "Life Cycle Based Sustainability Metrics." In *Sustainability Science and Engineering Defining Principles* (Volume 1), edited by Martin A. Abraham, 127–159. Oxford: Elsevier, 2006.

[7] Robert U. Ayres, personal communication.

[8] Personal communication.

[9] Hoffmann, Stephen R., L. Alice Pawley, L. Ranjani Rao, E. Monica Cardella, and W. Matthew Ohland. "Defining 'Sustainable Engineering': a Comparative Analysis of Published Sustainability Principles and Existing Courses." 2011. <http://www.slideshare.net/RIFEGroup/defining-sustainable-engineering-a-comparative-analysis-of-published-sustainability-principles-and-existing-courses> Accessed June 4, 2012.

its commitment to "improving methods for identifying and considering all of a project's environmental, social, and economic costs and impacts throughout its life-cycle".[10] Members of the American Institute of Chemical Engineers have expressed similar levels of organizational commitment through the development of "life cycle tools and applications for sustainability."[11] And in June 2013 nine industry leading firms signed up for Envision™—the Institute of Sustainable Infrastructure (ISI) rating system—which uses LCA as a basis to evaluate and rate the social, environmental, and economic benefits of infrastructure.

Everybody from Wal-mart, to Seventh Generation, the solar project developer First Solar to the state of California is doing LCA. On the one hand, as a general rule, LCA studies remain disjoined from the sustainability controversies of the 1990s separating the idea of a fundamental disunion between technics and socio-political questions and the conception of a tangible continuum throughout these two realms. Buttressed by the dominant engineering ideology of sustainability, the proliferation of LCA studies have at their core, and within practitioners in that field, a quantitative and free-market bias. Many contemporary LCAs can be read as basically advocating that the free market is going to take care of everything—accepting that it is always economical to improve industrial processes and evading critique of the economic incentives and disincentives around achieving sustainability.

On the other hand, the developing field of LCA is a much more complex and ideologically hybridized locus where the legacy of technopolitics is already present within the identities of contemporary sustainability engineers. For example, at the 2014 ASCE International Conference on Sustainable Infrastructure, topics like "Social LCAs" and "Sustainability, Quality of Life and Culture" were featured under the same thematic category with "The business case for social responsibility" and "Using design charrettes on infrastructure projects." (Figure 8.1). This example illustrates an important point of this book: sustainability engineering cultures have passed from a limited expression of technopolitics perspectives to their

---

[10] ASCE. "Civil Engineering for a Sustainable Future." 2012.
<http://www.asce.org/Press-Releases/2012/Civil-Engineering-for-a-Sustainable-Future/>
Accessed July 5, 2012.

[11] AIChE. "AIChE Webinar: Life Cycle Analysis Tools and Applications for Sustainability." 2011.
<http://apps.aiche.org/ChemEOnDemand/Preview.
aspx?ID=aa9bc724-3c1e-4879-b231-918ddf21593a>
Accessed May 23, 2012.

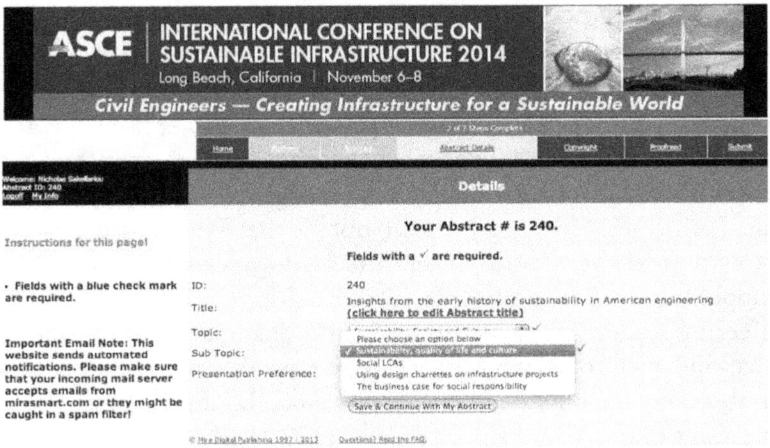

Figure 8.1 "Sustainability, quality of life and culture" features as potential topic of discussion in ASCE conferences alongside "Social LCAs," "The business case for social responsibility" and "Using design charrettes on infrastructure projects."

ongoing naturalization as legitimate subjects in dominant professional engineering discourse. Indeed the technopolitics axe shows that LCA *could* be developed and used in a much more diverse way.

A constant, then, of the engineering self in the 2000s and beyond is crystalized into an ontological view (life cycle thinking) and a technical methodology (LCA), which constitutes a medium for ideological and disciplinary convergence. Part II argued that SLCA is an exemplification of the continuing hybridization of engineering ideologies of sustainability. Yet to SLCA's ideological dimension must be added two others: one political, and one professional.

Politically, the question arises as to whether, and to what extent, SLCA discourse tends to produce conceptual and research opportunities to challenge the pathologies of social risks occurring across global product chains on a political basis. SLCA gives sustainability engineers a way of targeting where in design processes they can do what might be called politically driven engineering studies. First comes the identification of the problem with worker hours, unfair wages, etc., in a particular region or in several countries. Yet, what are the factors, what are the characteristics that lead to this? What is the role of militarism, or neoliberalism in an SLCA study? Contrary to the technological change orientation, SLCA could be more of a tool to search out and expose the root causes of environmental and social problems. In other words, via SLCA, practitioners may become more effective in learning how to stir political change in their profession. This is something that engineers typically do not think about because in traditional engineering mindsets, we learn to examine problems sector by sector.

Professionally, the technopolitics axe inherent in SLCA can help expand engineering identity. US engineering, for example, epitomizes a mainstream sensibility around, first of all, serving national, military, and corporate ends; whatever the mainstream is, engineering follows in terms of political priorities. Most engineers do not see that as a political choice, they see it as part of the natural order. The biggest potential contribution of technopolitics for future sustainability practitioners is that there might be engineering students who would read this book and say: 'I do not have to be such-and such to be an engineer. I can have different views or different ideas about the viability of market capitalism, about neoliberalism, justice or questions of peace and *still be an engineer.' Technopolitics offers budding engineering professionals a chance to pursue different trajectories than they thought they had to pursue.*

For all their social-condition improvement rhetoric, Part II of this book demonstrated that SLCA proponents and sustainability engineering practitioners currently rarely question why global corporations conduct business with "risky" suppliers in the first place. This has real implications for the case studies examined in Part III: namely engineers do not yet think about why RE technologies are being developed at all, or who is financing them, or how these technologies take on particular configurations.

Part III questioned how sustainability engineering is changing RE project development, if at all, as LCA boundaries and sustainability engineering identities are contested. I demonstrated that LCAs in RE may serve multiple roles: First as a way to exert control over engineering projects and to influence non-engineers into accepting how a particular solar or wind project is being configured. Second, as a means to form and express engineering identity. Third, as a site of contestation and reshaping the boundaries of what could or should be part of an LCA.

To date, LCAs of solar and wind technologies are comparative studies of systems (polysilicon, monosilicon, thin-film, etc.) that capture life cycle greenhouse gas emissions, criteria pollutant emissions and heavy metal emissions—few studies investigate financial and toxicity impacts.[12] In the

---

[12] See, for example, Hsu, David D., Patrick O'Donoughue, Vasilis Fthenakis, A. Garvin Heath, Hyung Chul Kim, Pamala Sawyer, Jun-Ki Choi, and E. Damon Turney. "Life Cycle Greenhouse Gas Emissions of Crystalline Silicon Photovoltaic Electricity Generation: Systematic Review and Harmonization. *Journal of Industrial Ecology* 16, no. 1 (2012): S122–S135; Padey, Pierryves, Isabelle Blanc, Denis Le Boulch, and Zhao Xiusheng.

case of the West Antelope Valley, major developers (First Solar), regulators (California Energy Commission) and investors relied on life cycle thinking to anticipate how a particular project like Solar Ranch One will perform in sustainability terms. Life cycle thinking, in other words, was used by developers to defuse opposition and to win approvals from regulators and investors. Such thinking did not adapt to different project sizes, to determine, for instance, if community-scale LCA-type of studies must include different concerns compared to industrial-scale LCAs. LCA thinking in RE thus forced projects in the WAV into an industrial template by favoring higher scales of economy and larger production facilities. At the same time, RE engineers built their perspective on specific assumptions (e.g., short-term profit), professional ideologies (quantification of impacts) and sets of evidence (technical expert knowledge). The extent to which, and for what purposes, RE developers in WAV used life cycle thinking to design and evaluate their potential projects underlines connections between sustainability engineering and participatory environmental policy processes. Similarly, my research offers a skeptical view of the ethical and value claims typical of much corporate sustainability discourses. In the WAV, corporate engineers failed to meaningfully operationalize both the involvement of multiple stakeholder perspectives and a perpetual process of decision, monitoring, reconsideration of assumptions and adjustment.

The case studies of utility-scale RE project development in California presented in this book may offer valuable insights into the politics of shaping and reshaping engineering identity. The political components are both cognitive (e.g., what an engineer knows, does not know, or does not care to know) and social/professional (bestowed by the actual fact of belonging to a community of practitioners). No less than their colleagues working in oil or coal industries, RE engineers ascribe to specific features of group membership, which, in turn, speak to what each community values or surmises. As a result, I found, RE engineers working on solar and wind projects were less likely to reshape their identities by embracing more site-specific and localized/variable engineering methods over other, more dominant engineering methods.

Yet this book shows that technical experts can actually make a difference in the social sustainability of RE as long as they know what questions to ask. This may be where the disconnect lies. A typical engineering course will consist of problem sets that have a right or wrong type of answer that students are expected to solve. In contrast, RE project development poses

"A Simplified Life Cycle Approach for Assessing Greenhouse Gas Emissions of Wind Electricity." *Journal of Industrial Ecology* 16, no. 1 (2012): S28–S38.

a very different situation. As we saw in the case studies of the WAV, a practitioner can face different scenarios and solutions based on the local context; but there is not one right answer. There are many different ways that a developer can calculate the cost of energy, or how much does it matter to a member of the local community if they have to sacrifice their rural vistas for red light-blinking industrial scale wind turbines. In that regard, the unthinking use of LCA methods in RE project development prompts deeper questions regarding the linkages between sustainability theory and epistemological discussions of "wicked problems"—problems for which no definitive formulations exist and for which "the planner has no right to be wrong."[13] This book challenges the idea that LCAs—with their current focus on "objective" and quantitative data—are the best way to support RE engineering policy.

In sum, to understand how engineering identities are formed and adapted in relation to RE we must thoroughly study a) the forms and trajectories that have characterized sustainability engineering to date; b) how solar and wind technologies fit into these forms and trajectories; and c) how conventional engineering cultures continue to influence RE engineering cultures and projects.

In conclusion, the RE case studies in Part III illustrated that although LCAs have now moved beyond one single discipline, community of practitioners, or area of application, they have continued to be an artificially bounded measurement tool (Figure 8.2). In the RE field a wide variety of actors now use life cycle thinking—communities, NGOs, investors, etc. This pervasive use illustrates the power that LCAs have gained in contemporary sustainability engineering. Yet this widening use also means that the suppositions of LCA are now under greater societal challenge. The local community in the WAV responded critically to RE engineers' assumptions: residents called for the inclusion of lay expertise, community ownership, the decommissioning of facilities and a fair distribution of benefits within the community. "Lay experts" too, not just engineers, are involved in making and pushing LCA boundaries. As a result, engineers' identities are being challenged to expand from quantitative analysis to include community/worker health and safety issues, as well as other kinds of project parameters such as community ownership and participation.[14]

---

[13] Thompson, Paul B, and P. Kyle Whyte. "What happens to Environmental Philosophy in a Wicked World?" *Journal of Agricultural and Environmental Ethics* 25 (2012): 485-498.

[14] For example, a comparative LCA study of solar modules that focused on end-of-life toxicity risks prompted some progressive engineers (First Solar) to reformulate the role

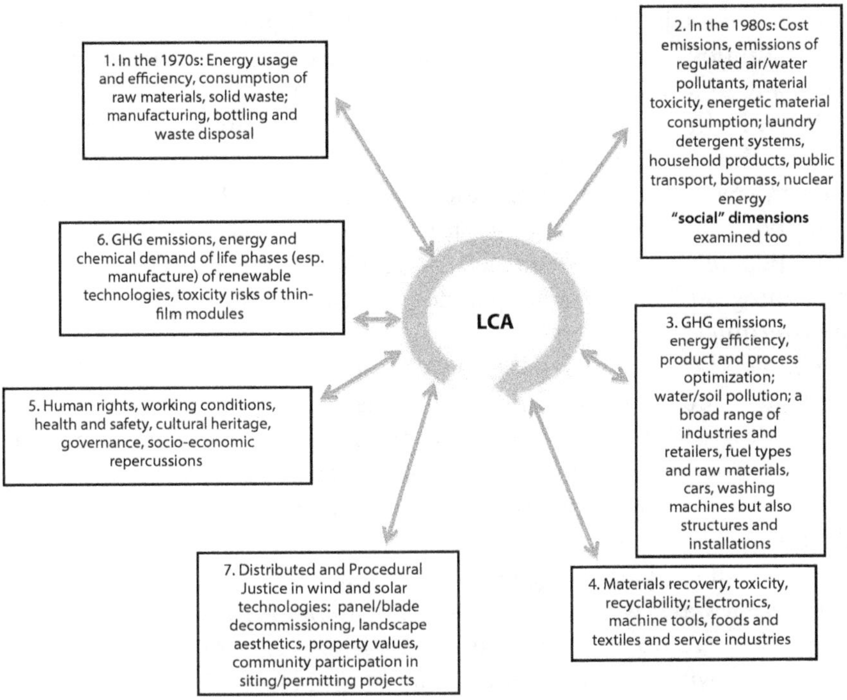

Figure 8.2 Bounding Life Cycle Assessment Measurement.

Both LCAs and engineering communities are altered in the process of deciding what counts as legitimate LCA boundaries. For example, on January 24, 2014, a total of thirty engineering practitioners, architects and planners had the opportunity to participate in a ASCE "continuing education" seminar—part of the society's series on best practices in sustainable engineering—designated "Community Participation: Effective Stakeholder Involvement Throughout the Project Lifecycle."[15] Likewise, as these concluding lines were being written, the ASCE began soliciting

of LCA in solar PV technology development and design for recyclability. Fthenakis Vasilis, C. Samuel Morris, D. Paul Moskowitz, and L. Daniel Morgan. "Toxicity of cadmium telluride, copper indium diselenide, and copper gallium diselenide." *Progress in Photovoltaics*, 7, no. 6 (1999): 489–497.

[15] Scott, Doug. "Learning to Listen: Seminar Connects the Design of Sustainable Projects with Community Involvement." Blogs.asce.org, January 31, 2014. <http://blogs.asce.org/learning-to-listen-seminar-connects-the-design-of-sustainable-projects-with-community-involvement/>

potential contributors to a book entitled "Engineering for Sustainable Communities," arguing that "current [engineering] approaches, practices and standards *do not address the full range of societal needs.*"[16]

What is, then, missing from engineering? One thing that is missing is *who* are the people that engineers serve. Who we serve is narrowly defined. We serve people who can pay. We serve governments. We serve private corporations. Can we broaden the groups that engineering serves to include other people who cannot afford those services? Can engineers be involved in framing problems and not just do what they are told by their corporate or government employers?

I contend that there needs to be a broadening of what engineering is—the field is too narrowly defined. There is a professional comfort level in just dealing with numbers, alleged facts, and objectivity. Any time there is any introduction of feelings or values into engineering discourses, it just does not fit with what practitioners perceive an engineering profession is all about.

Who is missing from engineering? People are missing from engineering in different ways: we see, for instance, that engineers are put in positions where they are not as interested in people, or something is missing with interacting with people, or practitioners simply do not want to interact with people, and that creates all kinds of problems.

Discussions of power are missing from engineering practice. Most engineers are not sensitive to that. Social workers think about these issues all the time. Social workers think about who they serve. Similarly in law, we have public interest law. There could be engineering firms that look like some of the law centers that work on social justice issues. It is my hope to see different models of engineering practice that include an awareness of power on the one hand, and the epistemological issue on the other. To have students confronting issues of power, being aware of what is missing from engineering, and why, and what these power structures are. At the same time, have students understand how that power relates to epistemology and how they are taught to think as an engineer and how they might think otherwise, and ultimately be able to perceive engineering reality in different frames.

Importantly, I think what is missing is a recognition of the heart. As engineers, we are so locked into our heads that there seems to be no place for feelings or empathy. There is no place for compassion. I hope that someday we could unlock the heart of engineering. It seems to be so far removed

[16] Richard N. Wright, personal communication, emphasis added.

from what we tend to do and what we tend to talk about. I think what does not belong in our profession is the notion that you have to have a specific political, economic, or political ideology *to be an engineer*. We have to thus open LCAs up for other ideologies and other economic systems. And for people's lives! Is this not, after all, what *Life* Cycle Assessment is about?

Bibliography

AAES, AIChE, ANS, ASME, NAE, NSPE. "A Declaration by the US Engineering Community to the World Summit on Sustainable Development." June 24, 2002.

AAES. "Statement of the American Association of Engineering Societies on the Role of the Engineer in Sustainable Development." Washington, DC: AAES, 1993.

AIChE. "AIChE Webinar: Life Cycle Analysis Tools and Applications for Sustainability." 2011.

AIChE. "Statement on Sustainable Engineering Forum of AIChE Ideas and Scope," November 2002.

Alan Bernheimer. "LA County Approves First Solar to Start Module Installation at AV Solar Ranch One." Investor.firstsolar.com, June 22, 2012.

Allen, David, *et al.* "Benchmarking sustainable engineering education: Final report." Austin: University of Texas at Austin, Carnegie Mellon University, Arizona State University, 2009.

Allen, Don, M. Carla, and D. Mark Webster. "Introduction." In *Sustainability Guidelines for the Structural Engineer*, edited by Dirk Kestner M., Jennifer Goupil, and Emily Lorenz, 1–7. Reston, VA: ASCE, 2010.

Anderson, Dennis. "Poppies and Panels." *Antelope Valley Press*, June 8, 2011.

Andersson, Karin, Merete Høgaas Eide, Ulrika Lundqvist, Berit Mattsson. "The feasibility of including sustainability in LCA for product development" *Journal of Cleaner Production* 6, (1998): 289–298.

Andrews, Evan, Pascal Lesage, Catherine Benoît, Julie Parent, Gregory Norris, and Jean-Pierre Revéret. "Life Cycle Attribute Assessment: Case Study of Quebec Greenhouse Tomatoes." *Journal of Industrial Ecology* 13, no. 4 (2009): 565–578.

Antelope Valley Area Plan (AVAP), Community Recognition, Section III-1. Undated.

Antelope Valley Conservancy (AVC). "Re: Preliminary Draft Significant Ecological Area and Hillside Management Area Conditional Use Permit Ordinance." February 25, 2012.

Antonsson, Ann-Beth, and Helene Carlsson. "The basis for a method to integrate work environment in life cycle assessments." *Journal of Cleaner Production* 3, no. 4 (1995): 215–220.

ASCE Steering Committee to Plan a Summit on the Future of the Civil Engineering Profession in 2015. *The Vision for Civil Engineering in 2025*. Reston, VA: ASCE, 2007.

ASCE. "Assuring a Desirable Quality of Life." ASCE Policy Statement no. 120. New York: ASCE, 1980.

ASCE. "Civil Engineering for a Sustainable Future." 2012.

ASCE. "Code of Ethics." Undated.

ASCE. *Sustainable Engineering Practice: An Introduction*. Reston, VA: American Society of Civil Engineers, 2004.

Atkisson, Alan. "Green Engineering and National Security: The US Army Corps of Engineers looks to the future and embraces the concept of sustainable development, an interview with Lt. General Henry J. Hatch." *In Context* 32 (1992): 40.

Avhidesert Forum.

Baldassare, Mark, Dean Bonner, Sonja Petek, and Jui Shrestha. "Californians & the environment." Public Policy Institute of California, July 2012.

Baumann, Henrikke, Rickard Arvidsson, Hui Tong, and Ying Wang. "Does the Production of an Airbag Injure More People than the Airbag Saves in Traffic?" *Journal of Industrial Ecology* 17, no. 4 (2013): 1–11.

Bazan-Arias, Cathy. "Guidance on ethical dilemmas and the history of sustainable development." *CENews*, December 30, 2008.

Bazan-Arias, Cathy. "Readers respond: Ethics and sustainable development." *CENews* January 6, 2009.

The Beacon Hill Institute at Suffolk University. "The Economic Impact of Arizona's Renewable Energy Standard and Tariff." Boston, MA: The Beacon Hill Institute at Suffolk University, 2013.

Bell, Derek, Tim Gray, and Claire Haggett. "The 'Social Gap' in Wind Farm Siting Decisions: Explanations and Policy Responses." *Environmental Politics* 14, no. 4 (2005): 460–477.

Benoît-Norris, Catherine, Deana Aulisio Cavan, and Gregory Norris. "Identifying Social Impacts in Product Supply Chains: Overview and Application of the Social Hotspot Database." *Sustainability* 4 no. 9 (2012): 1946–1965.

Benoît, Catherine, A. Gregory Norris, Sonia Valdivia, Andreas Ciroth, Asa Moberg, Ulrike Bos, Siddarth Prakash, Cassia Ugaya, and Tabea Beck. "The Guidelines for Social Life Cycle Assessment of Products: Just in Time." *International Journal of Life Cycle Assessment* 15, no. 2 (2010): 156–163.

Benoît, Catherine, and Gina Vickery-Niederman. "Social Sustainability Assessment Literature Review." Tempe, Arizona, and Fayetteville, Arkansas: The Sustainability Consortium, 2010.

Bernow, Stephen S., and D. Paul Raskin. "Ecology of Scientific Consciousness." *Telos* 28 (1976): 55–80.

Black and Veatch. "Renewable Energy Transmission Initiative Phase 1A Work Group Meeting 6." March 6, 2008.

The Board of Supervisors. "The Meeting Transcript of the Los Angeles County Board of Supervisors." November 23, 2010.

Boulding, Kenneth E. "The Economics of the Coming Spaceship Earth." In *Environmental Quality in a Growing Economy*, edited by Henry Jarrett, 3–14. Baltimore, MD: Resources for the Future/Johns Hopkins University Press, 1966.

Boulding, Kenneth E. *The Meaning of the 20th Century*. New York: Harper and Row, 1964.

Boustead Consulting & Associates, Ltd. "Analysis of Environmental and Energy Benefits of 'AVA Solar manufacturing Facilities' project." 2009.

Boustead, Ian. "LCA—How it Came About The Beginning in the U.K" *International Journal of Life Cycle Assessment* 1, no. 3 (1996): 147–150.

Boyle, Godfrey, and Peter Harper, eds. *Radical Technology*. London and New York: Pantheon/Random House, 1976.

Breault, Rich. "Solar, wind: Learn plan, enjoy early Halloween." *Antelope Valley Press*, October 25, 2011.

Britzius, Charles W. "Discussion of 'Engineers, Ethics, and Nuclear Weapons.'" *Journal of Professional Issues in Engineering* 115, no. 1 (1989): 100.

Burkhardt, Helmut, and H. Willem Vanderburg. "Preparing for a Sustainable Society." *IEEE Technology and Society* 10, no. 4 (1991): 6–8.

Burningham, Kate. "Using the language of NIMBY: a topic for research, not an activity for researchers." *Local environment* 5, no. 1 (2000): 55–67.

BWEA. "Wind farm planning approvals by local councils slumped to a record new low of 25%." 2009.

California Energy Commission, California Department of Fish and Game, Bureau of Land Management, and U.S. Fish and Wildlife Service (CEC, CDFG, BLM, and USFWS). "Planning Agreement by and among California Department of Fish and Game, California Energy Commission, United States Bureau of Land Management, and United States Fish and Wildlife Service for the Desert Renewable Energy Conservation Plan." May 2010.

California Energy Commission. "2013 Renewable Energy Action Team (REAT) Generation Tracking Projects." 2013.

California Hometown Locator. "Lancaster, CA 93536 ZIP Code Profile." Undated.

California Public Utilities Commission. "California Renewables Portfolio Standard (RPS)." Undated.

Calvert, John. "Email from John Calvert." AVHIDESERT, March 18, 2011.

CANWEA. "Best Practices for Community Engagement and Public Consultation." December 2013.

CANWEA. "Best Practices for Community Engagement and Public Consultation." December 2013.

Carew, Anna L., and A. Cynthia Mitchell. "Characterizing Undergraduate Engineering Students Understanding of Sustainability." *European Journal of Engineering Education* 27, no. 4 (2002): 349–361.

Carroll, William J. "Environmental Challenge for Engineers." *Journal of Professional Issues in Engineering Education and Practice* 121, no. 2 (1995): 126–129, 127.

Carroll, William J. "World Engineering Partnership for Sustainable Development." *Journal of Professional Issues in Engineering Education and Practice* 119, no. 3 (1993): 238–240.

Carter, Archie N. "Editorial." *Journal of Professional Issues in Engineering Education and Practice* 119, no. 3 (1993): 213.

CEC, CDFG, BLM, and USFWS (California Energy Commission, California Department of Fish and Game, Bureau of Land Management, and U.S. Fish and Wildlife Service. "Description and Comparative Evaluation of Draft DRECP Alternatives." December 17, 2012.

Center for Community Action and Environmental Justice (CCAE). "RE: Black and Veatch Renewable Energy Transmission Initiative Phase 1A." March 29, 2008.

Centre for Good Governance. "A Comprehensive Guide for Social Impact Assessment." Hyderabad, India: Centre for Good Governance, 2006.

Cha, Ariana Eunjung. "Solar Firms Leave Waste Behind in China." *Washington Post* March 9 (2008): A1.

Chadderton, Ronald A. "Should Engineers Counteract Environmental Extremism?" *Journal of Professional Issues in Engineering Education and Practice* 121, no. 2 (1995): 79–84.

Charles, Michael. "Sustainable growth: administration pursues 'livability agenda.'" *Civil Engineering* 69, no. 3 (1999): 10.

Ciroth, Andreas and Juliane Franze. "Conducting a Social LCA." Presented at the Workshop: Social Aspects of Products Over the Whole Life Cycle, Berlin, May 30, 2011.

Civil Engineering Research Foundation (CERF). "Engineering and Construction for Sustainable Development in the 21st Century: Assessing Global Research Needs." Washington, DC: American Society of Civil Engineers, 1996.

Clarke, Chris. "Group Calls for Strict Limits on Solar Power Near National Parks." *Rewire*, September 18, 2012.

Clarke, Chris. "What Happens When a Developer Abandons a Desert Solar Project." rewire, February 11, 2013.

Clift. Ronald, and Neil Morris. "Sustainable Development: Engineering with a human face." *Engineering Management Journal* 12, no. 5 (2002): 226–230.

Coates, Geoffrey H. "Facilitating Sustainable Development: Role of Engineer." *Journal of Professional Issues in Engineering Education and Practice* 119, no. 3 (1993): 225–229.

Colborn, Theo E., *et al.* "Great Lakes, Great Legacy?" Washington, DC, and Ottawa, Ontario: The Conservation Foundation and the Institute for Research on Public Policy, 1990.

Conlon, Edward. "The new engineer: Between employability and social responsibility," *European Journal of Engineering Education* 3, no. 2 (2008): 151–159.

Cook, Joan. "Carroll L. Wilson, Science and Energy Expert." *New York Times*, January 13, 1983.

Cottell, Michael N. T. "Facilitating Sustainable Development: Is Our Approach Correct?" *Journal of Professional Issues in Engineering Education and Practice* 119, no. 3 (1993): 220–224.

Council of Academies of Engineering and Technological Sciences (CAETS). *The Role of Technology in Environmentally Sustainable Development: A Declaration of the Council of Academies of Engineering and Technological Sciences.* Washington, DC: CAETS, 1995.

County of Los Angeles Public Library. "Antelope Valley Community History." Undated.

Currier, Craig. "Developer again seeks energy plant near poppy reserve." *Antelope Valley Press*, August 23, 2012.

Currier, Craig. "Locals battling windmills." Antelope Valley Press, November 8, 2011.

Currier, Craig. "Money, power tangle in west Valley." *Antelope Valley Press*, November 19, 2011.

Currier, Craig. "Solar firm ups community gift to $280,000." *Antelope Valley Press*, January 16, 2011.

Currier, Craig. "The question: Turbine or not turbine?" *Antelope Valley Press*, Wednesday January 25, 2012.

Dale, Alexander. "Philosophy for a Sustainable World." www.eswusa.org, January 13, 2012.

Daly, Herman E. "Toward a Stationary-State Economy." In *The Patient Earth*, edited by John Harte and Robert Socolow, 236–237. New York: Holt, Rinehart, and Winston, 1971.

De Coninck, Pierre, and Carmela Cucuzzella. "A Participatory Approach Seeking Consensus in a Context of Uncertainty. A Sustainable Development Perspective of Life Cycle Analysis. Paper presented at the CALCAS Workshop, Brussels, 27 September, 2007.

DeMers, Gerald L. "Discussion: Evaluating Technological Alternatives on Basis of Sustainability." *Journal of Professional Issues in Engineering Education and Practice* 123, no. 1 (1997): 43.

Derbyshire, David. "The wind turbine backlash: Growing public opposition thwarts green energy drive." *Mail Online*, July 11, 2011.

Derbyshire, Katherine. "Public policy for engineers: solar industry depends on policymakers' goodwill." ELECTROIQ, August 2008.

Devon, Richard. "Towards a Social Ethics of Engineering: The Norms of Engagement." *Journal of Engineering Education* 88, 1 (1999): 87–92.

DMCA. "Draft Environmental Impact Report Comments for the Fairmont Butte Motorsports Park Project, Parcel Map 26805 (SCH No. 2005031170)." July 17, 2009.

DMCA. "Memorandum, July 17, 2009." 2009. Courtesy of the DMCA.

Downey, Gary L., Arthur Donovan, and J. Timothy Elliott. "The Invisible Engineer: How Engineering Ceased to Be a Problem in Science and Technology Studies." *Knowledge and Society* 8, (1989): 189–216.

DRECP Independent Science Advisors (ISA). "Recommendations of Independent Science Advisors for The California Desert Renewable Energy Conservation Plan (DRECP)." October 2010.

Dreyer, Louise C. "A Site Specific Approach to Life Cycle Management of Labour Rights Issues: Using the Social LCA Toolbox." Presented at the International Seminar on Social LCA, Montpellier, France, May 2011.

Dreyer, Louise C. "Inclusion of Social Aspects in Life Cycle Assessment of Products: Development of a Methodology for Social Life Cycle Assessment." PhD diss., Technical University of Denmark, 2009.

Dreyer, Louise Camilla, Z. Michael Hauschild, and Jens Schierbeck. "A Framework for Social Life Cycle Impact Assessment." *International Journal of Life Cycle Assessment* 11, no. 2 (2006): 88–97.

Duffell, Roger J. "Toward the Environment and Sustainability Ethic in Engineering Education and Practice." *Journal of Professional Issues in Engineering Education and Practice* 124, no. 3 (1998): 78–90.

Duncan, Richard C. "A unified theory of human history: Summary presentation." *AESR Newsletter* 3, no. 3 (1990).

Duncan, Richard C. "Evolution, technology and the natural environment." Paper presented at the Annual ASEE Conference, Binghamton, NY, 1989.

Duncan, Richard C. "Sustainability—Is there a middle road?" Moses Greeley Parker Lecture Series, Lowell, MA.

Duncan, Richard C. "The Evolution of Social Control: Is a World Society Governable?" In *Preparing for a sustainable society*, edited by Helmut Burkhardt and H. Willem Vanderburg, 256–263. Piscataway: Institute of Electrical and Electronic Engineers, 1991.

Duncan, Richard C. "The Olduvai Theory: Sliding Towards a Post-Industrial Stone Age." Institute on Energy and Man, June 27, 1996, manuscript.

Duncan, Richard C. "The Olduvai Theory. Energy, Population, and Industrial Civilization." *The Social Contract*, (2005–2006): 1–12.

Duncan, Richard C. "The Olduvai Theory. Terminal Decline Imminent." *The Social Contract*, (2007): 141–151.

Duncan, Richard C. "The Olduvai Theory. Toward Re-Equalizing the World Standard of Living." *The Social Contract*, (2009): 67–80.

Duncan, Richard C. "World Energy Production, Population Growth and the Road to the Olduvai Gorge." May 2001, manuscript.
Earth Charter in Action. "A short story of the Earth Chapter." Undated.
Edmiston, Joseph T. "Memorandum." April 6, 2011.
EIARC. "A Proposed Eighth Fundamental Canon for the ASCE Code of Ethics." *Journal of Professional Issues in Engineering* 110, no. 3 (1984): 118–122.
Element Power. Brochure." Undated.
Elkington, John. "Enter the Triple Bottom Line." In *The Triple Bottom Line, Does it All Add up? Assessing the Sustainability of Business and CSR*, edited by Adrian Henriques, Julie Richardson, 1–16. Sterling, VA, 2004.
Ellis, Monica D., ed. *The Role of Engineering in Sustainable Development*. Washington: AAES, 1994.
Ember, Lois R. "Economic Growth Viewed as Engine for Sustainable Development." *Chemical and Engineering News* 70, no. 20 (1992): 19–20.
Ember, Lois R. "Environment protection: Global companies set new endeavor." *Chemical and Engineering News* 69, no. 14 (1991): 4.
Engineers Forum on Sustainability (EFS). *EFS Newsletter*. October 1997-April 2009.
ENVTL. PROTECTION AGENCY, BIOFUELS AND THE ENVIRONMENT: FIRST TRIENNIAL REPORT TO CONGRESS (2011).
EPA. "Memorandum of Agreement Between the Department of the Army and the Environmental Protection Agency Concerning the Determination of the Section 404 Program and the Application of the Exemptions Under Section 404(F) of the Clean Water Act." January 1989.
EPA. "U.S. Army Corps of Engineers Superfund Partnership with EPA." Undated.
EuPD. "Economic Impact of Extending the Section 1603 Treasury Program." Miami, Florida: EuPD, 2011.
Evans, Jim. "A brief (and incomplete) history of social responsibility in engineering." *AESR Newsletter* 2, no. 1 (1989): 8–10.
Evans, Jim. "Editorial: Economic Conversion." *AESR Newsletter* 1, no. 2 (1988): 14–16. [1988, (a)].
Evans, Jim. "Engineers, Ethics and Nuclear Weapons." *AESR Newsletter* 1, no. 1 (1988): 4–5. [1988, (b)].
Evans, Jim. "Further Thoughts on Sustainability." *AESR Newsletter* 5, no. 1 (1992): 1–2.
Evans, Jim. "Seattle Happenings." *AESR Newsletter* 3, no. 1 (1990): 13.
Evans, Jim. "Sustainability and the Enhancement of Human Welfare." *AESR Newsletter* 3, no. 4 (1990): 4–5.
Evans, Roger J. "Commentary on the Code of Ethics." *Journal of Professional Issues in Engineering* 114, no. 2 (1988): 148–156. [1988, (c)].
Evans, Roger J., and E. Thomas Munsey. "Engineers, Ethics, and Nuclear Weapons. *Journal of Professional Issues in Engineering* 113, no. 3 (1987): 268–275.
Executive Office, County of Riverside. "Submittal to the Board of Supervisors County of Riverside, State of California." June 23, 2011.

Fairmont Town Council. "Reasons for Appeal." January 4, 2011.

Farrell, Alex, and Daniel Sperling. "Getting the carbon out." *San Francisco Chronicle* Friday, May 18, 2007.

Farrell, Alex. "Sustainability and the Design of Knowledge Tools." *IEEE Technology and Society* 15, no. 4 (1996): 11–20.

Farrell, Alexander E., and Daniel Sperling. "A Low-Carbon Fuel Standard for California, Part 2: Policy Analysis." Berkeley, CA: University of California at Berkeley, 2007.

Fava, James A. "Editorial. Life-Cycle Assessment: A New Way of Thinking." *Environmental Toxicology and Chemistry* 13, no. 6 (1994): 853–854.

FAVOS. "AVSR1." Undated.

FIDIC. "FIDIC on the Environment: Special Issue of Independent Consulting Engineer." *Journal of International Federation of Consulting Engineers*, Autumn, 1990. Lausanne, Switzerland.

FIDIC. *Consulting engineers and the environment: guide for actions*. Lausanne, Switzerland: FIDIC, 1994.

Finance.yahoo. "First Solar, Inc. (FSLR)." Undated.

Finkbeiner, Matthias, M. Erwin Schau, Annekatrin Lehmann, and Marzia Traverso. "Towards Life Cycle Sustainability Assessment." *Sustainability* 2, no. 10 (2010): 3309–3322.

Finnveden, Göran. "Valuation Methods Within LCA—Where are the Values?" *International Journal of Life Cycle Assessment* 2, no. 3 (1997): 163–169.

Fischer, Frank, and Michael Black, eds. *Greening Environmental Policy: The Politics of A Sustainable Future*. New York: St. Martin's Press, 1995.

Flaherty, Margaret and Ann Rappaport. "Essential Matter. Corporate Environmentalism: From Rhetoric to Results." United Nations Research Institute for Social Development, Sept. 1997.

Freeman, Edward R., C. Andrew Wicks, and Bidhan Parmar. "Stakeholder Theory and 'The Corporate Objective Revisited.'" *Organizational Science* 15, no. 3 (2004): 364–369.

Friends of Antelope Valley Open Space and Concerned Citizens of Westside Antelope Valley (FAVOS and CCWAV). "Letter." May 22, 2012.

Frosch, Robert. A. "Sustainability Engineering (editorial)." *The Bridge* 29, no. 1 (1999): 2–3.

Fthenakis Vasilis, C. Samuel Morris, D. Paul Moskowitz, and L. Daniel Morgan. "Toxicity of cadmium telluride, copper indium diselenide, and copper gallium diselenide." *Progress in Photovoltaics*, 7, no. 6 (1999): 489–497.

Fthenakis, Vasilis M. "Sustainability metrics for extending thin-film photovoltaics to terawatt levels." *Materials Research Society Bulletin* 37, (2012): 1–6.

Fthenakis, Vasilis M., Hyung Chul Kim, and Erik Alsema. "Emissions from Photovoltaic Life Cycles." *Environmental Science and Technology* 42, no. 6 (2008): 2168–2174.

Funtowicz, Silvio O. and R. Jerome Ravetz. "A New Scientific Methodology for Global Environmental Issues." In *Ecological Economics: The Science and*

Management of Sustainability, ed. Robert Costanza. New York: Columbia University Press, 1991, 137–152.

Garcia-Johnson, Ronie. *Exporting Environmentalism. U.S. Multinational Chemical Corporations in Brazil and Mexico*. Cambridge: The MIT Press, 2000.

Gary Lee Downey, Juan C. Lucena, Barbara M. Moskal, Rosamond Parkhurst, Thomas Bigely, Chris Hays, Brent K. Jesiek, Liam Kelly, Jonson Miller, Sharon Ruff, Jane L. Lehr, and Amy Nichols-Belo. "The Globally Competent Engineer: Working Effectively with People Who Define Problems Differently." *Journal of Engineering Education* 95 no. 2 (2006): 1–17.

Gauthier, Caroline. "Measuring Corporate Social and Environmental Performance: The Extended Life-Cycle Assessment." *Journal of Business Ethics* 59 (2005): 199–206.

Georgescu-Roegen, Nicholas. *The Entropy Law and the Economic Process*, Harvard, 1971.

Gerry Coates. "The Responsibility of Engineers." *New Zealand Engineering*, September 1983.

Gibney, Kate. "Cleaning up on cleaning up: sustainable development is a new way of doing business." *ASEE Prism* 8, no. 5 (1999): 20–25.

Global Reporting Initiative. "G4 Sustainability Reporting Guidelines: Reporting Principles and Standard Disclosures." Amsterdam, The Netherlands: GRI, 2013.

Goldenberg, Suzanne. "Free-market research group's climate proposal denounced by host university." theguardian.com, December 5, 2013.

Gordon, Uri. *Anarchy Alive!: Anti-Authoritarian Politics from Practice to Theory*. London: Pluto Press, 2008.

Gram, Dave. "Vermont loves renewable energy, except when it arrives." reformer.com, March 2, 2014.

Grant. Albert A. "The Ethics of Sustainability: An Engineering Perspective." *Renewable Resources Journal*, Spring (1995): 23–25.

Great Lakes Wind Collaborative. "Best Practices for Sustainable Wind Energy Wind Development in the Great Lakes Region." Ann Arbor, MI: Great Lakes Wind Collaborative, 2011.

Grieβhammer, Rainer, Catherine Benoît, Louise Camilla Dreyer, Anna Flysjö, Andreas Manhart, Bernard Mazjin, Andrée-Lise Méthot, and Bo Weidema. "Feasibility Study: Integration of social aspects into LCA." Freiburg, 2006.

Haimes, Yacob Y. "Sustainable Development: A Holistic Approach to Natural Resource Management." *IEEE Transactions on Systems, Man, and Cybernetics* 22, no. 3 (1992): 413–417.

Halog, Anthony, and Yosef Manik. "Advancing Integrated Systems Modelling Framework for Life Cycle Sustainability Assessment." *Sustainability* 3, no. 2 (2011): 469–499.

Harding, Jim. *Social policy and social justice: the NDP government in Saskatchewan during the Blakeney years*. Waterloo, Ont: Wilfrid Laurier University Press, 1995.

Hargreaves, Steve. "First Solar shares plunge after project delay." Money.cnn.com, February 10, 2012.

Harrabin, Roger. "Local communities offered more say over wind farms." bbc.com, June 6, 2013.

Hatch, Hank. "Keynote speech at the ASCE Convention on October 8, 1989," manuscript, courtesy of Hank Hatch.

Hatch, Henry J. "Relevant Engineering in the 21st Century." *Journal of Professional Issues in Engineering Education and Practice* 119, no. 3 (1993): 216–219.

Hatch, Henry J. "Sustainable Development." Manuscript from talk given at during a mini-symposium on sustainable development, Tufts University, March 8, 2000.

Hatch, Henry. "Panelist Remarks." In *Partnerships for Global Ecosystem Management Science, Economics and Law. Proceedings and Reference Readings from the Fifth Annual World Bank Conference on Environmentally and Socially Sustainable Development*, edited by Ismail Serageldin and Joan Martin-Brown, 70–73. Washington DC: The World Bank, 1998.

Hauschild, Michael Z., C. Louise Dreyer, and Andreas Jørgensen. "Assessing Social Impacts in Life Cycle Perspective—Lessons Learned." *CIRP Annals Manufacturing Technology* 57 (2008): 21–24.

Hauschild, Michael, and Henrik Wenzel. *Environmental assessment of products. Vols. 1 and 2.* London: Chapman and Hall, 1998.

Hauschild, Michael, Jeswiet, Jack, and Leo Alting. "From Life Cycle Assessment to Sustainable Production: Status and Perspectives." *CIRP Annals Manufacturing Technology* 54, no. 2 (2005): 1–21.

Heaton, George R. Jr., Robert Repetto, and Rodney Sobin. *Backs to the Future: U.S. Government Policy Toward Environmentally Critical Technology*. Baltimore: World Resources Institute Publications, 1992.

Hedlund, Patric. "'Where the Trees Grow Sideways and the Sun Burns Bright.'" *The Mountain Enterprise*, April 29, 2011.

Hedlund, Patric. "Black gold and renewable energy make Kern County a recovery star." *The Mountain Enterprise*, February 8, 2013.

Hedlund, Patric. "Change on Display in the Antelope Valley. Solar Begins to Wind Hearts and Minds." *The Mountain Enterprise*, March 9, 2012.

Hedlund, Patric. "Company Says Solar Panels of Cadmium-Telluride Harmless." *The Mountain Enterprise*, March 9, 2012.

Hedlund, Patric. "Fairmont Town Council Asks for Emergency Planning Meeting on First Solar Intentions." *The Mountain Enterprise*, June 28, 2011.

Hedlund, Patric. "Fairmont Town Council's Solar Merry-Go-Round Ride." *The Mountain Enterprise*, June 24, 2011.

Hedlund, Patric. "First Solar and Rural Councils Inch Toward 'Good Neighbors' Goal." *The Mountain Enterprise*, February 10, 2012.

Hedlund, Patric. "First Solar Lays Off 165." *The Mountain Enterprise*, May 25, 2012.

Hedlund, Patric. "First Solar Lays off about 240." *The Mountain Enterprise*, June 1, 2012.

Hedlund, Patric. "First Solar Offers Jobs to Neenach Area—From Fire to 'Hire.'" *The Mountain Enterprise*, July 15, 2011.

Hedlund, Patric. "First Solar Seeks 'New Beginning' for 2012." *The Mountain Enterprise*, January 27, 2012.

Hedlund, Patric. "Meeting Held to Explain First Solar's Future Goals." *The Mountain Enterprise*, January 13, 2012.

Hedlund, Patric. "Neenach Area Files Two Appeals to L.A. County Solar Farm Permit." *The Mountain Enterprise*, January 7, 2011.

Hedlund, Patric. "Neenach-Fairmont Area Meet Energy Developers-Part Three." *The Mountain Enterprise*, February 11, 2011.

Hedlund, Patric. "Sabers Rattle as Fairmont Chides First Solar for 'Bullying.'" *The Mountain Enterprise*, July 1, 2011.

Hedlund, Patric. "Valley Voices. News from the Western Antelope Valley." *The Mountain Enterprise*, December 14, 2012.

Held, Michael. "Life Cycle Assessment of CdTe module recycling." Paper presented at the 24th European Photovoltaic Solar Energy Conference, Hamburg, Germany, 2009.

Henderson, Hazel. *Creating Alternative Futures: The End of Economics*. New York: The Berkeley Windhover, 1978.

Herkert, Joseph R. "Sustainable development and engineering: Ethical and public policy implications." Paper presented at the 1997 International Symposium on Technology and Society, Glasgow, UK, 20–21 June 1997.

Herkert, Joseph R., Alex Farrell, and J. James Winebrake. "Technology Choice for Sustainable Development." *IEEE Technology and Society* 15, no. 2 (1996): 12–20.

Herkert. Joseph. "ABET's Engineering Criteria 2000 and Engineering Ethics: Where do we go from Here?" Paper Presented at the OEC International Conference on Ethics in Engineering and Computer Science, March 1999.

Heylin, Michael. "Earth Day Revisited." *Chemical and Engineering News* 68, no. 16 (1990): 3.

Hileman, Bette. "Clinton forms sustainable development council." *Chemical and Engineering News* 71, no. 25 (1993): 7–8.

Hiles, Cheryl. "Regen SW response to the Community Energy Strategy call for evidence." Undated.

Hoffmann, Stephen R., L. Alice Pawley, L. Ranjani Rao, E. Monica Cardella, and W. Matthew Ohland. "Defining 'Sustainable Engineering': a Comparative Analysis of Published Sustainability Principles and Existing Courses." 2011.

Hofstetter, Patrick. Perspectives in Life Cycle Impact Assessment. A Structured Approach to Combine Models of the Technosphere, Ecosphere and Valuesphere." PhD diss., Swiss Federal Institute of Technology Zurich, 1998.

Holliday, Charles O., Stephan Schmidheiny, and Philip Watts. *Walking the talk: The business case for sustainable development*. San Francisco: Berrett-Koehler, 2002.

Hsu, David D., Patrick O'Donoughue, Vasilis Fthenakis, A. Garvin Heath, Hyung Chul Kim, Pamala Sawyer, Jun-Ki Choi, and E. Damon Turney. "Life Cycle Greenhouse Gas Emissions of Crystalline Silicon Photovoltaic Electricity Generation: Systematic Review and Harmonization. *Journal of Industrial Ecology* 16, no. 1 (2012): S122–S135.

Hsu, Tiffany. "Wind farms multiply, fueling clashes with nearby residents." *Los Angeles Times*, July 24, 2011.

Hudspith, Robert. "A teaching tool that exposes the non-neutrality of technology as it relates to sustainable development." In *Preparing for a sustainable society*, edited by Helmut Burkhardt and H. Willem Vanderburg, 294–301. Piscataway: Institute of Electrical and Electronic Engineers, 1991.

Hunkeler, David and Gerald Rebitzer. "The Future of Life Cycle Assessment." *International Journal of Life Cycle Assessment* 10, no. 5 (2005): 305–308.

Hunt, Robert G., and E. William Franklin. "LCA—How it Came About—Personal Reflections on the Origin and Development of LCA in the USA." *International Journal of Life Cycle Assessment* 1, no. 1 (1996): 4–7.

IChemE. "The Sustainability Metrics: Sustainable Development Progress Metrics Recommended for Use in the Process Industries." Warwickshire: Institution of Chemical Engineers, 2002.

ICSU and WFEO. "Addendum No. 8: Dialogue Paper by Scientific and Technological Communities. Role and Contributions of the Scientific and Technological Community (S&TC) to Sustainable Development." United Nations Economic and Social Council E/CN.17/2002/PC.2/6.Add.8. Advance Copy, 28 January 2002.

IEEE. "White Paper on Sustainable Development and Industrial Ecology." Hoboken, NJ: IEEE Electronics and the Environment Committee, 1995.

Independent Science Panel (ISP). "Initial Recommendations of the Desert Renewable Energy Conservation Plan (DRECP) Independent Science Panel Based on Review of Draft DRECP Materials." August 2012.

Institute for Local Self-Reliance. "Renewable Energy Mandate—Iowa." January 20, 2009.

Institute of Social and Ethical AccountAbility. *AA1000 Framework*. London: ISEA, 1999.

International Finance Corporation. "Stakeholder Engagement: A Good Practice Handbook for Companies Doing Business in Emerging Markets." Washington, DC: International Finance Corporation, 2007.

Jackson, Rosalind. "California desert communities say yes to solar, no to climate change." votesolar.org, February 9, 2012.

Jesse Ausubel H., and E. Hedy Sladovich, eds. *Technology and Environment*: National Academy of Engineering, 1989.

Jeswani, Harish K., Adisa Azapagic, Philipp, Schepelmann, and Michael Ritthoff. "Options for Broadening and Deepening the LCA Approaches." *Journal of Cleaner Production* 18, no. 2 (2010): 120–127.

Johnson, Darcie B. "What are the ethical responsibilities towards sustainable development?" *Journal of Professional Issues in Engineering Education and Practice* 125, no. 2 (1998): 35–37.

Johnson, Jeff. "Environment and the bottom line: Rand study of four leading companies shows environmental concerns are being joined to overall corporate planning." *Chemical and Engineering News* 77, no. 25 (1999): 25–26.

Johnson, Richard R. "Sustainability is key to engineering relevance." *Civil Engineering* 65, no. 12 (1995): 28.

Jones, Lorie. "Acceptance Speech." Engineers Australia, WA Division, AGM December 10, 2007.

Jørgensen, Andreas, Agathe Le Bocq, Liudmila Nazarkina, and Michael Hauschild. "Methodologies for Social Life Cycle Assessment." *International Journal of Life Cycle Assessment* 13, no. 2 (2008): 96–103.

Jørgensen, Andreas, C. H. Lufanna Lai, and Z. Michael Hauschild. "Assessing the Validity of Impact Pathways for Child Labour and Well-Being in Social Life Cycle Assessment." *International Journal of Life Cycle Assessment* 15, no. 1 (2010): 5–16.

Jørgensen, Andreas, T. Ivan Hermann, and Anders Bjørn. "Sketching a Life Cycle Impact Assessment for a Life Cycle Sustainability Assessment," unpublished manuscript.

Jørgensen, Andreas. "Developing the Social Life Cycle Assessment—addressing issues of validity and usability." PhD diss., DTU Management Engineering, 2010.

Jørgensen, Andreas. "Social LCA-A Way Ahead?" *International Journal of Life Cycle Assessment* 18, no. 2 (2013): 296–299.

Kahn, Debra. "Calif. Governor Vows to 'Crush' Foes of Renewable Energy." *New York Times*, July 26, 2011.

Kan Usuda, Koichi Kono, Keiko Ohnishi, Shin Nakayama, Yumiko Sugiura, Yasuhiro Kitamura, Akihiro Kurita, Yuko Tsuda, Motoshi Kimura and Yasuhisa Yoshida. "Toxicological aspects of cadmium and occupational health activities to prevent workplace exposure in Japan: A narrative review." *Toxicology and Industrial Health* 27, no. 3 (2011): 225–233.

Kaplan, Tracey. "Town Councils Give Communities a Chance to Be Heard : Government: The advisory boards reflect a 'grass-roots uprising' created to empower and safeguard rural northern L.A. County." *Los Angeles Times*, June 14, 1992.

Kaufmann, K. "Riverside County may levy fee on solar firms." Mydesert.com, June 17, 2011.

Kelly, William E. "Sustainability in Engineering Education." *EFS Newsletter*, March (2003): 1–3.

Keniston, Kenneth. "The crisis of the engineering algorithm." Paper presented at the Institute of Advanced Studies in Humanities, Politecnico di Torino, Torino, Italy, 1996, manuscript.

Kennard, Robert M. "Environmental Challenge for Engineers: Hong Kong Perspective." *Journal of Professional Issues in Engineering Education and Practice* 121, no. 2 (1995): 140–142.

Keoleian, Gregory A., and V. David Spitzley. "Life Cycle Based Sustainability Metrics." In *Sustainability Science and Engineering Defining Principles* (Volume 1), edited by Martin A. Abraham, 127–159. Oxford: Elsevier, 2006.

Kern County Planning and Community Development Department, Board of Supervisors. "Staff Report." October 30, 2012.
Khanna, Jiti. "Holistic Engineering: Beyond the Code of Ethics." *Engineering-ASCE* 63, no. 10 (1993): 6.
Kingsnorth, Paul. "Nimbys are the true democratic heroes." *The New Statesman* 17, (2004): 22–24.
Klöpffer, Walter. "Editorial." *International Journal of Life Cycle Assessment* 1 no. 1 (1996): 3.
Klöpffer, Walter. "Life Cycle Sustainability Assessment of Products." *International Journal of Life Cycle Assessment* 13, no. 2 (2008): 89–95.
Klöpffer, Walter. "Life-Cycle Based Methods for Sustainable Product Development." *International Journal of Life Cycle Assessment* 8, no. 3 (2003): 157–159.
Klöpffer, Walter. "Reductionism versus Expansionism." *International Journal of Life Cycle Assessment* 1 no. 2 (1996): 61.
Koren, James Rufus. "Cooling Off? Solar ranch nears launch as large projects lose sizzle." *Los Angeles Business Journal* 34, no. 35 (2012): 1 and 50.
Kroeger, Heidelore I., and P. Simonovic Slobodan. "Risk as a sustainable development Criteria." In *North American Water and Environment Congress & Destructive Water*, edited by Chenchayya Bathala, 1531–1536. New York: ASCE, 1996.
Kuhn, Thomas. *The Structure of the Scientific Revolutions*. Chicago: University of Chicago Press, 1996.
Kuo, Iris. "DOE loan chief: Where we'll invest in 2011." Venturebeat.com, December 15, 2010.
LA County Department of Regional Planning. "Notice of Preparation, Blue Sky Wind Project." September 20, 2011.
Labuschagne, Carin, C. Alan Brent, and Ron P.G. van Erck. "Assessing the Sustainability Performances of Industries." *Journal of Cleaner Production* 13 (2005): 373–385.
LADRP. "Chapter 4. Conservation and Open Space Element." Undated.
LADRP. "NOTICE OF PREPARATION AND NOTICE OF SCOPING MEETING. WILDFLOWER GREEN ENERGY FARM. Project No. R2010-00256-(05)/ Conditional Use Permit. 201000121/Environmental Impact Assessment 201000063. November 4, 2011.
LADRP. "Notice of Preparation, Blue Sky Wind Project." September 20, 2011.
LADRP. "Project No. R2009-02239-(5) AV SOLAR RANCH ONE, LLC. Vesting Tentative Tract Map No. 071035 Conditional Use Permit No. 200900026 Environmental Assessment No. 200900027 Agenda Item No. 7." Regional Planning Commission Public Hearing September 15, 2010.
LADRP. "Renewable Energy Meeting." Presentation, October 26, 2013.
Lantz, Eric, and Larry Flowers. "Social Acceptance of Wind Energy Projects, 'Winning Hearts and Minds' State-of-the-art report. Country report of: United States." The International Energy Agency, 2009.

Layton, Edwin T. Jr. *The Revolt of the Engineers: Social Responsibility and the American Engineering Profession*. Cleveland, PA: Press of Case Western Reserve University, 1971.

Lefevre, Clement, Damien Pellé, Shabnam Abedi, Raul Martinez, and Pierre-Francois Thaler. "Value of Sustainable Procurement Practices." Pricewaterhouse Coopers, 2010.

Lehmann, Annekatrin, Daniela Russi, Alba Bala, Matthias Finkbeiner, and Pere Fullana-i-Palmer. "Integration of Social Aspects in Decision Support, Based on Life Cycle Thinking." *Sustainability* 3, no. 4 (2011): 562–577.

Lehmann, Annekatrin, Eva Zschieschang, Marzia Traverso, Matthias Finkbeiner Matthias and Liselotte Schebek. "Social Aspects for Sustainability Assessment of Technologies—Challenges for Social Life Cycle Assessment (SLCA)." *International Journal of Life Cycle Assessment* 18, no. 8 (2013): 1581–1592.

Leonard, Raymond S. "Information Systems for Engineering Sustainable Development." Paper presented at the Workshop on Engineering Partnership for Sustainable Development: A Workshop in Conjunction with Prep Con IV of the United Nations Conference on Environment and Development. March 1–3, 1992, New York, NY.

Lepkowski, Will. "The social thermodynamics of Ilya Prigogine." *Chemical and Engineering News*, April 16 (1979): 30–33.

Levy, David L., Halina Szejnald Brown, and Martin de Jong. "The Contested Politics of Corporate Governance: The Case of the Global Reporting Initiative." *Business & Society* 49, no. 1 (2010): 88–115.

Life Cycle Initiative. "Guidelines for Social Life Cycle Assessment of Products." Paris, France: UNEP, 2009.

Life Cycle Initiative. "Why Take a Life Cycle Approach." Paris, France: UNEP, 2004.

Lippiatt, Barbara C. "Selecting Cost Effective Green Building Products: BEES Approach." *Journal of Construction Engineering and Management* 125, no. 6 (1999): 448–455.

Lipton, Eric, and Clifford Krauss. "A Gold Rush of Subsidies in Clean Energy Search." *New York Times*, November 11, 2011.

Los Angeles County Fire Department, Health Hazardous Materials Division (HHMD), Special Operations Section, Summary of Events, February 1, 2013, Log Number 12–38.

Los Angeles Department of Regional Planning (LADRP). "Findings of fact regarding the final environmental impact report for the AV SOLAR RANCH ONE PROJECT COUNTY PROJECT NO. R2009-02239 VESTING TENTATIVE TRACT MAP NO. TR071035 CONDITIONAL USE PERMIT NO. RCUPT200900026 ENVIRONMENTAL REVIEW NO. RENVT200900027 STATE CLEARINGHOUSE NO. 2009041145." December 2010.

Lovins, Amory B. *Soft energy paths: toward a durable peace*. San Francisco, CA: Friends of the Earth International, 1977.

Macadams, Lewis. "King of the hills: environmentalist Joe Edmiston might be the most powerful unelected official in Los Angeles." *Los Angeles Magazine*, July 1, 2005.

MacCorquodale, Patricia L., W. Martha Gilliland, P. Jeffrey Kash, and Andrew Jameton, eds. *Engineers and Economic Conversion: From the Military to the Marketplace*: Springer-Verlag, 1993.

Macombe, Catherine, Pauline Feschet, Michel Garrabé, and Denis Loeillet. "2nd International Seminar in Social Life Cycle Assessment—Recent Developments in Assessing the Social Impacts of Product Life Cycles." *International Journal of Life Cycle Assessment* 16, no. 9 (2011): 940–943.

Macombe, Catherine, Pekka Leskinen, Pauline Feschet, and Riina Antikainen. "Social Life Cycle Assessment of Biodiesel Production at Three Levels: A Literature Review and Development Needs." *Journal of Cleaner Production* 52 (2013): 205–216.

Manion, Mark. "Ethics, Engineering, and Sustainable Development." *IEEE Technology and Society* 21, no. 3 (2002): 39–48.

Mann, Margaret K., Michael Whitaker, and Todd Driver. "Life Cycle Assessment of Existing and Emerging Distributed Generation Technologies in California." Golden, Colorado, July 2011.

Massie, Robert K. "Poetry of the Possible." Manuscript, 1997.

Mazijn, Bernard. "Methodology of the Social LCA." Presented at the Workshop: Life Cycle Initiative, Brussels, 7 October 2010, emphasis added.

McIsaac, Gregory F., and C., Nancy Morey. "Engineers' Role in Sustainable Development: Considering Cultural Dynamics." *Journal of Professional Issues in Engineering Education and Practice* 124, no. 4 (1998): 110–119.

McLaren Loring, Joyce. "Wind energy planning in England, Wales, and Denmark: Factors influencing project success." *Energy Policy* 35, no. 4 (2007): 2648–2660.

McWhorter, Ron. "What are the ethical responsibilities towards sustainable development?" *Journal of Professional Issues in Engineering Education and Practice* 125, no. 2 (1998): 37–39.

Meadows Donella H., L. Dennis Meadows, Jørgen, Randers, and W. William Behrens III. *The Limits to Growth: A Report for the Club of Rome's Project on the Predicament of Mankind*. New York: Universe Books, 1972.

Mehle Michael. "Greenpeace Upstages Corporations At Earthtech—Environment Fair Produces Protests." *Seattle Times*, Thursday, April 5, 1990.

Méthot, Andrée-Lise. "FIDD: A Green and Socially Responsible Venture Capital Fund." Paper presented at the conference for Life Cycle Approaches for Green Investment, École Polytechnique Fédérale de Lausanne, Switzerland, 2005.

Meyer, Gary. "1,000 seek work at solar job fair." *The Mountain Enterprise*, March 8, 2013.

Meyer, Gary. "Unemployment Dips in Kern County." *The Mountain Enterprise*, March 30, 2012.

Michaelsen, Joel. "Mojave Desert Region Physical Geography." Undated.

Mitcham, Carl. "Responsibility and technology: the expanding relationship." In *Technology and Responsibility*, edited by Carl Mitcham, 3–39. The Netherlands: Springer, 1987.

Mitcham, Carl. "The Importance of Philosophy to Engineering." *Teorema: Revista internacional de filosofía* 17, no. 3 (1998): 27–47.

Morse, Paige Marie. "Taking a measure of sustainability: Chemical companies are using a variety of resources to determine effective ways to chart progress in sustainable business." *Chemical and Engineering News* 77, no. 30 (1999): 19–22.

Multiple authors. "Letters Responding to Richard Duncan's 'Olduvai Theory' Essay." Social Contract, 2006.

Mulvaney, Dustin. "Are Green Jobs just jobs? Cadmium narratives in the life cycle of photovoltaics." *Geoforum* 54, (2014): 178–186.

Munsey, Thomas E. "Sociological risk assessment. Likely confrontations and conflicts in the war for sustainability." In *Preparing for a sustainable society*, edited by Helmut Burkhardt and H. Willem Vanderburg, 402–409. Piscataway: Institute of Electrical and Electronic Engineers, 1991.

Nath, Ishan. "Cleaning-up After Clean Energy: Hazardous Waste in the Solar Industry." *Stanford Journal of International Relations* 11 no. 2 (2010): 6–15.

National Council of Examiners for Engineering and Surveying, 2011.

National Research Council. *Our Common Journey: A Transition Toward Sustainability*. Washington: National Academy Press, 1999.

Naudascher, Eduard. "Discussion of Richard C. Duncan's 'A Unified Theory of Human History: Summary.'" *AESR Newsletter* 3, no. 2 (1990): 10–11.

Nazarkina, Liudmila, and Agathe Le Bocq. "Social aspects of sustainability assessments: Feasibility of social life cycle assessment (SLCA)." Paper presented at EDF 2006, Moretsur-Loing, France, 2006.

Nelson, Jill Barnes. "Wind Company to Sue." *Mojave Desert News*, Friday April 6, 2012.

Nicholson-Lord, David. "Stake-out at the Body Shop. Social auditing: the stakeholder is alive and living in South America as Anita Roddick's empire undergoes an ethical inspection." *The Independent*, Sunday 21 January 1996.

Nicolis, Gregoire, and Ilya Prigogine. *Self-Organization in Non-equilibrium Systems*. New York: John Wiley & Sons, 1977.

Nivargikar, Rao and Debra Reinhart, eds. *Quality Assurance-A National Commitment*: ASCE, 1997.

Nordic Council of Ministers. *Product Life-Cycle Assessment-Principles and Methodology*. Copenhagen, Denmark: Nord 1992, 9th Nordic Council of Ministers, 1992.

Norris, Gregory A. "Revisions to LCA Needed to Address Sustainable Consumption." Paper presented at the conference of The Society of Non-Traditional Technology (AIST) Life cycle approach to sustainable consumption: conceptual design of a methodological framework, Tokyo, Japan, 2003.

Norris, Gregory A. "Social Impacts in Product Life Cycles." *International Journal of Life Cycle Assessment* 11, no. 1 (2006): 97–104.

NSF. "The Impact of Defense Downsizing on Technical Employment." Undated.

NSTC. *Bridge to a Sustainable Future: A National Environmental Technology Strategy*. Washington, DC: Executive Office of the President, 1995.

NWCC. "Siting Issues for Wind Power Plants." *Wind Energy Series*, no. 3 (1997).

O'Brien, Martin, Alison Doig, and Roland Clift. "Social and Environmental Life Cycle Assessment (SELCA): Approach and Methodological Development." *International Journal of Life Cycle Assessment* 1, no. 4 (1996): 231–237.

Office of the Governor of the State of California, Press Release, *Gov. Schwarzenegger Urges Local Counties to Expedite Large-Scale Renewable Energy Projects*. February 18, 2010.

Öko-Institut. *Produktlinienanalyse*. Cologne, Germany: Kölner Volksblatt Verlag, 1987.

Online Ethics Center. "Results from APPE Mini-Conference: Engineering Towards a More Just and Sustainable World." National Academy of Engineering, 2010.

Osborne, Mark. "First Solar fails to gain US DOE loan guarantee for 550 Topaz solar farm." www.pv-tech.org, September 22, 2011.

Overby, Chuck M. "Product Design for a Sustainable Future: A Matter of Ethics?" Paper presented at the Annual Conference of the American Society for Engineering Education, University of Massachusetts, 1980.

Padey, Pierryves, Isabelle Blanc, Denis Le Boulch, and Zhao Xiusheng. "A Simplified Life Cycle Approach for Assessing Greenhouse Gas Emissions of Wind Electricity." *Journal of Industrial Ecology* 16, no. 1 (2012): S28-S38.

Parenti, Christian. "'The Limits to Growth': A Book that Launched a Movement." www.thenation.com, December 5, 2012.

Parry, Simon and Ed Douglas. "In China, the true cost of Britain's clean, green wind power experiment: Pollution on a disastrous scale." *Daily Mail*, January 26, 2011.

Peet, John. "Being Fully Human and Creating a Better Future: Sustainable Development from an Integrated Systems Perspective." Paper Presented at the Workshop on Sustainable Development, Sigtuna Foundation, Sweden, 8–9 June 2000.

Peet, John. "Chemical Engineering & Sustainability: Is Green Processing Enough?" Paper presented at the APCChE (Asia-Pacific Conference on Chemical Engineering) Conference, Christchurch, NZ, September 30 - October 3, 2002, Proceedings paper #235.

Peet, John. *Energy and the Ecological Economics of Sustainability*. Washington, DC: Island Press, 1992.

Petroski, Henry. "The Invisible Engineer." *Civil Engineering—ASCE*, 60, no. 11 (1990): 46–49.

Petti, Luigia, K. Paola Sanchez Ramirez, L. Cássia Maria Ugaya, R.V. Vicoli, and S. Vaira. " A tomato case study using subcategory assessment method for social life cycle assessment." Presented at the SETAC 18th LCA Case Study Symposium, Sustainability Assessment in the 21st Century Tools, Trends & Applications, Copenhagen, 26–28 November 2012.

Phadke, Roopali. "Resisting and Reconciling Big Wind: Middle Landscape Politics in the New American West." *Antipode* 43 no. 3 (2011): 754–776.
Piper, Lennart, Sven-Olof Ryding, and Curt Henricson. *Continual Improvement with ISO 14000*. Stockholm, Sweden: SINF miljö&kvalitet AB, 2003.
Pletka, Ryan. "RETI Phase 2 Update Workgroup." December 10, 2009.
Poirot, James W. "Consulting engineers need to go public. Heightened public awareness will help profession embark upon a new age of innovation." *Engineering News Record*, January-March (1989): 3–6, special advertising section.
Poirot, James W. "The Earth Charter, Engineering Roles and Responsibilities." The Earth Charter Initiative, Virtual Library, 1999.
Poirot, Jim. "Engineering Implications in the Earth Charter." Paper Presented at the World Congress on Sustainable Development: Engineering & Technological Challenges of the 21st Century, Calcutta, India, January 21, 2000.
Pollak, Daniel. "Natural Community Conservation Planning (NCCP), The Origins of an Ambitious Experiment to Protect Ecosystems, Part 1 of a Series." Sacramento, CA: California Research Bureau, 2001.
Prelogar, Brittany, Laura Ardito, and Michael Navarre. "New Human Trafficking Laws and US Government Initiatives Make Anti-Trafficking a Compliance Priority for Businesses in 2013." Steptoe & Johnson LLP. February 2013.
Prendergast, John. "Engineering Sustainable Development." *Civil Engineering-ASCE* 63, no. 10 (1993): 39–42.
Pub. L. No. 109-58, Title XVII (Aug. 8, 2005).
Pub. L. No. 111-5 (Feb. 17, 2009).
Pvthin.org. "Sustainability. Sustainable energy from beginning to end." Undated.
Quinn, Larry. "Sustainability: Another New Paradigm." *Civil Engineering* 66, no. 10 (1996): 6.
Rada, Ramón J., Armando Caldeira-Pires, Catherine Benoit, and A. Claudia Pena. "Environmental Indicators Align with Integral Sustainability: The Chilean Copper Ongoing Process." *International Journal for Sustainable Innovations* 1, no. 1 (2011): 16–21.
Rajagopalan, Visvanathan. "The Engineer's role in sustainable development." *Civil Engineering* 62, no. 8 (1992): 6.
Ranganathan, Janet. "Sustainability Rulers: Measuring Corporate Environmental and Social Performance." Washington, DC: World Resources Institute, 1998.
Rappaport, Ann B. "Global Industry and Hazardous Waste Practice: Heading Toward Sustainability with Risk Management." *Practice Periodical of Hazardous, Toxic, and Radioactive Waste Management* 1, no. 4 (1997): 144–147.
Raskin, Paul, Tariq Banuri, Gilberto Gallopín, Pablo Gutman, Al Hammond, Robert Kates, and Rob Swart. "Great Transition. The Promise and Lure and the Times Ahead." Boston, Massachusetts: Stockholm Environment Institute-Boston, 2002.
Reilly, William K. "A perspective from EPA." *EPA Journal* 16, no. 4 (1990): 12.
Reilly, William K. "Generational Change in Environmental Law." *Natural Resources Journal* 31, no. 45 (1991): 1–6.

Reilly, William K. "Sustainable Development: From Paradox to Paradigm." Keynote Address—UN Economic Commission for Europe & US Environmental Protection Agency Workshop, The Economics of Sustainable Development. Washington, DC: Smithsonian Institution, January 23, 1990.

Reisch, Mark S. "Striving for Sustainability: Chemical industry leaders wrestle with sustainable development, Responsible Care." *Chemical and Engineering News* 79, no. 36 (2001): 17–22.

Reitinger, Claudia, Matthias Dumke, Mario Barosevcic, and Rafaela Hillerbrand. "A Conceptual Framework for Impact Assessment within SLCA." *International Journal of Life Cycle Assessment* 16, no. 4 (2011): 380–388.

Renewable Energy Transmission Initiative (RETI). "Phase 2B Maps." Undated.

Repetto, Robert. "Economic Aid and the Environment." *EPA Journal* 16, no. 4 (1990): 20–23. [Repetto (1990) (a)].

Repetto, Robert. *Promoting Environmentally Sound Economic Progress: What the North Can Do*. Washington, DC: World Resources Institute, 1990. [Repetto (1990) (b)].

Rhyne, Margaret. "Blue Sky Wind Energy Project NO. R2011-00408 EIR Scoping Meeting." Letter to the Los Angeles Department of Regional Planning, October 6, 2011.

Rhyne, Margaret. "June 2010 SEA Ordinance Summary Draft SEA Ordinance Change Comparison Chart 2012." July 30, 2012.

Rhyne, Margaret. "PRMDIA Response to Element Power's NOP." December 31, 2011.

Richards, Deanna J. "Harnessing Ingenuity for Sustainable Outcomes." *The Bridge* 29, no. 1 (1999): 16–22.

Risk Reduction Engineering Laboratory, US EPA and Pollution Prevention Office. *The Environmental Challenge of the 1990s: Proceedings of the International Conference on Pollution Prevention, Clean Technologies and Clean Products, Washington, DC, June 10–13, 1990*. Cincinnati, OH: Risk Reduction Engineering Laboratory, 1990.

Roberts, Billy J. "Photovoltaic Solar Resource of the United States." September 19, 2012.

Roberts, Don V. "Improving Engineers' Response to Agenda 21, Chapter 34." 2001, manuscript.

Roberts, Don V. "Sustainable development and the use of underground space." *Tunneling and Underground Space Technology* 11, no. 4 (1996): 383–390.

Roberts, Don V. "Sustainable Development-A Challenge for the Engineering Profession." Paper presented at the FIDIC Annual Conference, Oslo, Norway, June 18, 1990.

Roberts, Don V. "Sustainable Development—A Challenge for the Engineering Profession." In *The Role of Engineering in Sustainable Development*, edited by Monica D. Ellis, 44–61. Washington, DC: AAES, 1994.

Romanik, Clare. "A conversation with Joachim Spangenberg on sustainable development and social change," October 2012.

Rourke, William J. "A vessel for engineering education in the developing world." Paper presented at the 2nd World Congress, Sarawak, Malaysia, July 22–25, 2002.

Rubin, Debra K. "FIDIC delegates debate new ecological activism." *Engineering News Record* June 28 (1990): 15–16.

Sack, Fabian, and Louise Pastro. "The story behind the page-social Life Cycle Assessment of Focus Press." Paper presented at the 8th Life Cycle Conference Pathways to Greening Global Markets, Sydney, Australia, 2001.

Sakellariou, Nicholas. "A framework for social justice in Renewable Energy Engineering." In *Just Engineering: How Can Engineering Education Educate Engineers Who are Socially Just?*, edited by Juan C. Lucena, 243–267. New York: Springer, 2013.

Sala, Serenella, Francesca Farioli, and Alessandra Zamagni. "Life Cycle Sustainability Assessment in the Context of Sustainability Science Progress (Part 1)." *International Journal of Life Cycle Assessment* 18, no. 9 (2013): 1653–1672.

Sala, Serenella, Francesca Farioli, and Alessandra Zamagni. "Life Cycle Sustainability Assessment in the Context of Sustainability Science Progress (Part 2)." *International Journal of Life Cycle Assessment* 18, no. 9 (2013): 1686–1697.

Sanio, Michael R. "Engineering and Technology Program for International Development." (Revised Draft Proposal), April 7, 2003.

Sanio, Michael R., David Burack, and Sadaf Siddiqui. "Reuse of Urban Waste for Agriculture: An Investment Program for Progressive Action." Alexandria, VA: World Engineering Partnership for Sustainable Development, 1998.

Sanio, Michael. "The Role of Engineers in Sustainable Development." Paper presented at the 1997 AAAS CAIP Meeting, AAAS Headquarters, Washington, DC, 1997.

Schleicher, Ursula. "The Uses of Life Cycle Assessment for European Legislation." *International Journal of Life Cycle Assessment* 1, no. 1 (1996): 42–44.

Schmidt, Isabell, Manfred Meurer, Peter Saling, Andreas Kicherer, Wolfgang Reuter, and Carl-Otto Gensch. "Managing Sustainability of Products and Processes with the Socio-Eco-Efficiency Analysis by BASF." *Greener Management International* 45 (2004): 79–94.

Schmidt, Kati. "National Parks Conservation Association, San Bernardino County Convene Desert Renewable EnergyPlan Workshop with CA Energy Commission, BLM." www.prweb.com, September 5, 2013.

Schumacher, Ernst F. *Small in Beautiful: A study of economics as if people mattered.* London: Abacus, 1974.

Schwarzenegger, Arnold. "Letter to LA County." February 2010.

Scott, Doug. "Learning to Listen: Seminar Connects the Design of Sustainable Projects with Community Involvement." Blogs.asce.org, January 31, 2014.

Seely, Bruce E. *Building the American Highway System: Engineers as Policy Makers.* Philadelphia: Temple University Press, 1987.

Serageldin, Ismail. 1996. *Sustainability and the wealth of nations: first steps in an ongoing journey.* Environmentally sustainable development studies and monographs series; no. 5*ESSD Environmentally & Socially Sustainable Development Work in Progress. Washington, DC: The World Bank.

SETAC. *A conceptual framework for lifecycle impact assessment.* Edited by J. Fava, F. Consoli, R. Denison, K. Dickson, T. Mohin, and B. Vigon. Proceedings of a workshop in Sandestin, Florida. Society of Environmental Toxicology and Chemistry, February 1–7, 1992, in Pensacola, Florida, 1993.

SETAC. *A technological framework for life-cycle assessment.* Edited by J. Fava, R. Denison, B. Jones, M. Curran, B. Vigon, S. Selke, and J. Barnum. Proceedings of a workshop in Smugglers Notch, Vermont. Society of Environmental Toxicology and Chemistry, August 18–23, 1990, in Pensacola, Florida, 1991.

SETAC. *Guidelines for life-cycle assessment: A 'Code of Practice'* Edited by E Consoli, D. Allen, I. Boustead, J. Fava, W. Franklin, A. A. Jensen, N. de Oude, R. Parrish, R. Perriman, D. Postlethwaite, B. Quay, J. Skguin, and B. Vigon. Pensacola: Society of Environmental Toxicology and Chemistry Press, 1993.

Shapiro, Karen G., and L. Allen White. "Life-Cycle Design Practices at Three Multi-National Companies." Paper presented at the 1999 International Symposium on Electronics and the Environment, Danvers, MA, United States, 1999.

Sharon Beder. "Engineers, Ethics and Sustainable Development." Paper presented at the 10th International Congress of Logic, Methodology and Philosophy of Science, Florence, Italy, 1995.

Silicon Valley Toxics Coalition (SVTC). "2013 SOLAR SCORECARD." 2013.

Simmons, Ann M. "Antelope Valley residents not fired up over green energy projects." *Los Angeles Times*, July 4, 2011.

Simmons, Ann. M. "Wildflower reserve concerned about proposed racetrack." *Los Angeles Times*, July 27, 2009.

Smith Owen, and Mathias Bell. "Renewables' Bird Problem. It's not what you think. Meet the Duck Chart." Blog.rmi.org, October 29, 2013.

Sonnen, Michael B. "Sustainability Canon Is Not Justified." *Civil Engineering* 65, no. 9 (1995): 35–36.

Spangenberg, Joachim H. and Odile Bonniot. "Sustainability Indicators—A Compass on the Road Towards Sustainability." Wuppertal Paper No 81, February 1998.

Spillemaeckers, Sophie. "The Belgian social label: The first governmental social label based on social life cycle analysis." Paper presented at the CALCAS Workshop Governance and Life-Cycle Analysis. Opportunities for Going Beyond ISO-LCA, Brussels, Belgium, 2007.

Stallworth, Leo. "University of Antelope Valley pays employers for jobs." *Los Angeles News*, Friday September 9, 2011.

Stellar.in. "Frequently Asked Questions (FAQ's). Undated.

Stiles, Andrew. "First Solar Furlough." Freebeacon.com, May 30, 2012.

Strong Maurice, "Introduction." In Rowland, Wade. *The Plot to Save the World.* Toronto: Clarke, Irwin & Company Limited, 1972.

Strong, Maurice F. "The Road from Rio." *The Bridge* 23, no. 2 (1993): 3–7.

Strong, Maurice. "The Engineer as Agent of Global Change." Speech delivered at the American Association of Engineering Societies and Engineering Foundation Conference, entitled "Sustainable Development: Creating Agents of Change," held at Snowbird, Utah, August 4, 1995.

Swarr, Thomas E. "Societal Life Cycle Assessment—Could You Repeat the Question?" *International Journal of Life Cycle Assessment* 14, no. 4 (2009): 285–289.

Sweet, Cassandra. "Los Angeles County Approvals Boost $1.36 Billion First Solar-Exelon Deal." *www.4-traders.com*, February 17, 2012.

Syndhia Mathé. "Integrating participatory approaches in social LCA: What about functional units and impacts choices? Fish Farming case." Presented at the Conference: Environmental & Integrated Assessment of Complex Systems, Montpellier, France, 2011.

Szejnwald Brown, Halina, Martin de Jong, and Teodorina Lessidrenska. "The Rise of the Global Reporting Initiative (GRI) as a Case of Institutional Entrepreneurship." Working Paper No. 36 of the Corporate Social Responsibility Initiative, May 2007.

Technical University of Denmark. "Invitation to Seminar: Addressing the social dimension of product sustainability with life cycle assessment," May 31st 2010, 14.00-17.00 Meeting room 1, Building 101A, DTU.

Tetra Tech EC, Inc. "Wind Energy Siting Handbook." American Wind Energy Association, 2008.

Thom, David. "Engineering Education and the New Industrial Revolution." *International Journal of Engineering Education* 14, no. 2 (1998): 89–94.

Thom, David. "Engineering to Sustain the Environment." In *The Role of Engineering in Sustainable Development*, edited by Monica D. Ellis, 62–79. Washington, DC: AAES, 1994.

Thom, David. "The Role of Engineers to Promote Cleaner Production." In *The Engineer's response to sustainable development*, 36–45. Washington: WFEO, 1997.

Thom, David. "The WFEO Code of Ethics." *New Zealand Engineering*, June 1, 1988.

Thompson, Frederick A. "Sustainable Development: 'Depoliticking' the Environment and Making It a Matter of Natural Economics." WESTON WAY April/May 1995.

Thompson, Paul B, and P. Kyle Whyte. "What happens to Environmental Philosophy in a Wicked World?" *Journal of Agricultural and Environmental Ethics* 25 (2012): 485–498.

Tilde Herrera, "Walmart, Seventh Gen Give Free LCA Tool a Trial Run." Greenbiz.com, November 9, 2010.

Timberlake Lloyd. *Catalyzing Change: A short history of the WBCSD*. Geneva: WBCSD, 2006.

Toporow, Chantal C.M. "Values-led Technologies." In *Preparing for a sustainable society*, edited by Helmut Burkhardt and H. Willem Vanderburg, 228–235. Piscataway: Institute of Electrical and Electronic Engineers, 1991.

Toporow, Chantal, McCagie Rogers, Nik Warren, and Justin Biddle. "A Delicate Balance: Technics, Culture, and Consequences." California State University, Los Angeles, October 20–21, 1989. Torrance, CA: Los Angeles Chapter IEEE SSIT-30, 1990.

Trabish, Herman K. "$100,000 Vandalism at the First Solar/Exelon Antelope Valley Solar Ranch One." *Greentech media*, July 18, 2012.

Trabish, Herman K. "Construction Halted at First Solar's 230 MW Antelope Valley Site." *Greentech media*, April 22, 2013.

Trabish, Herman K. "Element Power, First Solar, and Intense Local Politics." *Greentech media*, November 29, 2011.

Trabish, Herman K. "How does a solar power plant developer win over the community?" *Greentech media*, October 11, 2011.

Trabish, Herman K. "Is the World's Biggest PV Project About to Get Stopped Cold?" *Greentech media*, June 29, 2011.

Trabish, Herman K. "Is this the new California Gold Rush?" *Greentech media*, July 1, 2011.

Trabish, Herman K. "Is Wind Becoming a Victim of Ecosabotage?" *Greentech media*, August 18, 2011.

Trabish, Herman K. "More Furloughs at First Solar while Dispute with Los Angeles County Drags On." *Greentech media*, June 18, 2012.

Trabish, Herman K. "The biggest US wind-solar hybrid project." *Greentech media*, October 18, 2011.

Trabish, Herman K. "The Lowdown on the Safety of First Solar's CdTe Thin Film." *Greentech media*, March 19, 2012.

Trabish, Herman K. "Update: First Solar 'Furloughs' Half its Solar Ranch One Workforce." *Greentech media*, May 30, 2012.

Trabish, Herman K. "Was the First Solar Fire at Solar Ranch One a Blessing in Disguise?" *Greentech media*, July 11, 2011.

Trabish, Herman K. "Will NextEra Energy's Wind be Part of the New California Gold Rush?" *Greentech media*, July 26, 2011.

Tracey, Dennis. "Sustainability and Environment: Art vs. Science." *EFS Newsletter*, December 2001.

Traverso, Marzia, Matthias Finkbeiner, Andreas Jørgensen, and Laura Schneider. "Life Cycle Benoît Sustainability Dashboard." *Journal of Industrial Ecology* 16, no. 5 (2012): 680–688.

Troge, Andreas. "Preface by Andreas Troge President of the Federal Environmental Protection Agency." *International Journal of Life Cycle Assessment* 1, no. 1 (1996): 1–2.

Tuerck, David G., Paul Bachman, and Michael Head. "RPS: A Recipe for Economic Decline, Scheme to cost Nevadans $2.275 billion over 12 years." Nevada Policy Research Institute, April 2013.

Tuszynski, Jack. "Cinderella of the Far West Antelope Valley." July 7, 2011.

Tuszynski, Jack. "OpEd: Richard the Fourth?" *The Mountain Enterprise*, September 23, 2011.

Udo de Haes, Helias A., Reinout Heijungs, Sangwon Suh, and Gjalt Huppes. "Three Strategies to Overcome the Limitations of Life-Cycle Assessment." *Journal of Industrial Ecology* 8, no. 3 (2004): 19–32.

UK Department of Energy and Climate Change. "Onshore wind: call for evidence." June 2013.

UN Economic Commission for Europe & U.S. Environmental Protection Agency Workshop, The Economics of Sustainable Development. Washington, DC: Smithsonian Institution, January 23, 1990.

UN General Assembly, *Draft World Charter for Nature.*, October 30, 1980, A/RES/35/7.

UNEP/ICC/FIDIC. *Environmental Management System Training Resource Kit.* UNEP, 1997.

UNESCO Secretariat. "Engineering and Technology for International Development: 'Engineering for a Better World,'" August 2003.
UNESCO Secretariat. "Engineering and Technology for International Development: 'Engineering for a Better World,'" August 2003.
UNISA. "Dr. Pityana's speech at the inauguration of global reporting initiative, UN headquarters, new york." Thursday 4 April 2002.

United Nations (UN). "Plan of Implementation of the World Summit on Sustainable Development." World Summit on Sustainable Development, 2002.

United Nations Conference on Environment and Development (UNCED). *Agenda 21: Programme of Action for Sustainable Development.* United Nations, 1992.

Unknown. "A question of Ethics: Development of Sustainability Provisions in ASCE Code of Ethics." Undated.

Unknown. "ASCE leads world's engineers in advocating sustainable development." *ASCE News* 20, no. 1 (1995): 1 and 6.

Unknown. "ASCE Names 10 Distinguished Members for 2009. *Engineering News Record*, ENR.com, July 23, 2009.

Unknown. "Auriga Maintains a 'Buy' on First Solar (FSLR); The Antelope Raises its Head Again." Streetinsider.com, February 10, 2012.

Unknown. "Beautiful Earth Group breaks ground on large California solar plant." CleanEnergyAuthority.com, January 3, 2011.

Unknown. "Engaging the wind community." ecogeneration.com, September/October 2011.

Unknown. "Firms show ecological awareness at tech fair." *Chemical and Engineering News* 68, no. 16 (1990): 5.

Unknown. "If we could just channel that passion elsewhere." *Engineering News Record* June 28 (1990): 130.

Unknown. "Momentum Gathers for Sustainable Development." *ASCE News* 18, no. 6 (1993): 6.

Unknown. "Nextera told by 400 at meeting to Leave Town!" Ontario-wind-resistance.org, February 5, 2010.

Unknown. "Synthesis Report: The Social Value of LCA." *International Journal of Life Cycle Assessment* 1, no. 2 (1996): 65.

Unknown. "Tapping Maple Ridge. A Film About Maple Syrup, Wind and Community." Tappingmapleridge.com, undated.

Unknown. "The Ethic of Environmental Engineering." *New Zealand Engineering* 39, no. 8 (1984): 19–21.

Unknown. "Wind Energy planning in England, Wales and Denmark: Factors influencing Project Success (McLaren 2007)." Undated.

US Department of Energy (DOE). "Final Environmental Assessment, for Department of Energy Loan Guarantee for the AV Solar Ranch One Project in Los Angeles and Kern Counties, California." Washington, DC: DOE, 2011.

US EPA. "Life-cycle assessment: Inventory guidelines and principles." EPA/600/R-92/245. Prepared by Battelle and Franklin Associates for the US EPA. Cincinnati: Risk Reduction Laboratory, Office of Research and Development, 1993.

Vanderburg, Willem H. "Sustainable Development and the Practice of Engineering." In *Preparing for a sustainable society*, edited by Helmut Burkhardt and H. Willem Vanderburg, 9–16. Piscataway: Institute of Electrical and Electronic Engineers, 1991.

Vanegas, Jorge A., R. Jennifer DuBose, and R. Annie Pearce. "Sustainable Technologies for the Building Construction Industry." In *Designing for the global environment: a symposium*, edited by Saeid Lohbani Sadri [16 pages]. Atlanta: College of Architecture, Georgia Institute of Technology, 1995.

Vara.freeforums.org, August 13, 2009.

Vesilind, Aarne P., and S. Alastair Gunn. "Sustainable Development and the ASCE Code of Ethics." *Journal of Professional Issues in Engineering Education and Practice* 124, no. 3 (1998): 72–74.

Waddell, James. "Engineering for Sustainable Development" manuscript, 1991.

Walker, Gordon P., Patrick Devine-Wright, Sue Hunter, Helen High, and Bob Evans. "Trust and community: exploring the meanings, contexts and dynamics of community renewable energy." *Energy Policy* 38 no. 6 (2010): 2655–2633.

Weidema, Bo P., Miguel Brandao, and Jenna M. Watson. "The International Life Cycle Academy: A Quality Network for Teaching in Quantitative Sustainability Assessment: The Why and How." Presented at the SETAC Europe LCA Case Study Symposium, Copenhagen, 27 November 2012.

Weidema, Bo. "Quantifying Corporate Social Responsibility." Undated manuscript.

Weil, Vivian. *Engineering Ethics in Engineering Education: Report on a Conference, June 12–13, 1990*. Chicago: Center for the Study of Ethics in the Professions, Illinois Institute of Technology, 1992.

Weiner, Terry, and Miriam Raftery. "Advocates of rooftop solar raise voices in protest over industrial-scale desert solar and wind projects." *East County Magazine*, March 2012.

Weiner, Terry, and Miriam Raftery. "San Bernardino Board of Supervisors OK's 45-day solar moratorium." *East County Magazine*, March 15, 2012.

Weldon, David G. "One Environmental Canon Is Enough." *Civil Engineering* 65, no. 9 (1995): 35–36.

Wenzel, Henrik, Michael Hauschild, and Leo Alting. *Environmental Assessment of Products. Vol. 1 – Methodology, tools and case studies in product development.* United Kingdom: Kluwer Academic Publishers, 1997.

Wesoff, Eric. "Sharp Solar and Solar Frontier Ready to Reclaim Japan's Leadership in Solar." *Greentech media*, October 19, 2010.

Weston, Roy F. "Sustainable Development: The Economic Model of the Future." Paper presented at the New Mexico Conference on the Environment, Albuquerque, NM, United States, April 25, 1994.

Weston, Roy F. "Sustainable Development: To Better Understand the Concept." WESTON WAY April/May 1995.

Weston, Roy F. "Towards Understanding the Idea of Sustainable Development." Roy F. Weston, Inc., 1992. Manuscript.

WFEO Executive Council. "Earth Charter Resolution." September 20, 2000.

WFEO-ComTech. "Engineers and Sustainable Development." August 2002. Manuscript.

WFEO. "WFEO Commitment to Implementing Sustainable Development." 1997.

WFEO. *Arusha Declaration*—The UNCED Conference 1992 Statement by the World Federation of Engineering Organizations on Environment and Development, 1991.

WFEO. *The Engineer's response to sustainable development.* Washington: WFEO, 1997.

WFEO/FMOI (UNESCO). "Code of Ethics." 2001.

White, Allen L. "Time for Social Contract 2.0?" Undated manuscript.

White, Allen L., Monica Becker, and E. Deborah Savage. "Environmentally Smart Accounting: Using Total Cost Assessment To Advance Pollution Prevention." *Pollution Prevention Review* (1993): 247–259.

The White House, The President's Council for Sustainable Development. "Sustainable America: A New Consensus for Prosperity, Opportunity, and A Healthy Environment for the Future." Washington, DC, The White House, 1996.

The White House. *The National Strategy of the United States*. Washington, DC, The White House, 1991.

Wiggins, John H. "Challenge for Engineers." *Journal of Professional Issues in Engineering Education and Practice* 121, no. 3 (1995): 199.

Winner, Langdon. *The whale and the reactor: a search for limits in an age of high technology*. Chicago: University of Chicago Press, 1986.

Wisnioski, Matthew H. "How Engineers Contextualize Themselves." In *Engineering in Context*, edited by Steen Hyldgaard Christensen, Bernard Delahousse, and Martin Meganck, 403–416. Aarhus: Academica, 2009.

Wisnioski, Matthew H. *Engineers for Change: Competing Visions of Technology in 1960's America*. Cambridge, MA: The MIT Press, 2012.

Wolsink, Maarten. "Entanglement of interests and motives: Assumptions behind the NIMBY-theory on facility siting." *Urban Studies* 31, no. 6 (1994): 857–866.

Wolsink, Maarten. "Invalid theory impedes our understanding: A critique on the persistence of the language of NIMBY." *Transactions Institute of British Geographers* 31, no. 1 (2006): 85–91.

Wolsink, Maarten. "Wind Power and the NIMBY-myth: Institutional Capacity and the Limited Significance of Public Support." *Renewable Energy* 21, no. 1 (2000): 49–64.

Woolard, Edgar S. "Creating Corporate Environmental Change." *The Bridge* 29, no. 1 (1999): 8–11.

The World Bank. *WDR 1992: Development and the Environment.* Oxford: World Bank and Oxford University Press, 1992.

World Commission on Environment and Development (WCED). *Our Common Future.* Oxford: Oxford University Press, 1987.

Young, Steven B. "The Materials Life-Cycle Approach." In *Preparing for a sustainable society*, edited by Helmut Burkhardt and H. Willem Vanderburg, 318–325. Piscataway: Institute of Electrical and Electronic Engineers, 1991.

Zadek, Simon. "Towards a Progressive Buddhist Economics." 1997.

Zahnter, Susan. "Re: Element Power Wildflower Green Energy Farm, Project Number R2010-00256-(5), RENVT201000063, RCUPT201000121." January 5, 2012.

Zahnter, Susan. "Re: Met Tower Appeals, NextEra Blue Sky Windfarm, Project No. 201001402-(5), CUP No.201000136; Wildflower Green Energy Farm (County Project No. R2010-00256-(5),Conditional Use Permit No. 201000121, Environmental Review No. 201000063." January 23, 2012.

Zahnter, Susan. "RE: West Antelope Solar Project, R2012-01589-(5) / CUP 201200086 / ENV 201200158." November 20, 2013.

Zamagni, Alessandra, Oscar Amerighi, and Patrizia Buttol. "Strengths or Bias in Social LCA?" *International Journal of Life Cycle Assessment* 16, no. 7 (2011): 596–598.

Zamagni, Alessandra. "Inclusion of Economic Mechanisms into Life Cycle Analysis: Start with 'Framing the Question.'" *Integrated Environmental Assessment and Management* 6, no. 4 (2010): 777–785.

2.0 LCA Consultants. "What is (was) SPOLD." Undated.

Appendix 1: Solar Ranch One Timeline (Siting, Permitting, Financing and Pre-Construction)

2008 Renewable Portfolio Standard (RPS) solicitation between PG&E and AV Solar Ranch LLC

The property for Solar Ranch One was bought in 2008.

On February 18, 2009, the Antelope Acres Town Council voted through the endorsement of Solar Ranch One after a town council meeting presentation given on that same day by NextLight.

San Francisco based NextLight Renewable Power files an application with LA County's Regional Planning Department for Solar Ranch One on March 18, 2009.

On March 23, 2009, the Antelope Acres Town Council sends a letter to LA County's Regional Planning Department to confirm its support for Solar Ranch One.

Notice of Preparation (NOP) for Solar Ranch One's EIR and a Scoping Meeting Notification were issued on April 12, 2009.

Copies of the NOP are available for public review from May 1, 2009, to June 1, 2009.

A Scoping Meeting takes place in Antelope Acres—fifteen miles away from the project's location— on May 14, 2009, to consider oral and written input regarding the scope and content of Solar Ranch One's EIR.

NextLight representatives attend a June 25, 2009, Association of Rural Town Councils (ARTC) meeting to discuss the Solar Ranch One. Approximately twenty-five people attend the meeting.

On April 28, 2010, First Solar and NextLight Renewable Power, announced they have entered into an agreement for First Solar to acquire NextLight—Solar Ranch One becomes property of First Solar.

LA County released Solar Ranch One's Draft EIR (DEIR) on June 16, 2010.

A 45-day public comment period for the DEIR was set from June 16 to July 30, 2010.

The first hearing for Solar Ranch One at the Department of Regional Planning was scheduled for June 30, 2010.

Solar Ranch One's Final EIR was published on August 31, 2010.

On September 14, 2010, Antelope Acres Town Council President Vickie Nelson writes in an email that "the original support position of…[her] Town Council [for Solar Ranch One] has not changed."

The mitigation agreement between Solar Ranch One and the Desert Mountain Conservation Authority (DMCA) was also executed on September 14, 2010.

The second hearing for Solar Ranch One at Department of Regional Planning took place on September 15, 2010.

LA County certified the EIR, and approved the Project, including the CUP on December 7, 2010.

On June 6, 2011, Solar Ranch One holds a reception with guests from Antelope Acres Town Council and the Poppy Reserve Association.

On June 30, 2011, the Department of Energy (DOE) publicized an offer including a potential $680 million loan guarantee for Solar Ranch One.

On July 2, 2011, three days before construction was supposed to begin in Solar Ranch One, a fire erupted on site, allegedly caused by a non-local employee engaged in soil sampling.

On September 30, 2011, Exelon Corporation announced its acquisition of Solar Ranch One from First Solar—First Solar remains the developer and operator of the project.

On October 25, 2011, First Solar ousted CEO Rob Gillete amid more than 25% drop in its stock price to the company's lowest level since 2007.

On April 5, 2012, Exelon announces that Solar Ranch One received the first advance of its loan guarantee from DOE.

Appendix 2: Sample of Community Concerns as Regards to the Solar Ranch One, Alpine, Wildflower and Blue Sky RE projects

Name of project:	Community concern:	Concern expressed as:*
1. Wildflower Green Energy Farm (Element Power)	Preservation of Antelope Valley's poppy reserve	
2. Wildflower Green Energy Farm (Element Power)	Procedural justice	"We have to wake up at 4am to be [in downtown LA] at 9 to be heard by planning [department] folks for 2 minutes. These two minutes determine our way of life."

Name of project:	Community concern:	Concern expressed as:*
3. Wildflower Green Energy Farm (Element Power)	Distributional justice	"Not only will we lose the opportunity to benefit from the tax revenue, but we will also witness the power from this project ending up in urban areas..."
4. Wildflower Green Energy Farm (Element Power)	What counts as "local jobs"?	
5. Wildflower Green Energy Farm (Element Power)	Industrial turbines offend rural landscape aesthetic sensibilities	
6. Wildflower Green Energy Farm (Element Power)	Urban Rural divide	"In rural WAV the only form of democracy at a level lower than the county is city."
7. Wildflower Green Energy Farm (Element Power)	Role and negotiating power of town councils	"...stewardship of mitigation moneys and lands cannot be delegated to entities that lack due diligence..."
8. Wildflower Green Energy Farm (Element Power)	Misunderstanding of local needs/concerns	"[Wildflower] was canceled not so much because of community opposition, but because developers did not have a clue of what bothers the rural folks..."
9. Wildflower Green Energy Farm (Element Power)	Noise from turbines	
10. Wildflower Green Energy Farm (Element Power)	Property values	
11. Blue Sky Project (NextEra Energy)	Distributional justice	"Why do rural folks have to sacrifice themselves for climate change?"
12. Blue Sky Project (NextEra Energy)	Cumulative impacts	"These projects will soon make the WAV look like Tehachapi..."

Name of project:	Community concern:	Concern expressed as:*
13. Blue Sky Project (NextEra Energy)	Projects' effects on local flora and fauna	
14. Solar Ranch One (Exelon)	What counts as "local jobs"?	"One small job fair was done locally [in Fairmont]. The large [fair, however] was organized 30 miles away in a city [Lancaster] of lower class working people."
15. Solar Ranch One (Exelon)	Urban rural divide	
16. Solar Ranch One (Exelon)	Local knowledge ignored	"The trees used in landscaping were pathetic. They won't survive three years. Plus First Solar used a San Diego landscaper [Recon]."
17. Solar Ranch One (Exelon)	Developer violates safety rules	"...construction vehicles were using small dirt roads; blocking Highway 138 and risking collisions."
18. Solar Ranch One (Exelon)	Misunderstanding of local needs/wants	"These engineers never felt the need to properly explore the unique dynamics of our region."
19. Solar Ranch One (Exelon)	Planning department failed to facilitate cooperation between developers and community	"The meetings at the airport were a joke. The County did not show any meaningful leadership."
20. Solar Ranch One (Exelon)	Water use	
21. Solar Ranch One (Exelon)	Developers are late in addressing, or completely failed to address, community concerns	"To make sure we monitored their impact on our air, we suggested that they performed air quality testing, but we were blatantly ignored."
22. Solar Ranch One (Exelon)	Developer is involved in non-transparent permitting	

Name of project:	Community concern:	Concern expressed as:*
23. Solar Ranch One (Exelon)	PV LCAs' focus is problematic	"Life cycles quantify emissions and leave everything else out"
24. Solar Ranch One (Exelon)	Projects' effects on local flora and fauna	
25. Solar Ranch One (Exelon)	Developer does not keep their promises	"When I supported Solar Ranch One I did not imagine that their word had no value…" "They would lie to AQMD [Air Quality Management District]; they would say we shut that job down. But I was there on 170th and could see forklifts driving up and down, the dust pouring of their wheels."
26. Solar Ranch One (Exelon)	Policing of the project's site	"I was literally chased by a security guard on 170th st by Solar Ranch One's project site."
27. Solar Ranch One (Exelon)	Developers possessing a distorted notion of "community engagement"	"Events like an obligatory open house where no direct answers are given and no collaboration is decided upon do nothing to engage the community."
28. Solar Ranch One (Exelon)	Landscaping	
29. Solar Ranch One (Exelon)	Distributional justice	
30. Solar Ranch One (Exelon)	Dust mitigation	You claim the "palliatives" being used are "benign". Wonderful! You were asked to identify them. You have failed to do so. (David Jefferies P.128)

Name of project:	**Community concern:**	**Concern expressed as:***
31. Solar Ranch One (Exelon)	Extending construction activity	"Without asking us, in early 2013 they started having four work shifts, seven days a week."
32. Solar Ranch One (Exelon)	Poor planning	
33. Solar Ranch One (Exelon)	What counts as "local jobs"	My son worked at Solar Ranch One for eighty-nine days… And then they laid him and others off and brought in a new crew…[So,] outsiders come and they rent homes here so they use the local addresses. 'So look,' they say, 'we're local.' No, you're not. These are guys from Utah, Nevada, Arizona…"
32. Alpine Solar (NRG Energy)	Procedural justice/Poor planning	"Alpine did not do an EIR and thought they would get away with it."
33. Alpine Solar (NRG Energy)	Urban rural divide	
34. Alpine Solar (NRG Energy)	Cumulative effects	
35. Alpine Solar (NRG Energy)	Poor planning	"Believe it or not planning folks placed Alpine's materials at the Rosamond Library for review. Rosamond belongs to another County [Kern]…"
36. Alpine Solar (NRG Energy)	Developer does not keep their promises	"I could pull the meeting minutes where they gave the presentations and give you specifics. Like, 'we're going to hire x number of employees.' NRG was lying through their teeth."

* Interview material was used to describe local concerns, where appropriate.

Index

www.ingramcontent.com/pod-product-compliance
Lightning Source LLC
Chambersburg PA
CBHW080505040125
19783CB00084B/31

9 781119 418160